T0312693

Principles of Ad Hoc Networking

Principles of Ad Hoc Networking

Michel Barbeau and Evangelos Kranakis
Carleton University, Canada

John Wiley & Sons, Ltd

Telephone (+44) 1243 779777

Email (for orders and customer service enquiries): cs-books@wiley.co.uk
Visit our Home Page on www.wileyeurope.com or www.wiley.com

Other Wiley Editorial Offices

John Wiley & Sons Inc., 111 River Street, Hoboken, NJ 07030, USA

Jossey-Bass, 989 Market Street, San Francisco, CA 94103-1741, USA

Wiley-VCH Verlag GmbH, Boschstr. 12, D-69469 Weinheim, Germany

John Wiley & Sons Australia Ltd, 42 McDougall Street, Milton, Queensland 4064, Australia

John Wiley & Sons (Asia) Pte Ltd, 2 Clementi Loop #02-01, Jin Xing Distripark, Singapore 129809

John Wiley & Sons Canada Ltd, 6045 Freemont Blvd, Mississauga, Ontario, L5R 4J3, Canada

Wiley also publishes its books in a variety of electronic formats. Some content that appears in print may not be available in electronic books.

Anniversary Logo Design: Richard J. Pacifico

British Library Cataloguing in Publication Data

A catalogue record for this book is available from the British Library

ISBN: 978-0-470-03290-9

Typeset in 10/12pt Times by Laserwords Private Limited, Chennai, India
Printed and bound in Great Britain by Antony Rowe Ltd, Chippenham, Wiltshire
This book is printed on acid-free paper responsibly manufactured from sustainable forestry in which at least two trees are planted for each one used for paper production.

To Eda and Esther for their love,
compassion and understanding (EK, MB)

Contents

PREFACE

...The shore of the morning sea and the cloudless sky brilliant blue and yellow all illuminated lovely and large. Let me stand here. Let me delude myself that I see these things...
Cavafy (1976)[Morning Sea, page 54]

Exchanging and sharing information has been a vital human activity since ancient times. Communication, in its simplest form, involves a sender who wants to communicate messages to an intended receiver. The word *communication* is derived from Latin and refers to the social need for direct contact, sharing information and promoting mutual understanding. The word *telecommunication* adds the prefix *tele* (meaning distance) and was first used by Edouard Estaunié in his 1904 book *Traité Pratique de Télecommunication Electrique* (see Huurdeman (2003)[page 3]). It is a technology that eliminates distance in communication.

Networks are formed by a collection of interconnected entities that can exchange information with each other. Simple systems consisting of a combination of runners, calling posts, mirrors, smoke and fire, pigeons, heliographs and flags have been used since ancient times. Efficiency in communication (i.e. the amount of information transmitted per time unit) has always been a driving force in developing new technologies. This led to the creation of permanent networked systems that could maintain consistent communication capability over large geographic areas. This gave rise to Claude Chappe's semaphore system in France (1791) and Abraham Edelcrantz' beacon system in Sweden (1794). The successful implementation of the telegraph with Samuel Morse's code (1832) and Alexander Graham Bell's telephone (1876) in the United States (see Holzmann and Pehrson (1995)) created the seeds of a telecommunication revolution that has continued with ever-increasing intensity till the present day.

The growing popularity of time-sharing systems created the need for combining communication lines and computers. Ever since the development of ARPANET that led to the invention of packet switching networks in the early 1960s, there has been no shortage of paradigms in computer networking. Ad hoc networking, which is the subject of the current book, is the latest. Despite the fact that it is less than a decade old, it is already becoming the foremost communication paradigm in wireless systems. According to "The New Dictionary of Cultural Literacy", *ad hoc* comes from Latin and means *toward this (matter)*. It is a phrase describing something created especially for a particular occasion. It may be improvised and often impromptu but it is meant to address a situation at hand.

Networks have been around for sometime. They have been the object of numerous, sophisticated graph theoretic studies by mathematicians ever since Euler proposed the celebrated *Königsberg bridge* problem in 1736. Starting with ARPANET, engineers have continued to provide a plethora of inventions that enable dynamic networking solutions. So one may wonder why we need the new term ad hoc networks. An ad hoc network is an assembly of wireless devices that can quickly self-configure to form a networked topology. In traditional networking, nodes had specific, well-defined roles, usually as routers, switches, clients, servers, and so on. In contrast, nodes in ad hoc networks have no preassigned roles and quick deployment makes them suitable for monitoring in emergency situations.

In a way, the design of ad hoc networks needs to abstract simplicity from the midst of a meaningless complexity since topology formation has to take advantage of the physical connectivity characteristics of the environment. Often, studies are interdisciplinary and bring forth a paradigm shift in that they encompass a research approach to networking problems that combines ideas from many diverse disciplines like communication, control, geometry, graph theory and networking, probability, and protocol design that have given rise to many interesting new ideas. Nothing could describe more graphically the vitality of computer science than Alan Perlis' exuberant statement quoted from his 1966 Turing award lecture on "The Synthesis of Algorithmic Systems" (see Perlis (1987)[page 15]).

> Computer science is a restless infant and its progress depends as much on shifts in point of view as on the orderly development of our current concepts.

Computer science is often inspired by combining the sophistication of mathematical abstractions with the practicality of engineering design. In the sequel, we provide a discussion of some of the important developments of ad hoc networks with applications and provide a road map to the contents of the book.

Development of Ad Hoc Networking

An ad hoc network consists of nodes that may be mobile and have wireless communications capability without the benefit of a mediating infrastructure. Every node can become aware of the presence of other nodes within its range. Such nodes are called *neighbors* because direct wireless communications links can be established with them. Links established in the ad hoc mode do not rely on the use of an access point or base station. Neighbors can communicate directly with each other. The nodes and links form a graph. Any pair of nodes, not directly connected, can communicate if there is a path, consisting of individual links, connecting them. Data units are routed through the path from the origin to the destination. Routing in the ad hoc mode means that there is no need for an address configuration server such as DHCP or routers. Every node autonomously configures its network address and can resolve the way to reach a destination, using help from other nodes. Every node also plays an active role in forwarding data units for other nodes.

Ad hoc network applications

Here are three kinds of applications of the ad hoc concept: ad hoc linking, ad hoc networking and ad hoc association. Ad hoc linking is a feature present in a number of infrastructure-based systems. The D-STAR system is illustrative of an application using the ad hoc linking mode. The developer of the D-STAR protocols is the Japan Amateur Radio League (2005). D-STAR provides digital voice and digital data capability for fixed users, pedestrians and vehicles. It is intended mainly for emergency communications and is TCP/IP based. The data rate is 128 kbps (with a 130 kHz bandwidth). A D-STAR radio may have an Ethernet port for interfacing a computer to the radio. D-STAR radios can communicate directly without any access point, base station or repeater. They do, however, have backbone and Internet interconnection capability. D-STAR radios are currently available. There are a number of other technologies available with ad hoc linking capability such as Bluetooth, WiFi/802.11 and WiMAX/802.16 (mesh topology).

Ad hoc networking refers to the capability that the members of a network have to build routing information and forward data units from one location to another in the network. The Dedicated Short Range Communications (DSRC) is a radio service allocated in the 5.850 GHz-5.925 GHz range for vehicular ad hoc networks developed by the IEEE (2005). Such networks support roadside-to-vehicle and vehicle-to-vehicle communications. Envisioned applications include safety and traffic information dissemination and collision avoidance.

An ad hoc association is a relation established between two applications that find each other. Clients and servers of location-based services are examples of such applications. The goal of location-based services is to link a node's location to other useful information, resource or service. Applications include location of health services and goods. Discovery protocols with awareness of location are required to support location-based services. Protocols have been defined for the purpose of service discovery such as the Bluetooth Service Discovery Protocol and IETF Service Discovery Protocol. They can be combined with location information attributes. Location-based services are also envisioned in the context of DSRC.

Impact on protocol architecture

As depicted in Figure 1, operation in the ad hoc mode has an impact on a network architecture at all levels of the layered hierarchy. The physical layer is responsible for the transmission and reception of frames of bits as radio signals, also known as *framing*. Nodes in the ad hoc mode need their own mechanism to synchronize the start and end of their physical layer frames. Multiple access modes are preferred over modes relying on the use of the signal of a base station to synchronize physical layer frames, such as in time division multiplexing. The roles of the link layer are the control of the access to radio channels, hardware addressing and error control of frames. The link layer in the ad hoc mode needs to address neighbor discovery and the establishment of links, which may be unidirectional or asymmetric. The task of the network layer is to build routing information and forward data units from their origin node to their destination node using available paths. The network layer needs to deal with address configuration, address conflict detection and routing. The purpose of the transport layer is to provide to processes, on a node, process-to-process communication channels. Higher error rates and long interruptions characterize the network level service. These issues need to be addressed by the transport layer. The application layer consists of protocols for supporting the needs of applications. Over ad hoc networks, protocols are required to dynamically discover resources and services.

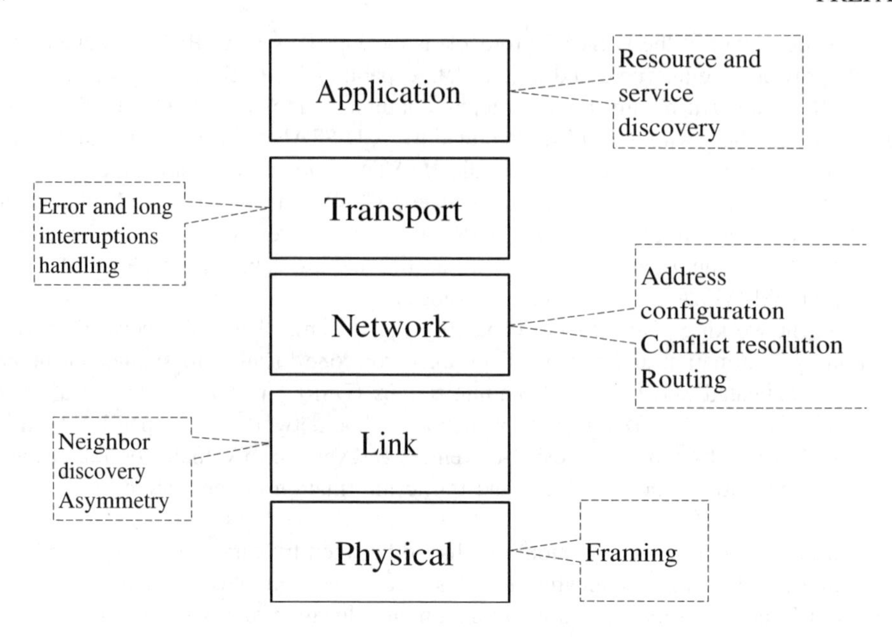

Figure 1 Impact of the ad hoc mode on a protocol architecture.

Roadmap and Style of Presentation

This book grew out of courses that each of us taught over the past few years at both the senior undergraduate level and junior graduate level. We have tried to present issues and topics in as orderly a manner as possible. Generally, chapters are relatively independent of each other and if you are already familiar with the subject you can read it in any order. Figure 2 provides a simple chart of dependencies that the reader may want to follow. Following is a brief outline of the contents of the chapters.

Chapter 1, on *Wireless Data Communications*, looks at the physical layer characteristics of ad hoc networks. Highlights of this chapter include signal representation, analog to digital conversion and digital to analog conversion, architecture of an SDR application, quadrature modulation and demodulation, spread spectrum, antennas and signal propagation.

Chapter 2, on *Medium Access Control*, addresses how wireless media are shared with distributed access. Control mechanisms are discussed, which insure non interfering access. After introducing the fundamentals of probability and statistics, this chapter presents some of the medium access protocols used in ad hoc networks. Highlights of this chapter include traffic modeling, multiple (uncoordinated, contention based, and demand assigned) access, CSMA/CD and CSMA/CA.

Chapter 3, on *Ad Hoc Wireless Access*, goes into the deeper details of a particular technology and discusses the principles of Bluetooth network formation. Highlights of this chapter include architecture of Bluetooth, access control, protocols for node discovery and topology formation, and mesh mode of WiMAX/802.16.

Chapter 4, on *Wireless Network Programming*, describes how to use the packet socket API in order to access the WiFi/802.11 wireless interface in a Linux system and communicate

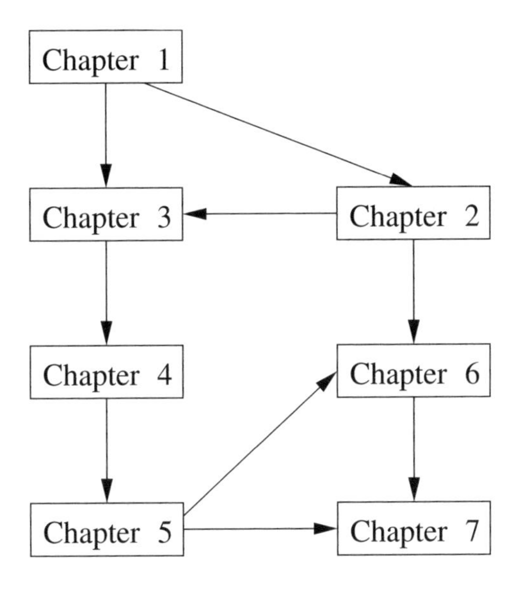

Figure 2 Dependency of chapters.

with other nodes in the ad hoc mode. Highlights of this chapter include ad hoc linking in WiFi/802.11, sockets, parameters and control and receiving/sending frames.

Chapter 5, discusses *Ad Hoc Network Protocols* thus focusing on the network layer. In particular, it addresses the issue of how packets should be forwarded and routed to their destination. Highlights of this chapter include reactive, proactive and hybrid approaches, clustering, ad hoc network model and cluster formation, quality of service, and sensor network protocols (flat routing, hierarchical routing and ZigBee).

Chapter 6, on *Location Awareness*, brings attention to simple geometric principles that enrich the infrastructureless character of ad hoc networks. It investigates how dynamic communication solutions (e.g. route discovery, geolocation) can be established taking advantage only of geographically local conditions. In addition, the guiding principle is that algorithms must terminate in constant time that is independent of the size of the input network. Highlights of this chapter include geographic proximity, neighborhood graphs, preprocessing the ad hoc network in order to construct spanners, radiolocation techniques and localization algorithms, information dissemination, geometric routing and traversal in (undirected) planar graphs, graph and geometric spanners and properties of random unit graphs, as well as coverage and connectivity in sensor networks.

Chapter 7, on *Ad Hoc Network Security*, discusses a variety of security problems arising in ad hoc networks. These include authentication and signatures, physical layer attacks, security of application protocols (WiFi/802.11, ZigBee), biometrics, routing and broadcast security as well as secure location verification and security of directional antenna systems.

Finally, each chapter concludes with several exercises. Some are rather routine and are meant to complement the text, many others are less so, while the more challenging ones are marked with ($\star$). The reader is advised to attempt them all and may occasionally have to refer to the original published source for additional details.

The book has a companion Web site at http://www.scs.carleton.ca/~barbeau/pahn/index.htm. The companion Web site for the book contains presentation slides and source code for the examples in the book.

Acknowledgements

We would like to thank our students and collaborators for the learning experience that led to this book. We are particularly thankful to Christine Laurendeau who read and commented extensively on portions of the manuscript. Many thanks also to Gustavo Alonso, Paul Boone, Prosenjit Bose, Mathieu Couture, Costis Georgiou, Jen Hall, Danny Krizanc, Pat Morin, Michel Paquette, Tao Wan, and Peter Widmayer for many stimulating conversations. The second author would also like to express his deepest appreciation to Jorge Urrutia for providing a stimulating environment at the Mathematics Institutes in Morelia and Guanajuato and the rest of the routing group Edgar Chavez, Jurek Czyzowicz, Stefan Dobrev, Hernan González-Aguilar, Rasto Kralovic, Jarda Opatrny, Gelasio Salazar, and Laco Stacho for interesting conversations at our annual summer meeting.

During the preparation of the book, the authors were supported in part by grants from MITACS (Mathematics of Information Technology and Complex Systems) and NSERC (Natural Sciences and Engineering Research Council of Canada).

Michel Barbeau and Evangelos Kranakis
Ottawa, Ontario, Canada

Glossary

AP	Answering Protocol, 75
CP	Conditional Protocol, 75
LP	Listening Protocol, 75
RP	Random Protocol, 75
SP	Sleep Protocol, 75
ACL	Asynchronous ConnectionLess, 70
ADC	Analog to Digital Conversion, 5
AES	Advanced Encryption Standard, 204, 205
AOA	Angle Of Arrival, 169
AODV	Ad Hoc On-Demand Distance Vector, 121
API	Application Programming Interface, 103
BASN	Body Area Sensor Network, 207
BER	Bit Error Rate, 20
BFS	Breadth First Search, 149
BiBa	Bins and Balls, 218
BPF	Band Pass Filter, 7
bps	Bit per second, 11
BS	Base Station, 87
CA	Certification Authority, 204
CCK	Complementary Code Keying, 11
CRC	Cyclic Redundancy Code, 89
CSMA	Carrier Sense Multiple Access, 52
CSMA/CA	Carrier Sense Multiple Access/Collision Avoidance, 52
CSMA/CD	Carrier Sense Multiple Access/Collision Detection, 58
CTS	Clear to Send, 54
CW	Contention Window, 53
DAC	Digital to Analog Converter, 8
dB	decibel, 17
dBi	dB isotropically, 17
DBPSK	Differential Binary Phase Shift Keying, 11
DCF	Distributed Coordination Function, 52

DDC	Direct Digital Conversion, 5
DDS	Direct Digital Synthesis, 9
DES	Data Encryption Standard, 204
DFS	Depth First Search, 149
DHCP	Dynamic Host Configuration Protocol, 88
DLTR	Distance Left To Right, 179
DMAC	Distributed and Mobility-Adaptive Clustering, 133
DQPSK	Differential Quadrature Phase-Shift Keying, 11
DS	Direct Sequence, 11
DSDV	Destination-Sequenced Distance Vector, 215
DSP	Digital Signal Processing, 1
DSR	Dynamic Source Routing Protocol, 116
DSSS	Direct Sequence Spread Spectrum, 56
DT	Delaunay Triangulation, 150
ECG	Electrocardiogram, 207
EIRP	Effective Isotropic Radiated Power, 24
ESSID	Extended Service Set ID, 104
FER	Frame Error Rate, 20
FH	Frequency Hopping, 11
FS	Frequency Synchronization, 68
FSK	Frequency Shift Keying, 12
GFSK	Gaussian-shape Frequency Shift Keying, 11
GG	Gabriel Graph, 148
GPS	Global Positioning System, 146
HAMA	Hybrid Activation Multiple Access, 57
HMAC	Hash Message Authentication Code, 91
HSP	Half Space Proximal, 155
IARP	Intrazone Routing Protocol, 125
IBSS	Independent Basic Service Set, 52
IERP	Interzone Routing Protocol, 125
IETF	The Internet Engineering Task Force, 121
IF	Intermediate Frequency, 5
IP	Internet Protocol, 70
IPI	InterPule Interval, 208
K	Kilo, 11
L2CAP	Logical Link Control and Adaptation Layer, 70
LAMA	Link Activation Multiple Access, 57
LAN	Local Area Network, 191
LF	Link Formation, 67

LMR	LMR Flexible Low Loss Cable (A trademark of Times Microwave Systems), 21
LocalMST	Local Minimum Spanning Tree, 157
LPF	Low Pass Filter, 5
M	Mega, 11
MAC	Medium Access Control, 46
MAC	Message Authentication Code, 193
MaxDLTR	Max Distance Left To Right, 179
MD	Message Digest, 194
MIC	Message Integrity Code, 205
MidDLTR	Mid Distance Left To Right, 179
MinDLTR	Min Distance Left To Right, 179
MPR	MultiPoint Relay, 122
MS	Mobile Station, 87
MSH-CSCF	Mesh centralized configuration, 93
MSH-CSCH	Mesh centralized scheduling, 92
MSH-DSCH	Mesh distributed scheduling, 92
MSH-NCFG	Mesh network configuration, 90
MSH-NENT	Mesh network entry, 91
MST	Minimum Spanning Tree, 149
MTU	Maximum Transmission Unit, 105
NA	Neighbor Algorithm, 170
NAMA	Node Activation Multiple Access, 56
NCR	Neighbor Aware Contention Resolution, 56
NDP	Neighbor Discovery Protocol, 125
NNG	Nearest Neighbor Graph, 147
ODRP	On-Demand Delay Constrained Unicast Routing Protocol, 136
OFDM	Orthogonal Frequency Division Multiplexing, 11
OLSR	Optimized Link-State Routing, 121
OSI	Open Systems Interconnection, 70
PAMA	Pairwise Link Activation Multiple Access, 57
PCF	Point Coordination Function, 52
PDU	Protocol Data Unit, 89
PER	Packet Error Rate, 20
PKI	Public-Key Infrastructure, 192
PPG	Photoplethysmogram, 207
PSK	Phase Shift Keying, 12
QAM-16	Quadrature Amplitude Modulation-16 states, 11
QPSK	Quadrature Phase shift Keying, 11

r.v.	Random Variable, 30
RB	Random Backed, 68
RC4	Route Coloniale 4, 203
RNG	Relative Neighbor Graph, 148
RoA	Region of Acceptance, 222
RTS	Request to Send, 54
SC	Single Carrier, 11
SCO	Synchronous Connection Oriented, 70
SDP	Service Discovery Protocol, 70
SDR	Software Defined Radio, 1
SEAD	Secure, Efficient, Ad Hoc, Distance vector, 215
SEAL	SElf Authenticating vaLue, 219
SHA	Secure Hash Algorithm, 195
SKKE	Symmetric-Key Key Establishment, 206
SNMP	Simple Network Management Protocol, 88
SS	Spread Spectrum, 11
TC	Topology Control, 124
TCP	Transmission Control Protocol, 23
TDMA	Time Division Multiple Access, 89
TDOA	Time Difference Of Arrival, 168
TKIP	Temporal Key Integrity Protocol, 204
TOA	Time Of Arrival, 168
TTL	Time To Live, 124
UDG	Unit Disk Graph, 146
UDP	User Datagram Protocol, 91
UWB	Ultrawideband, 24
WEP	Wired Equivalent Privacy, 203
WiFi	Wireless Fidelity, 103
WiMAX	Worldwide Interoperability for Microwave Access, 87
WPA	WiFi Protected Access, 204
ZRP	Zone Routing Protocol, 125

1

WIRELESS DATA COMMUNICATIONS

Ce n'est pas possible!...This thing speaks!
Dom Pedro II, 1876

The telecommunication revolution was launched when the great scientist and inventor Alexander Graham Bell was awarded in 1876, US patent no. 174,465 for his "speaking telegraph." His resolve to help the deaf led to his perseverance to make a device that would transform electrical impulses into sound. Bell's telephone ushered a revolution that changed the world forever by transmitting human voice over the wire. Dom Pedro II, the enlightened emperor of Brazil (1840 to 1889), could not be easily convinced of the telephone's ability to talk when Bell provided explanations at the Philadelphia centennial exhibition in 1876. But lifting his head from the receiver exclaimed "My God it talks" as Alexander Graham Bell spoke at the other end (see Huurdeman (2003), pages 163–164). Bell had been trying to improve the telegraph when he came upon the idea of sending sound waves by means of an electric wire in 1874. His first telephone was constructed in March of 1876 and he also filed a patent that month. Although the receiver was essentially a coil of wire wrapped around an iron pole at the end of a bar magnet (see Pierce (1980)), there is no doubt that Bell had a very accurate view of the place his invention would take in society. At the same time he stressed to British investors that "all other telegraphic machines produce signals that require to be translated by experts, and such instruments are therefore extremely limited in their application. But the telephone actually speaks" (see Standage (1998), pages 197–198). This chapter is dedicated to the important topic of signaling that makes communication possible in all wireless systems.

Nowadays, wireless communications are performed more and more using software and less and less using hardware. Functions are performed in software using digital signal processing (DSP). A very flexible setup is the one involving a general hardware platform whose functionality can be redefined by loading and running appropriate software. This is the idea of software defined radio (SDR). In the sequel, the focus is on the use of SDRs, rather than on their design. Essential DSP theory is presented, but only to a level of detail

Principles of Ad hoc Networking Michel Barbeau and Evangelos Kranakis

required to understand and use a SDR. The emphasis is on architectures and algorithms. Wireless data communications are addressed from a mathematic and software point of view. Detailed SDR hardware design can be found in the book by Mitola (2000) and papers from Youngblood (2002a,b,c, 2003). Detailed DSP theory can be found in the books by Smith (1999, 2001) and in a tutorial by Lyons (2000).

The overall architecture of an SDR is pictured in Figure 1.1. There is a transmitter and a receiver communicating using radio frequency electromagnetic waves, which are by nature analog. The transmitter takes in input data in the form of a bit stream. The bit stream is used to modulate a carrier. The carrier is translated from the digital domain to the analog domain using a digital to analog converter (DAC). The modulated carrier is radiated and intercepted using antennas. The receiver uses an analog to digital converter to translate the modulated radio frequency carrier from the analog domain to the digital domain. The demodulator recovers the data from the modulated carrier and outputs the resulting bit stream. This chapter explains the representation of signals, Analog to digital conversion (ADC), digital to analog conversion, the software architecture of an SDR, modulation, demodulation, antennas and propagation.

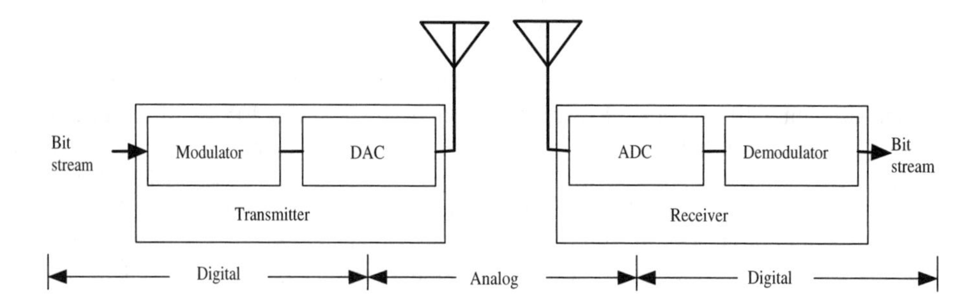

Figure 1.1 Overall architecture of an SDR.

1.1 Signal representation

This section describes two dual representations of signals, namely, the real domain representation and the complex domain representation.

A continuous-time signal is a signal whose curve is continuous in time and goes through an infinite number of voltages. A continuous-time signal is denoted as $x(t)$ where the variable t represents time. $x(t)$ denotes the amplitude of the signal at time t, expressed in volts or subunits of volts. A radio frequency signal is by nature a continuous-time signal. $\cos(2\pi ft)$ is a mathematic representation of a continuous-time periodic signal with frequency f Hertz. A signal $x(t)$ is periodic, with period $T = 1/f$, if for every value of t, we have $x(t) = x(t + T)$. A discrete-time signal is a signal defined only at discrete instants in time. A discrete-time signal is denoted as $x(n)$. The variable n represents discrete instants in time. A discrete-time sampled signal is characterized by a sampling frequency f_s, in samples per second, and stored into computer memory.

The aforementioned representations model signals in the real domain. Equivalently, signal can be represented in the complex domain. This representation captures and better explains the various phenomena that occur while a signal flows through the different functions of a radio. In addition, in the complex domain representation the amplitude, phase and frequency of a signal can all be directly derived. The complex domain representation of a signal consists of the in-phase original signal, denoted as $I(t)$, plus j times its quadrature, denoted as $Q(t)$, with $j = \sqrt{-1}$:

$$I(t) + jQ(t).$$

The quadrature is just the in-phase signal whose components are phase shifted by 90°. The representation of such a signal in the complex plane helps grasp the idea. A periodic complex signal of frequency f whose in-phase signal is defined as $\cos(2\pi ft)$ is pictured in Figure 1.2. Its quadrature is $\sin(2\pi ft)$, since it is the in-phase signal shifted by 90°. The evolution in time of the complex signal is captured by a vector, here of length 1, rotating counterclockwise at $2\pi f$ rad/s. The value of $I(t)$ corresponds to the length of the projection of the vector on the x- axis, also called the *Real axis*. This projection ends at position $\cos(2\pi ft)$ on the Real axis. The value of $Q(t)$ corresponds to the length of the projection of the vector on the y-axis, also called the *Imaginary axis*. This projection ends at position $\sin(2\pi ft)$ on the Imaginary axis.

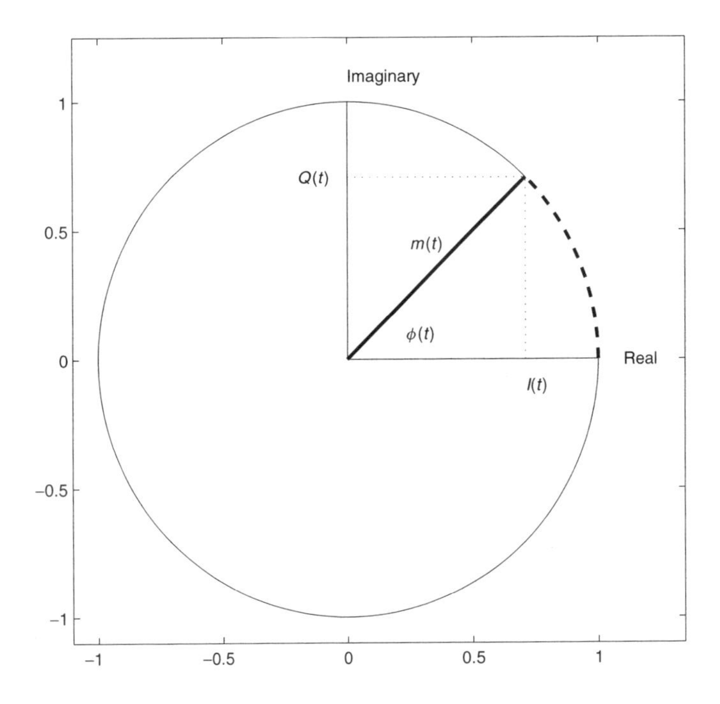

Figure 1.2 Complex signal.

Here is an interesting fact. Euler has originally uncovered the following identity:

$$\cos(2\pi ft) + j\sin(2\pi ft) = e^{j2\pi ft}.$$

The left side of the equality is called the *rectangular form* and the right side is called the *polar form*. Not intuitive at first sight! It is indeed true and a proof of the Euler's identity can be found in the book of Smith (1999) and a tutorial from Lyons (2000). Above all, it is a convenient and compact notation. Figure 1.3 depicts an eloquent representation of a signal in the polar form. The values of $I(t) = \cos(2\pi ft)$ and $Q(t) = \sin(2\pi ft)$ are plotted versus time as a three-dimensional helix. The projection of the helix on a Real-time plane yields the curve of $I(t)$. The projection of the helix on the Imaginary-time plane yields the curve of $Q(t)$. It is equally true that $\cos(2\pi ft) - j\sin(2\pi ft) = e^{-j2\pi ft}$. In this case, however, the vector is rotating clockwise.

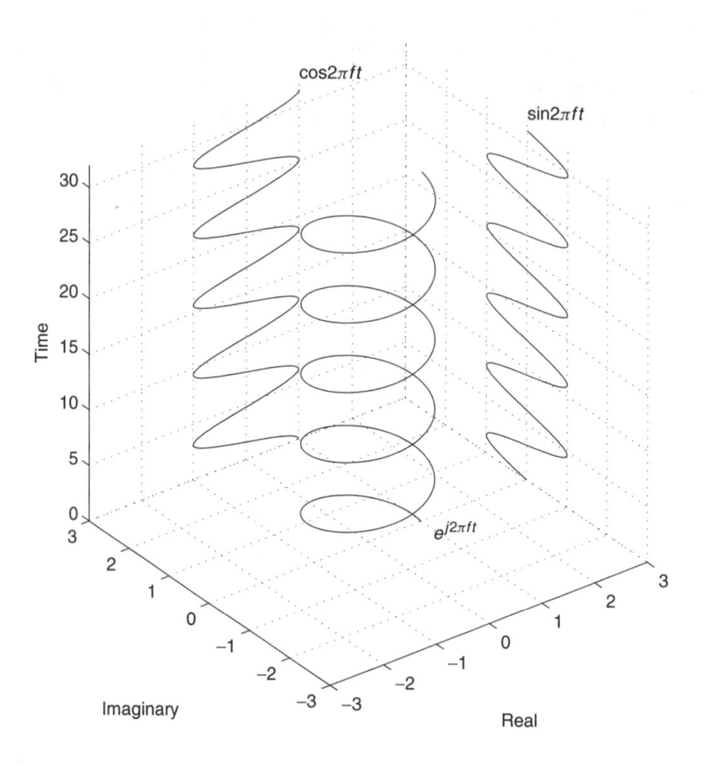

Figure 1.3 Complex signal in 3D.

The beauty of the representation is revealed as follows. Given its $I(t)$ and $Q(t)$, a signal can be demodulated in amplitude, frequency or phase. Its instantaneous amplitude is denoted as $m(t)$ (it is the length of the vector in Figure 1.2) and, according to Pythagoras' theorem, is defined as:

$$m(t) = \sqrt{I(t)^2 + Q(t)^2} \text{ Volts.} \tag{1.1}$$

Its instantaneous phase is denoted as $\phi(t)$ (it is the angle of the vector in Figure 1.2, measured counter clockwise) and is defined as:

$$\phi(t) = \arctan\left(\frac{Q(t)}{I(t)}\right) \text{ Radians.} \tag{1.2}$$

For discrete-time sampled signals, variable t is replaced by variable n in Equations 1.1 and 1.2. On the basis of the observation that the frequency determines the rate of change of the phase, the instantaneous frequency of a discrete-time sampled signal at instant n is derived as follows:

$$f(n) = \frac{f_s}{2\pi}\,[\phi(n) - \phi(n-1)] \text{ Hertz.} \quad (1.3)$$

The real domain representation and complex domain representation are dual to each other. It is always possible to map the representation in one model to another, as illustrated in Table 1.1. The column on the left lists real domain representations and their dual representations in the complex domain are listed in the right column. Note that no quantity is added nor subtracted during the translation from one domain to another.

Table 1.1 Equivalence of real and complex representations of signals.

Real domain	Complex domain
$\cos(2\pi ft)$	$\frac{1}{2}\left([\cos(2\pi ft) + j\sin(2\pi ft)] + [\cos(2\pi ft) - j\sin(2\pi ft)]\right) = \frac{1}{2}\left(e^{j2\pi ft} + e^{-j2\pi ft}\right)$
$\sin(2\pi ft)$	$\frac{1}{j2}\left([\cos(2\pi ft) + j\sin(2\pi ft)] - [\cos(2\pi ft) - j\sin(2\pi ft)]\right) = \frac{j}{2}\left(e^{-j2\pi ft} - e^{j2\pi ft}\right)$

1.2 Analog to digital conversion

ADC is the process of translating a continuous-time signal to a discrete-time sampled signal. According to Nyquist, if the bandwidth of a signal is f_{bw} Hertz then a lossless ADC can be achieved at a sampling frequency corresponding to twice f_{bw}. This is called the *Nyquist criterion* and is denoted as:

$$f_s \geq 2 f_{bw}.$$

The architecture of an analog to digital converter is pictured in Figure 1.4. On the left side, the input consists of a modulated radio signal. It goes through a low pass filter (LPF) whose role is to limit the bandwidth such that the Nyquist criterion is met. Signal components in the input at frequencies higher than f_{bw} are cut off. Otherwise, false signal images, called *aliases*, are introduced and cause distortion in ADC. On the right side, the output consists of a discrete-time sampled signals. This method is termed *direct digital conversion (DDC)*.

Direct conversion and sampling, as pictured in Figure 1.4, is limited by the advance of technology and the maximum sampling frequency that can be handled by top of the line processors. Actually to be able to reach the high end of the radio spectrum, an architecture such as the one pictured in Figure 1.5 is used. The modulated radio signal, whose carrier frequency is f_c, is down converted to an intermediate frequency (IF) or the baseband and LPF before ADC. Down conversion is done by mixing (represented by a crossed circle) the frequency of the modulated radio signal with a frequency f_{lo} generated by a local oscillator. If the frequency f_{lo} is equal to f_c, then the IF is 0 Hz and the signal is down converted to baseband. The cut off frequency of the LPF is the upper bound of the bandwidth of the baseband signal: f_{bw}.

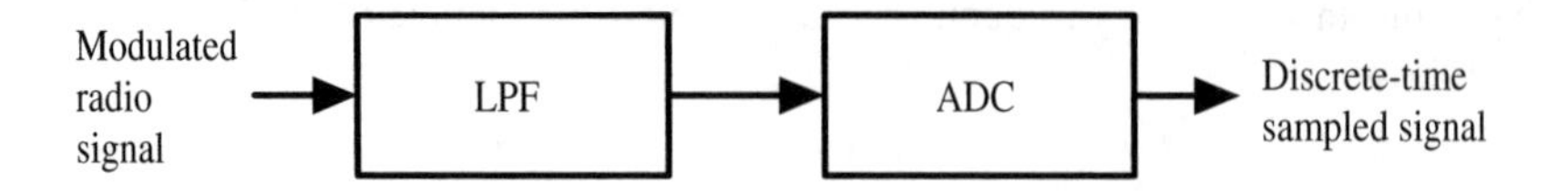

Figure 1.4 Architecture of ADC.

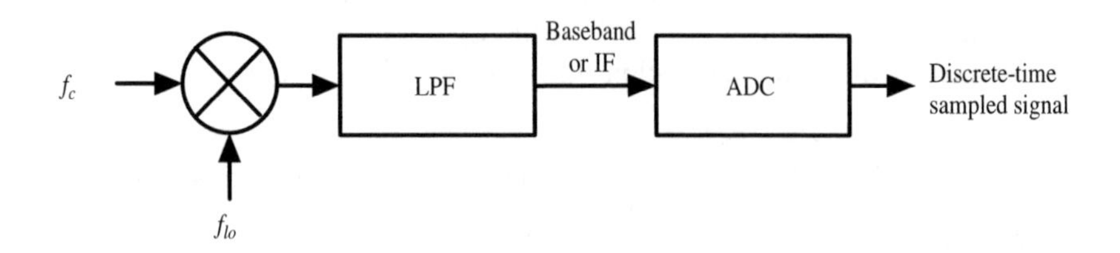

Figure 1.5 Architecture of down conversion and ADC.

When the signals of frequencies f_c and f_{lo} are mixed, their sum $(f_c + f_{lo})$ and differences are generated. The sum is not desired and eliminated by the LPF. One of the differences, that is, $f_c - f_{lo}$, is called the *primary frequency* and is desired because it falls within the baseband. The other one, that is, $-f_c + f_{lo}$, is called the *image frequency*. Note that we can dually choose $-f_c + f_{lo}$ as the primary frequency and $f_c - f_{lo}$ as the image frequency. The image frequency is introducing interference and noise from the lower side band of f_c, that literally *folds over the primary*, as explained momentarily. It is undesirable. The phenomenon is difficult to explain clearly with signals in the real domain representation, but becomes crystal clear when explained with signals in the complex domain. The mixing of two signals at frequencies f_c and f_{lo}, $\cos(2\pi f_c t)$ and $\cos(2\pi f_{lo} t)$ respectively, is defined as the product of their representation in the polar form:

$$\left(\frac{e^{j2\pi f_c t} + e^{-j2\pi f_c t}}{2}\right)\left(\frac{e^{j2\pi f_{lo} t} + e^{-j2\pi f_{lo} t}}{2}\right)$$

which is equal to:

$$\frac{e^{j2\pi(f_c+f_{lo})t} + e^{-j2\pi(f_c+f_{lo})t} + e^{j2\pi(f_c-f_{lo})t} + e^{-j2\pi(f_c-f_{lo})t}}{4}. \tag{1.4}$$

The term $e^{j2\pi(f_c+f_{lo})t} + e^{-j2\pi(f_c+f_{lo})t}$ is a signal in the complex domain whose translation in the real domain, $\cos(2\pi(f_c + f_{lo})t)$, is a signal corresponding to the sum of the two frequencies, cut off by the LPF. Without loss of generality, assume that $f_{lo} < f_c$. In the complex domain representation, the signal at the primary frequency $f_c - f_{lo}$, that is, the signal $e^{j2\pi(f_c-f_{lo})t}$, is a translation of the signal at frequency $f_{lo} + (f_c - f_{lo}) = f_c$. The signal at the image frequency $-(f_c - f_{lo}) = -f_c + f_{lo}$, that is, the signal $e^{-j2\pi(f_c-f_{lo})t}$, is a translation of the signal at frequency $f_{lo} - f_c + f_{lo} = 2f_{lo} - f_c$. Figure 1.6 illustrates the relative locations, on the frequency axis, of the different signals involved. The term $e^{j2\pi(f_c-f_{lo})t} + e^{-j2\pi(f_c-f_{lo})t}$ translated in the real domain representation is $\cos(2\pi(f_c - f_{lo})t)$, the difference of the two frequencies. The image frequency folds over the primary image.

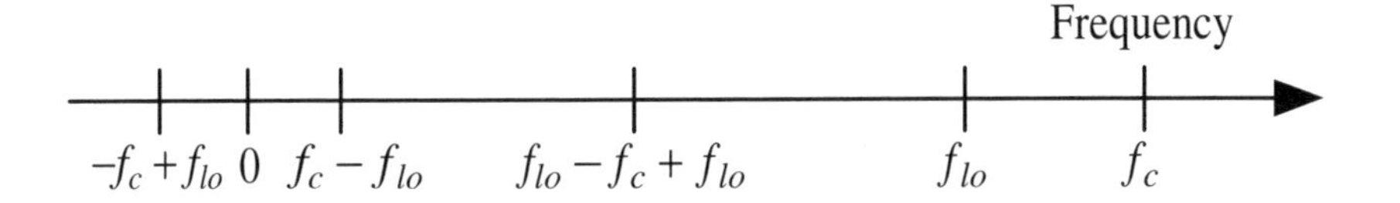

Figure 1.6 Frequencies involved in down conversion.

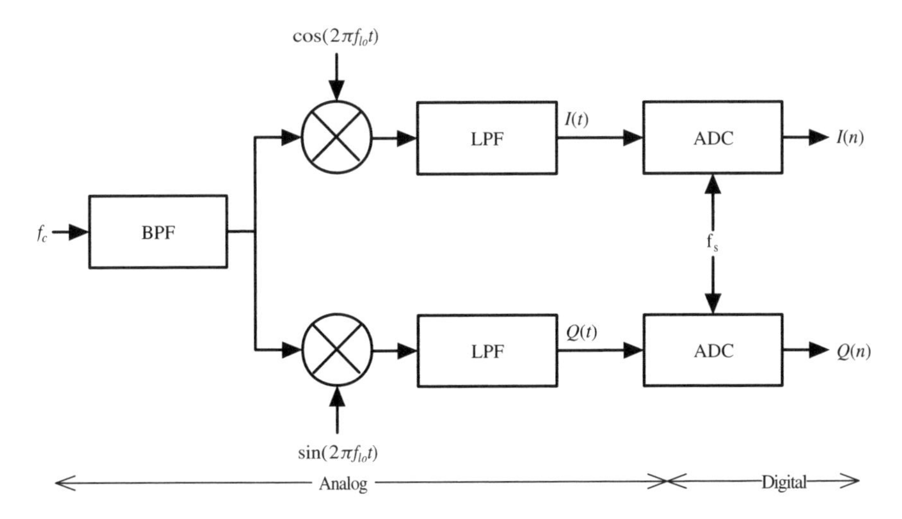

Figure 1.7 Architecture of quadrature mixing.

Image frequencies can be removed by using quadrature mixing, which is depicted in Figure 1.7. From the left side, the radio signal is first band pass filtered (BPF) . There is a low cut off frequency below which signals are strongly attenuated and a high cut off frequency above which signals are strongly attenuated. Signals between the low cut off frequency and high cut off frequency flow as is. The local oscillator consists of both a cosine signal and a sine signal. The top branch mixes the radio signal at frequency f_c with signal $\cos(2\pi f_{lo}t)$ while the bottom branch mixes it with signal $\sin(2\pi f_{lo}t)$. They are both individually low pass filtered to remove the signals at the sum frequencies. The two ADCs are synchronized by the same sampling frequency f_s. Modeled in the complex domain, the output of the top branch is:

$$I(n) = \frac{e^{j2\pi(f_c-f_{lo})n} + e^{-j2\pi(f_c-f_{lo})n}}{4} \quad (1.5)$$

while the output of the bottom branch is:

$$Q(n) = j\frac{e^{j2\pi(f_c-f_{lo})n} - e^{-j2\pi(f_c-f_{lo})n}}{4}. \quad (1.6)$$

Taking the sum,

$$I(n) - jQ(n) = \frac{e^{j2\pi(f_c-f_{lo})n}}{2}$$

cancels the image frequency in the complex domain representation.

Figure 1.8 pictures the signals that flow in quadrature mixing, assuming a baseband or intermediate signal at frequency 1 Hz, that is, $f_c - f_{lo} = 1$ Hertz, defined as $\cos(2\pi t)$. The phase evolves at a rate of 2π rad/s. In the top branch, between the LPF and ADC the signal, in-phase and continuous, is $I(t) = \cos(2\pi t)$. After the ADC, it is a discrete-time signal $I(n) = \cos(2\pi n)$. In the bottom branch, between the LPF and ADC the signal, phase shifted by 90° and continuous, is $Q(t) = \sin(2\pi t)$. After the ADC, it is a discrete-time signal $Q(n) = \sin(2\pi n)$. Note that, with this method, the quadrature of the signal is produced in an analog form, that is, the $Q(t)$, then it is digitized to form the imaginary part of the complex domain representation of a signal being processed. In other words, the quadrature is generated in hardware. An implementation of this method is described by Youngblood (2002a). Smith (2001) describes another method in which the quadrature is calculated after ADC using the Hilbert transform. In other words, the quadrature is generated in software.

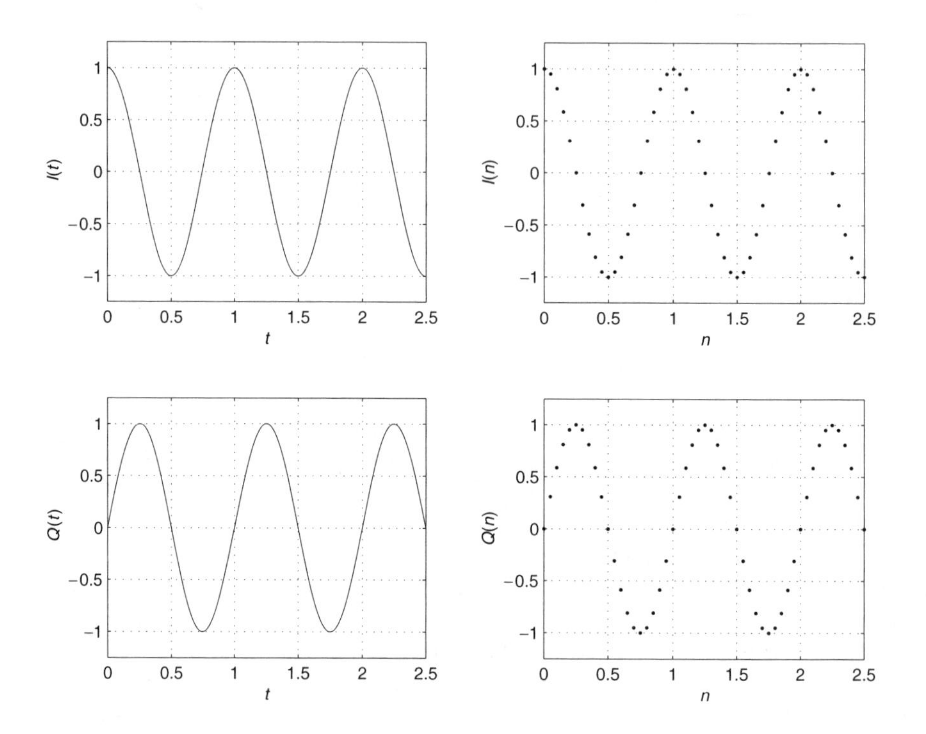

Figure 1.8 Flow of signals in quadrature mixing, with the assumption $f_c - f_{lo}$ Hertz.

1.3 Digital to analog conversion

Digital to analog conversion is an operation that converts a steam of binary values to a continuous-time signal. A digital to analog converter (DAC) converts binary values to voltages by holding the value corresponding to the voltage for the duration of a sample. Figure 1.9 pictures the architecture of a digital to analog conversion. Two signals are involved: a modulating signal and a carrier at frequency f_c. The modulating signal consists

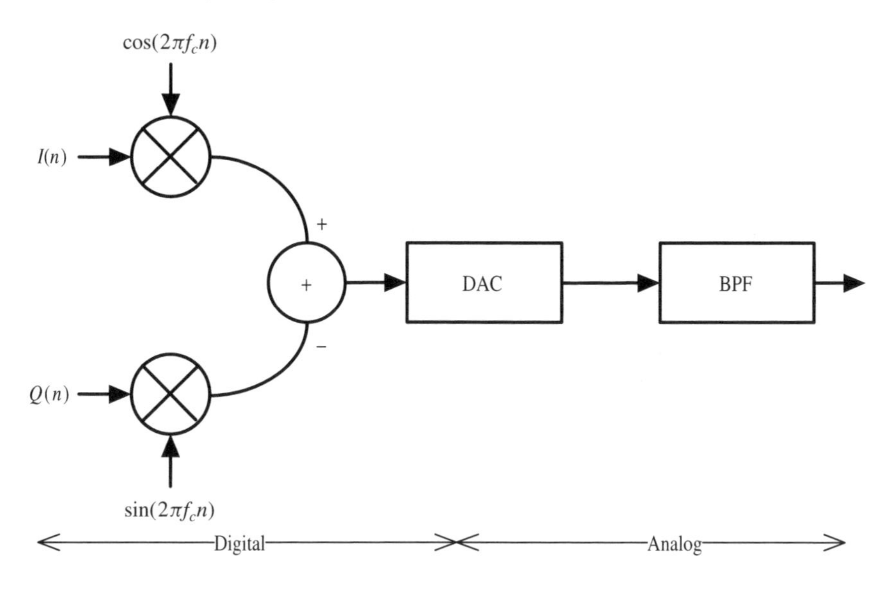

Figure 1.9 Digital to analog conversion.

of an in-phase signal $I(n)$ and its quadrature $Q(n)$. There are two digital mixers, represented by crossed circles. The top mixer computes the product $I(n)\cos(2\pi f_c n)$ while the bottom mixer does the product $Q(n)\sin(2\pi f_c n)$. The result of the first product is added to the value of the second product, with sign inverted. This sum is fed to a DAC then BPF. This method is called *direct digital synthesis* (*DDS*). With this method negative frequencies are eliminated. The DDS computes the real part of the product:

$$e^{j2\pi f_c n}(I(n) + jQ(n))$$

that is:

$$I(n)\cos(2\pi f_c n) - Q(n)\sin(2\pi f_c n)$$

the imaginary part:

$$j\left[I(n)\sin(2\pi f_c n) + Q(n)\cos(2\pi f_c n)\right]$$

does not need to be computed since it is not transmitted explicitly (it is in fact regenerated by the receiver, as discussed in Section 1.2).

1.4 Architecture of an SDR application

The good news is that given the $I(n)$ and $Q(n)$, theoretically, there is nothing that can be demodulated that cannot be demodulated in software. This section reviews the architecture of a software application, an SDR application, that does demodulation (as well as modulation), given a discrete-time signal represented in the complex domain as $I(n)$ and $Q(n)$. The application consists of a capture buffer, a playback buffer and an event handler, in which is embedded DSP.

The capture buffer is fed by the ADCs, in Figure 1.5, and stores the digital samples in the $I(n)$ and $Q(n)$ form. The capture buffer is conceptually circular and of size 2^k samples. Double buffering is used, hence the capture buffer is subdivided into two buffers of size 2^{k-1} samples. While one of the buffers is being filled by the ADCs, the other one is being processed by the SDR application.

The playback buffer is fed by the event handler. It stores the result of digital processing, eventually for digital to analog conversion if, for instance, the baseband signal consists of digitized voice. The playback buffer is also conceptually circular and of size 2^l samples. Quadruple buffering is used, hence the playback buffer is subdivided into four buffers of size 2^{l-2} samples. Among the four buffers, one is being written by the event handler while another is being played by the DAC. Playback starts only when the four buffers are filled. This mitigates the impact of processing time jittering, which arises if the SDR application shares a processor with other processes.

The SDR application is event driven. Whenever one of the capture buffers is filled, an event is generated and an event handler is activated. The event handler demodulates the samples in the buffer using DSP. The result is put in a playback buffer.

The initialization of the SDR application is detailed in Figure 1.10. There are three variables. Variable `i` is initially set to 0 and is the index of the last capture buffer that has been filled by the ADCs and is ready to be processed. The variable `j` is also initially set to 0 and is the index of the playback buffer in which the result of DSP is put when an event is being handled. Variable `first_round` is initially set to true and remains true until the four playback buffers have been filled.

```
Initialization:
   // index over capture buffer
   i = 0
   // index over playback buffer
   j = 0
   // true while in first round of playback buffer filling
   first_round = true
```

Figure 1.10 Initialization of the SDR application.

The event handler of the SDR application is described in Figure 1.11. Firstly, the samples in the current capture buffer are processed and the result is put in the current playback buffer. Then, if the value of the variable `first_round` is true and all four playback buffers are filled and ready, playback is started and variable `first_round` is unasserted. Finally, variable `i` is incremented modulo 2 and variable `j` is incremented modulo 4 for the next event handling instance.

If the sampling frequency is f_s, then the event rate is:

$$r = \frac{f_s}{2^{k-1}} \text{ events/second}$$

Each event should be processed within a delay of:

$$d = \frac{1}{r} \text{ seconds}$$

After initialization, the delay before playback is started is therefore at least $4d$ seconds.

```
Event handling:
   process capture buffer[i]
   put result in playback buffer[j]
   if first_round and j = 3 then
      start playback
      first_round = false
   i = (i + 1) mod 2
   j = (j + 1) mod 4
```

Figure 1.11 Event handler of the SDR application.

1.5 Quadrature modulation and demodulation

The goal of modulation is to represent a bit stream as a radio frequency signal. Table 1.2 reviews the modulation methods of radio systems employed for ad hoc wireless computer networks (note that WiMAX/802.16 does not strictly have an ad hoc mode, but its mesh mode has ad hoc features). Bluetooth uses Gaussian-shape frequency shift keying (GFSK) combined with frequency hopping (FH) spread spectrum (SS) transmission (Bluetooth, SIG, Inc., (2001a)). WiFi/802.11 uses two or four-level GFSK, for the data rates 1 Mbps and 2 Mbps respectively, combined with FH SS transmission (IEEE (1999a)). WiFi/802.11 also uses differential binary phase shift keying (DBPSK) and differential quadrature phase shift keying (DQPSK), for the data rates 1 Mbps and 2 Mbps respectively, combined with direct sequence (DS) SS. WiFi/802.11b uses complementary code keying (CCK) (IEEE (1999c)). WiFi/802.11a uses orthogonal frequency division multiplexing (OFDM) (IEEE (1999b)). WiMAX/802.16, with the single-carrier (SC) transmission and 25 MHz channel profile, uses quadrature phase shift keying (QPSK) (downlink or uplink) and quadrature amplitude modulation-16 states (QAM-16) (downlink only) (see IEEE et al. (2004)). WiMAX/802.16, with the OFDM transmission and 7 MHz channel profile, uses QAM-64 states (note that WiMAX/802.16 defines other transmission and modulation characteristics).

Table 1.2 Modulation schemes.

System	Bandwidth (MHz)	Modulation	Rate (Mbps)	Transmission
Bluetooth	1	GFSK	1	FH SS
802.11	1	GFSK	1 and 2	FH SS
	10	DBPSK	1	DS SS
	10	DQPSK	2	DS SS
802.11b	10	CCK	11	CCK
802.11a	16.6	OFDM	54	OFDM
802.16 SC-25	25	QPSK	40	SC
802.16 SC-25	25	QAM-16	60	SC
802.16 OFMD-7	7	QAM-64	120	OFDM

Frequency shift keying (FSK) and Phase shift keying (PSK) modulation are discussed in more detail in the sequel. SS transmission is presented in Section 1.6.

The frequency shifts according to the binary values are tabulated for WiFi/802.11 GFSK in Tables 1.3 and 1.4.

Table 1.3 Two-level GFSK modulation.

Symbol	Frequency shift (kHz)
0	−160
1	+160

Table 1.4 Four-level GFSK modulation.

Symbol	Frequency shift (kHz)
00	−216
01	−72
10	+216
11	+72

The phase shifts according to the binary values are tabulated for WiFi/802.11 DBPSK and DQPSK in Tables 1.5 and 1.6. For DBPSK, the phase of the carrier is shifted by 0°, for binary value 0, or 180°, for binary value 1.

Table 1.5 DBPSK modulation.

Symbol	Phase shift (°)
0	none
1	180

Generation of a discrete-time sampled FSK signal, for playback, can be done as follows. Let f_o be the frequency of a symbol to represent and T_b be the duration of a symbol. The modulating signal (used in Figure 1.9) for the period T_b is:

$$e^{j2\pi f_o n}.$$

Note that $I(n) = \cos(2\pi f_o n)$ and $Q(n) = \sin(2\pi f_o n)$. This modulating signal shifts the instantaneous frequency of the carrier f_c proportionally to the value of f_o:

$$e^{j2\pi f_c n} e^{j2\pi f_o n} = e^{j2\pi (f_c + f_o) n}.$$

As mentioned in Section 1.3, only the real part needs to be computed and transmitted.

Table 1.6 DQPSK modulation.

Symbol	Phase shift (°)
00	none
01	90
10	−90
11	180

Production of a PSK signal can be done as follows. Let ϕ be the phase of a symbol and T_b its duration. The modulating signal (usable in Figure 1.9) for the duration T_b is:

$$e^{j\phi}.$$

This modulating signal shifts the instantaneous phase of the carrier f_c proportionally to the value of ϕ:

$$e^{j2\pi f_c n} e^{j\phi} = e^{j(2\pi f_c n + \phi)}.$$

This is termed *exponential modulation*. The algorithm of a software exponential modulator is given in Figure 1.12. The bits being transmitted are given in an output buffer. The samples of the modulation, that is, the modulated carrier, are stored by the modulator in the playback buffer. Let r be the bit rate, the number of samples per bit corresponds to:

$$s = \frac{f_s}{r}.$$

The length of the playback buffer corresponds to the length of the output buffer times s. A for loop, that has as many instances as the length of the playback buffer, generates the samples. The sample at position i encodes the bit from output buffer at index $\lfloor \frac{i}{s} \rfloor$. The index of the first bit is 0 in both the output buffer and playback buffer. The function `fo()` returns the value of the frequency shift as a function of the value of the symbol. The time of the first sample is 0 and incremented by the value $\frac{1}{f_s}$ from sample to sample. The symbol `fc` represents the frequency of the carrier. The term `exp(j*x)` corresponds to the expression e^{jx}.

```
for i = 0 to length of playback buffer, minus one
   // determine value of symbol being transmitted
   symbol = output buffer[floor(i / s)]
   // determine the frequency shift
   shift = fo(symbol)
   // determine time
   n = i * 1/fs
   // Generate sample at position "i"
   playback buffer[i] =
      real part of exp(j*2*pi*(fc+shift)*n)
```

Figure 1.12 Algorithm of a software exponential modulator.

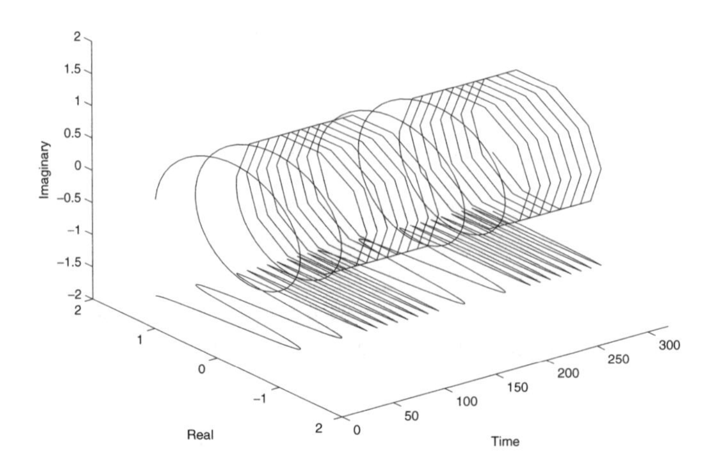

Figure 1.13 Exponential modulation of bits `1 0 1 0`.

An example modulation is plotted as a three-dimensional helix in Figure 1.13. The picture also contains a projection of the helix on a real-time plane, which is the signal effectively transmitted. For this example, the frequency of the carrier is 5 Hz, the frequencies of the modulating signal are $+3$ Hertz for bit 0 and -3 for bit 1, the rate is 1 bps and the sampling frequency is 80 samples/s. The picture corresponds to the modulation of bits `1 0 1 0`.

The pseudo code of a FSK demodulator is given in Figure 1.14. The algorithm consists of a for loop that has as many instances as there are samples in the capture buffer, parsed into the `Quadrature` buffer and `InPhase` buffer. Given an instance of the loop, the frequency of the previous sample is available in `prev_f` and phase of previous sample in `prev_p`. While a carrier is being detected and frequency of the carrier is invariant, the variable `count` counts the number of samples demodulated. The variable `count` is initialized to 0. When the variable `count` reaches the value s, a complete symbol has been received and its value is determined according to the frequency shift of the carrier. Then, the variable `count` is reset.

1.6 Spread spectrum

There are two forms of SS transmission in use: direct sequence (DS) and frequency hopping (FH).

With DS SS, before transmission the *exclusive or* of each data bit and a pseudorandom binary sequence is taken. The result is used to shift the carrier. DBPSK or DQPSK may be used for that purpose. The length of the resulting bit stream is the length of the original bit stream multiplied by a factor corresponding to the length of the pseudorandom binary sequence. This applies as well to the data rate and bandwidth occupied by the radio signal: the signal is spread over a larger bandwidth.

```
// Initialization
prev_f = 0
prev_p = 0
count = 0
// Demodulation loop
for i = 0 to length of capture buffer, minus one
   // Compute the instantaneous phase
   phase = atangent Quadrature(i) / InPhase(i)
   // Compute the instantaneous frequency
   freq = fs * ((phase - prev_p) / (2 * pi) )
   // Detection of carrier
   if freq == (fc + fo(1)) or freq == (fc + fo(2))
      if (count==0)
         // no bit is being demodulated, start demodulation
         count = 1
      else if freq==prev_f
         // continue demodulation while frequency is constant
         count = count + 1
      else
         count = 0
   // determine if a full bit has been demodulated
   if count==s
      if freq==fc+fo(1)
         symbol = 0
      else
         symbol = 1
      count = 0
   // save phase and frequency for the next loop instance
   prev_p = phase
   prev_f = freq
end
```

Figure 1.14 Algorithm of a software demodulator.

In data communications, the pseudorandom binary sequence is often of fixed length, the same from one data bit to another and shared by all the transmitters and receivers communicating using the same channel.

The 11-bit Baker pseudorandom binary sequence is very popular:

10110111000

Application of the Baker sequence to a sequence of data bits is illustrated in Figure 1.15.

The popularity of the Baker sequence is due to its self-synchronization ability. That is to say, bit frontiers can be determined. Indeed, as the bits that were transmitted are received they pass through an 11-bit window. The autocorrelation of the bits in the 11-bit window and bits of the 11-bit Baker sequence is calculated. The autocorrelation is the value of a counter, initialized to 0. From left to right and i from 1–11, if the bit

Data bits	0	1	0	0	0
Transmitted sequence	10110111000	01001000111	10110111000	10110111000	10110111000

Figure 1.15 Application of the Baker sequence.

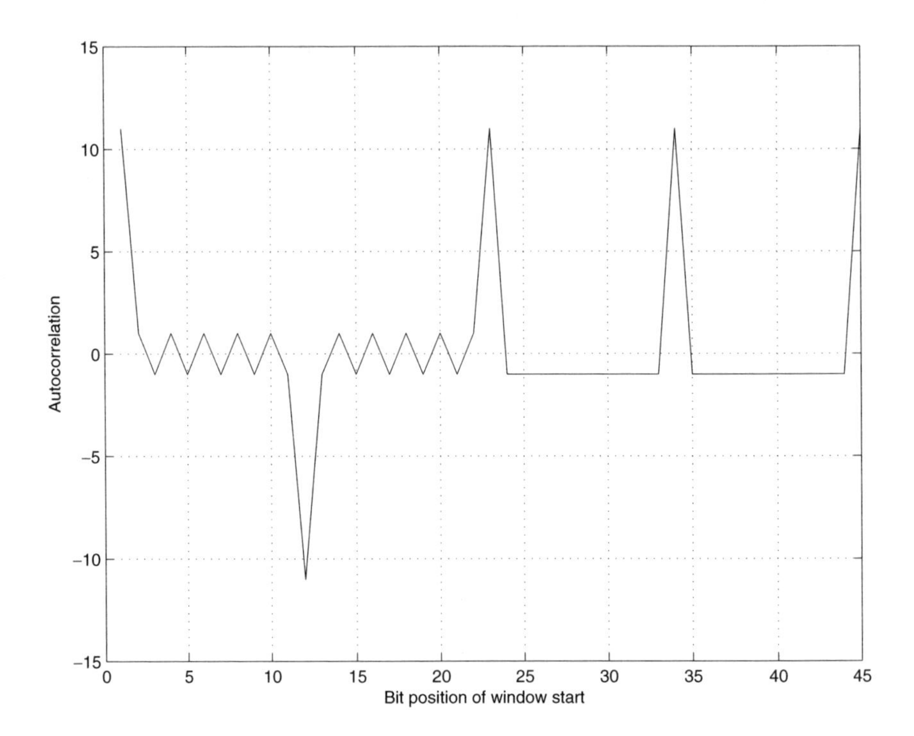

Figure 1.16 Autocorrelation with Baker sequence.

at position i in the window matches the bit at the same position in the Baker sequence, then the counter is incremented. Otherwise it is decremented. In the absence of errors, the count varies from minus 11 (lowest autocorrelation) to 11 (maximum autocorrelation). Maximum autocorrelation marks the starting positions of bits with value 0. Figure 1.16 plots the autocorrelation of the bit sequence of Figure 1.15 when the 11-bit window slides from left to right.

The 802.11 radio system uses DS SS (IEEE (1999a)). The carrier is modulated using DBPSK and DQPSK at respectively 1 Mbps and 2 Mbps. The 11-chip Baker pseudorandom binary sequence is used.

DS SS offers resilience to jamming and higher capacity, given a power of signal to power of noise ratio. An analysis by Costas (1959) shows that it is more difficult to jam a broad channel because the required power is directly proportional to the bandwidth.

Shannon (1949) has developed a model giving the capacity of a channel in the presence of noise:

$$C = W \log_2 \left(1 + \frac{S}{N}\right)$$

where C is the capacity in bps, W the bandwidth in Hertz, S the power of the signal in Watts and N the power of the noise in Watts. If S/N cannot be changed, then higher capacity can be achieved if the signal is using a broader bandwidth as in DS SS.

With FH SS, a radio frequency band is divided into segments of the same bandwidth. Each segment is characterized by its centre frequency. The transmitter jumps from one frequency to another according to a predetermined hopping pattern. The transmitter sits on each frequency for a duration called the *dwell time*. The receiver(s) synchronizes with the transmitter and jumps from one frequency to another in an identical manner. FSK is used to send data during dwell time.

The Bluetooth radio system uses FH SS (Bluetooth, SIG, Inc., (2001a)). In North America, the number of frequencies is 79 and they are defined as follows, for $i = 0 \ldots 78$:

$$f_i = 2402 + i \text{ Mega Hertz.} \tag{1.7}$$

Each segment has a bandwidth of 1 MHz. The carrier is modulated using GFSK. The hopping rate is 1600 hops/s. This is termed *slow FH* because the data rate is higher than the hopping rate. A short frame can be sent during one dwell time.

The WiFi/802.11 radio system uses FH SS as well (IEEE (1999a)). Three sets of hopping patterns are used, hence defining three logical channels. 802.11 uses 79 frequencies defined as in Equation 1.7. The carriers are also modulated using GFSK. At 2.5 hops/s, 802.11 is also slow FH.

FH offers resilience to narrow band interference. If a source of interference is limited to one frequency, the frequency can be retracted from the hopping pattern.

1.7 Antenna

Antennas are to wireless transmission systems what speakers are to sound systems. A sound system is not better than its speakers. A wireless transmission system is not better than its antennas. Antennas are devices that radiate and pick up electromagnetic power from free space. There are directional antennas and omni-directional antennas. *Directional antennas* radiate a focused electromagnetic power beam and pick up a focused source of energy. *Omni-directional antennas* spread and pick up electromagnetic power in all directions. This section discusses in a more formal manner the directivity and the maximum separation distance between two antennas.

The directivity of an antenna is captured mathematically using the notion of decibel (dB) and dB isotropically (dBi). The dBs translate the magnitude of a ratio of the quantities x and y as:

$$z \; dB = 10 \log_{10} \frac{x}{y}.$$

Directivity means that an antenna radiates more power in certain directions and less in others. An antenna that would radiate power uniformly in all directions is called an *isotropic antenna*. Such an antenna cannot be physically realized, but serves as a reference to characterize designs of real antennas.

Given an antenna and the power P_1 it radiates in the direction of strongest signal, a value in dBi translates the magnitude of the ratio of this power over the power P_2 radiated in a direction by an isotropic antenna in the same conditions. This ratio,

$$g\ dBi = 10 \log_{10} \frac{P_1}{P_2}$$

is called the *gain* of the antenna.

The *radiation pattern* defines the spatial distribution of the power radiated by an antenna. A radiation pattern can be graphically represented by a three-dimensional surface of equal power around the antenna. Figure 1.17 represents the radiation pattern of an omni-directional antenna. It is an antenna of type ground plane with a gain of 1.9 dBi. Figure 1.18 shows the radiation pattern of a directional antenna. It is an antenna of type Yagi with a gain of 12 dBi.

For terrestrial microwave frequencies (1 GHz–40 GHz), antennas shaped as parabolic dishes are commonly used. They are used for fixed, focused, line-of-sight transmissions.

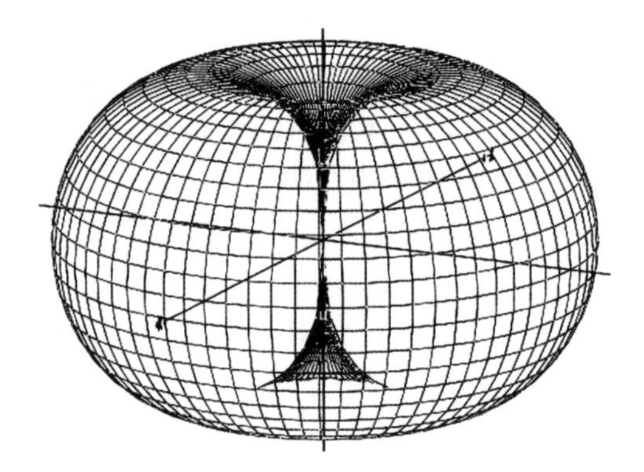

Figure 1.17 Radiation pattern of an omni-directional antenna.

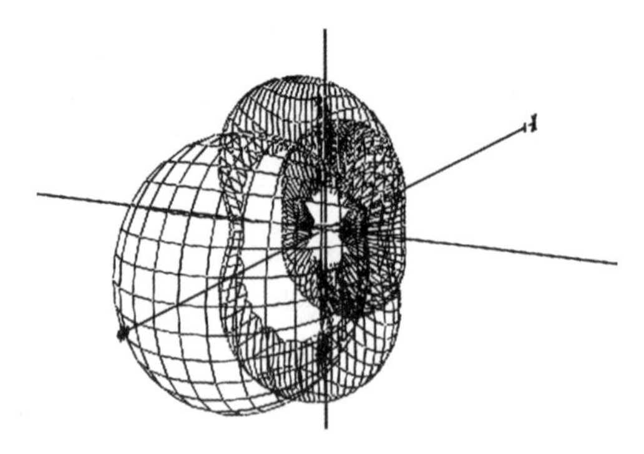

Figure 1.18 Radiation pattern of a directional antenna.

Line-of-sight transmission is achievable whenever two antennas can be connected by an imaginary non-obstructed straight line.

Let h be the height, in meters, of two antennas. The maximum distance between the antennas with line-of-sight transmission, in kilometers, is computed by the following formula:

$$d = 7.14\sqrt{Kh}.$$

K is an adjustment factor modeling refraction of waves with the curvature of the earth (hence waves propagate farther), typically $K = 4/3$. Maximum distance between antennas at heights up to 300 m is plotted in Figure 1.19.

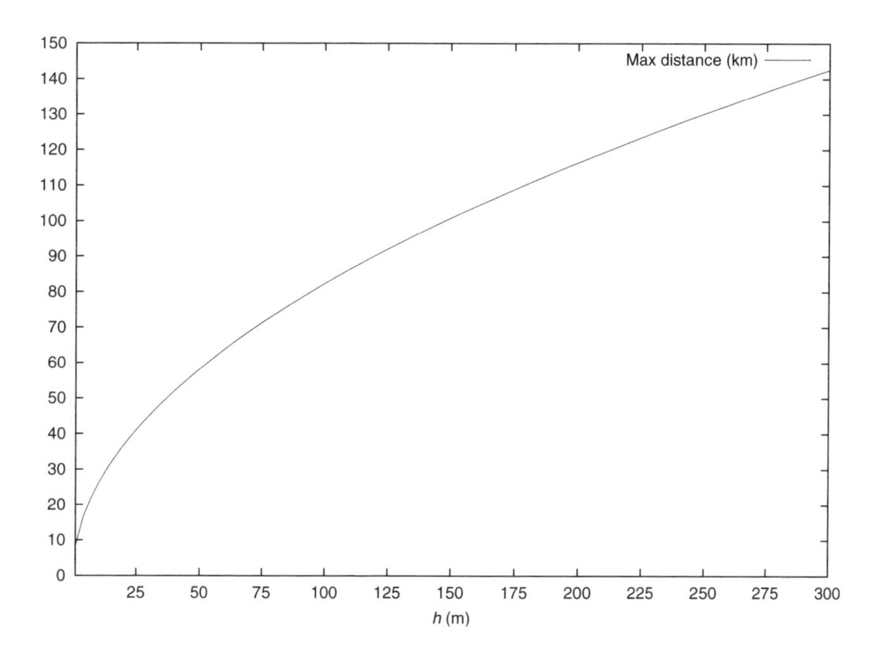

Figure 1.19 Maximum distance between antennas.

1.8 Propagation

Successful propagation in free space of a radio frequency signal from a transmitter to a receiver is subject to a number of parameters, namely, the:

1. output power of the transmitter (P_T),
2. attenuation of the cable connecting the transmitter and transmitting antenna (C_T),
3. transmitting antenna gain (A_T),
4. free space loss (L),
5. receiving antenna gain (A_R),

6. attenuation of the cable connecting the receiving antenna and receiver (C_R) and
7. receiver sensitivity (S_R).

Antenna gain is discussed in Section 1.7. The remaining six parameters are discussed in the following text.

In the gigahertz range, the output power of transmitters is often weak, in the order of milli Watts (mW) and conveniently expressed in decibel milli Watts (dBm). Given a power level P in mW, the corresponding value in dBm is defined as:

$$Q\ dBm = 10 \log_{10} PmW$$

Tables 1.7 and 1.8 gives the power level requirement of various Bluetooth, WiFi/802.11 and WiFi/802.16 radios. The data was extracted from product standards. For instance, the output power of an 802.11 radio is 100 mW or 20 $dBm = 10 \log_{10} 100\ mW$.

The sensitivity of a receiver is also expressed in dBm as a function of two performance parameters: data rate and error rate. The rate of error is given as a Bit Error Rate (BER), Frame Error Rate (FER) or Packet Error Rate (PER). Frames are assumed to be of length 1024 bytes while packets are assumed to be of length 1000 bytes. For instance, according to Table 1.8 a 802.11 radio should achieve a data rate of 1 Mbps with an expected FER of 3% if the strength of the signal at the receiver is −80 dBm or higher. If the hardware parameters are fixed, then the table also indicates that the data rate can be increased solely if the strength of the signal at the receiver can be increased. Note that the specifications of products from manufacturers often exhibit better numbers than the requirements. It often difficult to compare products from different manufacturers because the performance figures are often given under different constrains. that is, different BER, FER or PER. There is no

Table 1.7 Transmission performance parameters of 802.11 and Bluetooth radios.

Radio	Frequency (GHz)	Power (dBM)
Bluetooth Class 1	2.4–2.4835	20
Bluetooth Class 2		4
Bluetooth Class 3		0
801.11	2.4–2.4835	20
801.11b	2.4–2.4835	20
801.11a	5.15–5.35	16–29
802.16 SC-25 QPSK	10–66	≥ 15
802.16 SC-25 QPSK	10–66	≥ 15
802.16 SC-25 QAM-16	10–66	≥ 15
802.16 SC-25 QAM-16	10–66	≥ 15
802.16 OFMD-7	2–11	15–23

Table 1.8 Reception performance parameters of 802.11 and Bluetooth radios.

Radio	Rate (Mbps)	Error	Sensitivity (dBm)
Bluetooth Class 1	1	10^{-3} (BER)	−70
Bluetooth Class 2	1	10^{-3} (BER)	−70
Bluetooth Class 3	1	10^{-3} (BER)	−70
801.11	1	3 % (FER)	−80
	2	3 % (FER)	−75
801.11b	11	8 % (FER)	−83
801.11a	54	10 % (PER)	−65
802.16 SC-25 QPSK	40	10^{-3} (BER)	−80
802.16 SC-25 QPSK	40	10^{-6} (BER)	−76
802.16 SC-25 QAM-16	60	10^{-3} (BER)	−73
802.16 SC-25 QAM-16	60	10^{-6} (BER)	−67
802.16 OFMD-7	120	10^{-6} (BER)	−78 to −70

standard reference. Independent evaluations, when available, might be the best source for performance comparisons.

The strength of a signal, that is, its attenuation, decreases with distance. In cables or free space, what is important is the relative reduction of strength. It is normally expressed as a number of decibels (dB). A value in dB expressing a ratio of two powers P_1 and P_2 is given by the following formula:

$$10 \log_{10} \frac{P_2}{P_1}.$$

If P_1 expresses the power of a transmitted signal and P_2 the power of the signal at the receiver, which is normally lower, then a loss corresponds to a negative value in dB. Through an amplifier, P_1 may express the power of an input signal and P_2 the power of the signal at the output, which is normally higher, then the gain is expressed by a positive value in decibels.

The total gain or loss of a transmission system consisting of a sequence of interconnected transmission media and amplifiers can be computed by summing up the individual gain or loss of every element of the sequence. Equivalently, it could be expressed by a ratio obtained by multiplying together the gain or loss ratio of every element. Doing calculations in dB is much more convenient because additions are more convenient to calculate than multiplication.

If the frequency of the radio signal is fixed, then the cable attenuation, in dB, is a linear function of distance. Attenuation, per unit of length, can be extracted from manufacturer specifications (Table 1.9). For instance, a radio signal at frequency 2.5 GHz is subject to an attenuation of 4.4 dB per 100 feet over the LMR 600 cable.

Table 1.9 Cable attenuation per 100 feet.

Type	Frequency (GHz)	Attenuation (dB)
Belden 9913	0.4	2.6
	2.5	7.3
	4	9.5
LMR 600	0.4	1.6
	2.5	4.4
	4	5.8
	5	6.6

On the other hand, free space loss, in dB, is proportional to the square of distance. If d is the distance and λ the wavelength, both expressed in the same units, the loss is given by the following formula:

$$L = 10\log_{10}\left(\frac{4\pi d}{\lambda}\right)^2 \; dB. \tag{1.8}$$

Other factors are not taken into account. For instance, rainfall increases attenuation, in particular at frequency 10 GHz and above.

Figure 1.20 compares attenuation presented to a 1 MHz signal, over a wireless medium, with a Category 5 cable, where loss in dB is proportional to a factor times the distance. Loss over a wireless medium is significantly higher.

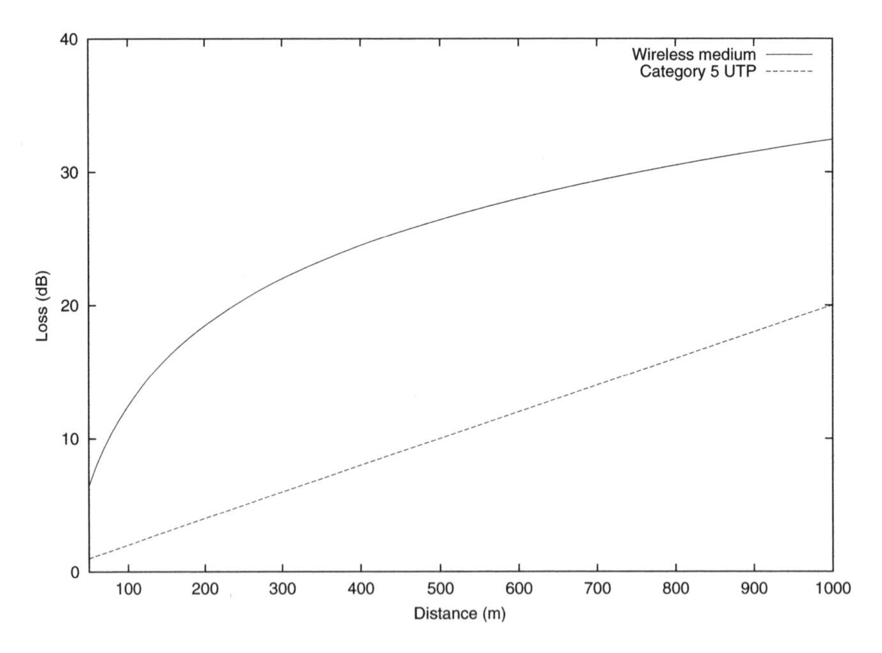

Figure 1.20 Comparison of attenuation of a 1 MHz signal over a wireless medium and a Category 5 cable.

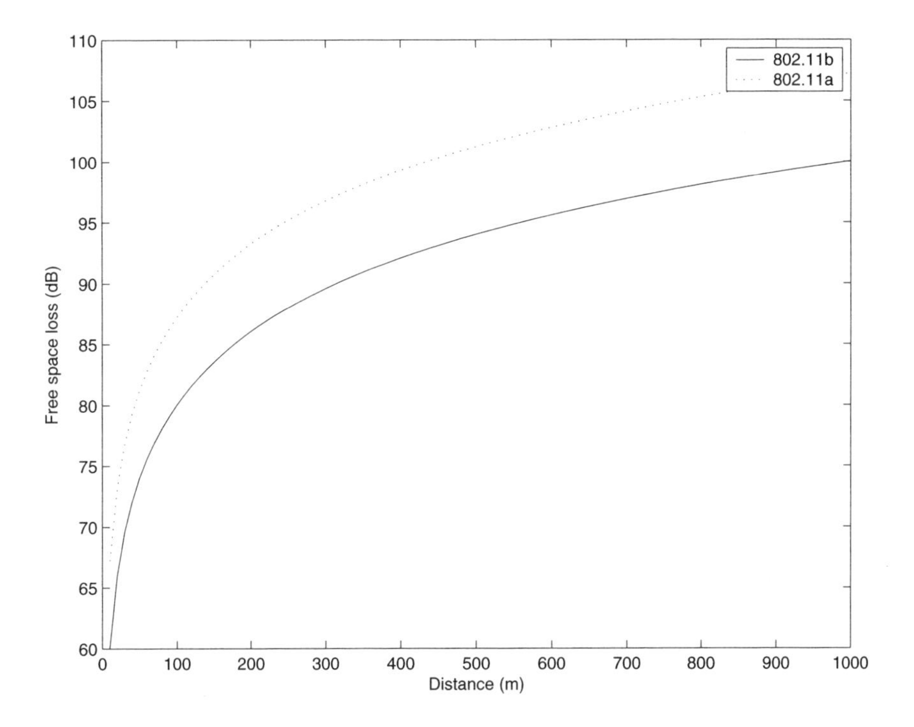

Figure 1.21 Comparison of attenuation of 802.11a and 802.11b.

Figure 1.21 compares free space loss for 802.11a (5.5 GHz) and 802.11b (2.4 GHz) as a function of distance. At equal distance, 802.11a has an additional cost of 7.2 dB in free space loss.

One can observe that, in contrast to cable medium, free space loss becomes quickly high and stays substantially higher. What is the consequence of that? Higher loss means that the signal at the receiver is weaker relative to noise. It means that the probability of bit errors is higher. For the sake of comparison, Table 1.10 gives BERs typically obtained with wireless, copper and fiber media. The relatively higher BERs over a wireless medium cause problems when a protocol such as Transmission ControlProtocol (TCP), devised for wired networks, is used for wireless communications. This has to with the way TCP handles network congestion. TCP interprets delayed acknowledgements of transmitted data segments as situations of congestion. In such cases, TCP decreases its throughput to relieve the

Table 1.10 Typical BERs as a function of the medium type.

Medium	BER
Wireless	10^{-6}–10^{-3}
Copper	10^{-7}–10^{-6}
Fiber	10^{-14}–10^{-12}

network by increasing the value of its retransmission timer, which means that it waits longer before retransmitting unacknowledged segments of data. On a wireless medium, however, absence of acknowledgements is mainly because of the high BER. Hence, reduction of this throughput is useless since it does not improve the BER and it delays delivery of data segments.

The performance of transmitting system (i.e. a transmitter, a cable and an antenna) is termed the *effective isotropic radiated power* (*EIRP*) and is defined using the following equation:

$$EIRP = P_T - C_T + A_T.$$

Assuming line of sight between a transmitting antenna and a receiving antenna and if directional, orientation towards each other, the maximum free space loss, given a data rate and a BER can be calculated as:

$$L = EIRP + A_R - C_R - S_R.$$

The maximum coverage of a system can be calculated by resolving the variable d which appears in Equation 1.8.

The performance (i.e. the data rate and BER) of a radio system can be improved either by:

- increasing the power of the transmitter,
- lowering the sensitivity of the receiver,
- increasing the gain of the transmitting antenna or receiving antenna,
- shortening lengths of cables or
- shortening the distance.

This model of performance is consistent with models used by manufacturers of wireless interfaces, for instance Breeze, Wireless Communications Ltd. (1999). Other factors do affect the performance of radio systems and limit their coverage. Some of these are multipath reception, obstruction of line of sight, rain, antenna motion due to wind, noise, interference and hardware or software faults. McLarnon (1997) reviews some of them in more detail in the context of microwave communications.

1.9 Ultrawideband

A transmission system is said to be an ultrawideband (UWB) if the bandwidth of the signal it generates is much larger than the center frequency of the signal. Since, according to Shannon's equation, the capacity of a channel is proportional to its bandwidth, UWB supports relatively much larger data rates. It is targeted for low power (1 mW or less) and short distance operation (i.e. few meters).

Parameters of a typical UWB system are given in Table 1.11 (Stroh (2003)). In UWB, the carrier consists of a stream of very short pulses, from 10–1000 ps each. By nature, pulsative signals occupy a very large amount of bandwidth. Neither up or down frequency conversion of signals is required.

Table 1.11 Parameters of an UWB system.

Bandwidth	500 MHz
Frequency range	3.1 GHz to 10.6 GHz
Data rate	100 Mbps to 500 Mbps
Range	10 m
Transmission power	1 mW

Table 1.12 Shape of modulated UWB pulses.

Modulation	1	0
Amplitude	Full	Half
Bipolar	Positive	Inverted
Position	Non delayed	Delayed

Modulation is done by changing the amplitude, direction or spacing of the pulses. Possible UWB modulation schemes are described in Table 1.12 (Leeper (2002)). An example is pictured in Figure 1.22. Binary values are represented in part (a). Part (b) illustrates pulse amplitude modulation. A full amplitude pulse represents the binary value 1 while a half amplitude pulse represents the binary value 0. With pulse bipolar modulation (c), a positive

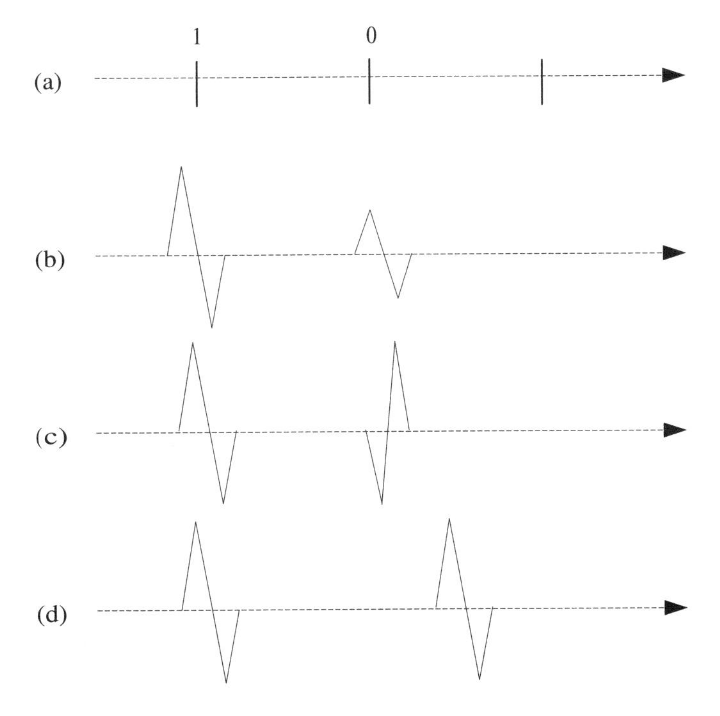

Figure 1.22 UWB modulation.

rising pulse represents the binary value 1. The binary value 0 is coded as an inverted falling pulse. Pulse position modulation (d) encodes the binary value 1 as a non-delayed pulse, whereas the pulse for the binary value 0 is delayed in time.

In few words, UWB is a very short range high data rate wireless data communications system. The signal being very large bandwidth, UWB has a natural resilience to noise. On the other hand, UWB may create interference to other systems.

1.10 Energy management

Energy management is an issue in ad hoc networks because devices are battery powered with limited capacity. Energy refers to the available capacity of a device for doing work, such as computing, listening, receiving, sleeping or transmitting. The amount of work is measured in Joules. The available capacity for doing work, the energy, is also measured in joules. The rate at which work is done is measured in Joules per second, or Watts.

The energy consumed by a wireless interface in the idle, receive, sleep and transmission modes has been studied by Feeney and Nilsson (2001). Table 1.13 lists representative energy consumption values. There is an important potential of saving energy when a device is put in sleep mode. This is an aspect that is studied in more detail in Section 3.4.4.

Table 1.13 Energy consumption.

State	Consumption (mW)
Idle	890
Receive	1020
Transmit	1400
Sleep	70

1.11 Exercises

1. Explain the math between the following phenomenon. In North America, the audio of the TV channel 13 is transmitted at frequency 215.75 MHz. A receiver that has the capability to handle signals within 216–240 MHz and an IF of 10.8 MHz, can actually receive the audio of the channel 13 at frequency 237.35 MHz.

2. Demonstrate that Equations 1.5 and 1.6 hold.

3. Given a sampling frequency f_s equal to 44,100 samples/s and a capture buffer of size 2^{12} samples, calculate the event rate, event processing delay and delay before start of playback.

4. Given a radio system with a maximum free space loss L operating at wavelength λ, develop an expression representing the maximum separation distance d (in meters) between a transmitter and receiver.

5. Compute the maximum coverage for each type of radio listed in Tables 1.7 and 1.8.

6. Given a distance d in Km, demonstrate that at 2.4 GHz, the expression of free space loss can be simplified as $100 + 20\log_{10} d$.

2

MEDIUM ACCESS CONTROL

> In the final analysis the vitality of mathematics arises from the fact that its concepts and results, for all their abstractness, originate, as we shall see, in the actual world...
> Aleksandrov (1963)

In his beautiful essay "A General View of Mathematics," Aleksandrov (1963, Vol. 1, page 3) argues that abstraction, proof and application are inseparable features that constitute the essential nature of mathematics. There is a platonic unifying cycle of abstract mathematical results with their proofs that make them convincing and lead to applications that forge even more new abstractions. Although this chapter is about medium access control (MAC) in ad hoc networks, it should not surprise the reader that it will follow a detour that will include probabilistic abstractions, queuing models and MAC models leading all the way from abstraction to applications and back.

Broadcast channels form the basis of communication in all wireless systems. In contrast to point-to-point systems, broadcast channels are shared by multiple users. This means that a packet transmitted by a single host will be received by all hosts participating in this channel. Therefore a control mechanism is required to insure noninterfering accesses. Depending on the protocol being used, access decisions may be made either in a centralized or a distributed manner. For example, in a centralized system a distinguished node (e.g. base station) becomes responsible for allocating channels and as such will also be responsible for quality of service considerations.

This is in contrast with ad hoc and sensor networks, where wireless media are shared with distributed access. In such systems, contention resolution protocols are required for scheduling hosts. For example, in neighbor aware contention, either deterministic or randomized priority may be used in order to elect one among multiple winners. There is a variety of techniques being used for providing power management, guaranteeing fair scheduling and quality of service support.

Section 2.1 outlines the fundamentals of probability and statistics, which are useful for the analysis of MAC. This is a useful section with particular emphasis on examples relating to ad hoc network analysis and performance. Section 2.2 discusses traffic modeling and

Principles of Ad hoc Networking Michel Barbeau and Evangelos Kranakis

queuing characteristics. Sections 2.3, 2.4 and 2.5 look at the fundamental issues in multiple access. However, the latter section is specific to the IEEE 802.11 protocol standard. The protocols addressed range from uncoordinated access to contention-based, carrier sense and collision avoidance. Finally, the chapter concludes with Section 2.6, which presents medium access protocols that are being used in ad hoc networks.

2.1 Fundamentals of probability and statistics

This section summarizes various concepts from probability theory that will be essential in understanding the forthcoming material.

2.1.1 General concepts

Let X be a random variable (also abbreviated r.v.) defined over a (typically finite) probability space. The probability mass function is defined by

$$p_X(a) = Pr[X = a].$$

This is also called the *probability density* of the r.v. X.

Example 2.1.1 *Let X denote the r.v. that is the product of two fair dice.*

$$\begin{aligned}
&Pr[X = 1] = Pr[(1, 1)] = \tfrac{1}{36}\\
&Pr[X = 2] = Pr[(1, 2), (2, 1)] = \tfrac{2}{36}\\
&Pr[X = 3] = Pr[(1, 3), (3, 1)] = \tfrac{2}{36}\\
&Pr[X = 4] = Pr[(1, 4), (4, 1), (2, 2)] = \tfrac{3}{36}\\
&Pr[X = 5] = Pr[(1, 5), (5, 1)] = \tfrac{2}{36}\\
&Pr[X = 6] = Pr[(1, 6), (6, 1), (2, 3), (3, 2)] = \tfrac{4}{36}\\
&Pr[X = 7] = Pr[\emptyset] = 0\\
&Pr[X = 8] = Pr[(2, 4), (4, 2)] = \tfrac{2}{36}\\
&Pr[X = 9] = Pr[(3, 3)] = \tfrac{1}{36}\\
&Pr[X = 10] = Pr[(2, 5), (5, 2)] = \tfrac{2}{36}
\end{aligned}$$

Random variables are described by their (cumulative) distribution function:

$$\begin{aligned}
F_X(a) &= \sum_{x \le a} p_X(x)\\
&= \sum_{x \le a} Pr[X = a]\\
&= Pr[X \le a].
\end{aligned}$$

Example 2.1.2 *Consider the r.v. X given in Example 2.1.1. It is easy to calculate the probability that the outcome of the two dice has product at most* 3*, since* $Pr[X \le 3] = \frac{1}{36} + \frac{2}{36} + \frac{2}{36} = \frac{5}{36}$.

If the random variable is continuous, then the derivative of $F_X(a)$ is also called *density*.

2.1.1.1 Expectation and variance

The *expectation* of a discrete r.v. X with non-negative integer values is defined by

$$E[X] = \sum_k k \cdot Pr[X = k].$$

For the random variables $X_1, X_2, \ldots, X_n$, the following fundamental identity holds:

$$E\left[\sum_{i=1}^{n} X_i\right] = \sum_{i=1}^{n} E[X_i].$$

Example 2.1.3 *A counter counts by adding tally marks on a sheet of paper for each object counted. Tallying is a r.v., in the sense that the tally is placed with probability p and is missed with probability* $1 - p$. *When finished, the number of tallies is equal to the number of objects. Let X be the number of tallies placed on the paper. Let N be the actual number of objects. Let*

$$X_i = \begin{cases} 1 & \text{if a tally is placed for the } i\text{th object} \\ 0 & \text{otherwise} \end{cases}.$$

The expected number of tallies on the paper is

$$E[X] = E\left[\sum_{i=1}^{N} X_i\right] = \sum_{i=1}^{N} E[X_i] = \sum_{i=1}^{N} Pr[X_i = 1] = Np.$$

A random process $\mathbf{X}(t) = \{X_n(t) : n \geq 0\}$ is a collection of random variables possibly dependent on t (the variable n is sometime interpreted as time). If $X_1, X_2, \ldots \geq 0$ are independent and identically distributed r.v.s with mean $E[X] < \infty$ and stopping time N, then

$$E\left[\sum_{i=1}^{N} X_i\right] = E[N] \cdot E[X]. \tag{2.1}$$

Equation 2.1 is one of the most useful identities for computing the *stopping time* and is also known as *Wald's identity*. A complete proof is outlined in Exercise 5 of Chapter 3.

Example 2.1.4 *A gambler is betting to win or lose* 1 *dollar depending on the outcome of a biased coin which is heads (resp., tails) with probability p (resp.,* $1 - p$*), where* $p > 1/2$. *Let N be the waiting time until the gambler has won k dollars, in which case he stops. What is the expected value of the waiting time? Let X be the r.v., which assumes the values* $+1, -1$ *depending on whether the gambler wins or loses, respectively. Clearly,* $E[X] = (+1) \cdot p + (-1) \cdot (1 - p) = 2p - 1$. *If* $X_1, X_2, \ldots$ *are the independent and identically distributed (equal to X) random variables representing the outcomes at times* $1, 2, \ldots$ *then the capital at time N is* $\sum_{i=1}^{N} X_i$. *Moreover, when he it stops* $\sum_{i=1}^{N} X_i = k$. *So applying Equation 2.1,* $E[N] = k/(1 - 2p)$.

The *covariance* of X, Y is defined by the formula

$$\mathrm{Cov}(X, Y) = E[(X - E[X]) \cdot (Y - E[Y])]$$

and the *variance* of X by

$$\mathrm{Var}(X) = \mathrm{Cov}(X, X).$$

Two random variables X, Y are *independent* if for all a, b,

$$Pr[X \leq a, Y \leq b] = Pr[X \leq a] \cdot Pr[Y \leq b].$$

If X, Y are independent, then for any functions f, g we have

$$E[f(X) \cdot g(Y)] = E[f(X)] \cdot E[g(Y)].$$

If X, Y are independent, then we have

$$\mathrm{Var}(X + Y) = \mathrm{Var}(X) + \mathrm{Var}(Y).$$

2.1.2 Random variables and distributions

There is a significant volume of literature on random variables and their properties. Here, we mention a few random variables that will be important for our investigations.

2.1.2.1 Bernoulli random variable

For a given experiment, a *Bernoulli random variable* X is defined by

$$X = \begin{cases} 1 & \text{outcome of experiment is success} \\ 0 & \text{outcome of experiment is failure} \end{cases}$$

Given that p (respectively, $1 - p$) is the probability of a positive (respectively, negative) outcome to the experiment, we have $Pr[X = 0] = 1 - p$ and $Pr[X = 1] = p$.

Example 2.1.5 *A random coin is tossed with outcome T (Tail) or H (Head): T (respectively, H) occurs with probability p (respectively, $1 - p$).*

2.1.2.2 Geometric random variable

In a *Geometric random variable* X, independent trials are performed. $p = Pr[\text{success}]$, and $1 - p = Pr[\text{failure}]$. X represents the number of trials required until the first success, that is $Pr[X = k] = (1 - p)^{k-1} p$.

Example 2.1.6 *A memoryless token moves on the nodes of a 4-node complete network. If the token is at a given node, it moves to any of the other three nodes with the same probability. Next time it moves, it does not remember its previous move or position. How many steps are needed until all nodes are visited? Let $N =$ # of steps until all nodes are visited. Let $X_i =$ # of steps required to visit one more node when i nodes have been visited, $i = 0, 1, 2, 3$. Clearly, $N = X_0 + X_1 + X_2 + X_3$, $X_0 = X_1 = 1$. Moreover, X_2, X_3 follow the geometric distribution with $p = 2/3$ and $p = 1/3$, respectively. Hence, $E[N] = E[X_0] + E[X_1] + E[X_2] + E[X_3] = 1 + 1 + \frac{3}{2} + 3 = \frac{13}{2}$.*

2.1.2.3 Poisson random variable

The basic setup in *Poisson* approximations is an experiment in which independent events happen with small probability. This is of course important for networks since packet arrivals resemble experiments of this type. A random variable X taking values $0, 1, 2, \ldots$ is called *Poisson* with parameter $\lambda > 0$, if for $k = 0, 1, 2, \ldots$

$$Pr[X = k] = \frac{\lambda^k}{k!} e^{-\lambda}.$$

This random variable will be used to model data traffic in communication networks as well as in random unit disk graphs. Packets arrive at a rate of λ packets per second. Interarrival times (i.e. times between packet arrivals) are random, independent of each other and have an exponential density function with mean $1/\lambda$.

Example 2.1.7 *A set of n sensors is dropped (say from an airplane) randomly and independently on a unit interval. This can be modeled by assuming the n points are generated using a Poisson process with a rate of arrival n (see Ross (1996, 2nd edition)). Given that the sensors have a range $r > 0$, one is interested to know for what value of r the resulting sensor network is connected, provides full coverage of the interval, and so on.*

2.1.2.4 Binomial random variable

A *Binomial random variable* X is the # of successes in n trials, where p is the probability of success, and $1 - p$ is the probability of failure. This random variable typically models lottery-ticket and radioactive decay problems, among others.

Example 2.1.8 *The binomial distribution approximates the Poisson, for n large and k small. Let $\lambda = np$. For $n \to \infty$ and k small,*

$$\begin{aligned} Pr[X = k] &= \binom{n}{k} p^k (1-p)^{n-k} \\ &= \frac{n!}{k!(n-k)!} \left(\frac{\lambda}{n}\right)^k \left(1 - \frac{\lambda}{n}\right)^{n-k} \\ &= \frac{\lambda^k}{k!} \left(1 - \frac{\lambda}{n}\right)^n \frac{n!}{n^k (n-k)!} \left(1 - \frac{\lambda}{n}\right)^{-k} \\ &\approx \frac{\lambda^k}{k!} e^{-\lambda}, \quad as\ n \to \infty. \end{aligned}$$

2.1.2.5 Exponential distribution

The *exponential* distribution is defined as follows:

$$Pr[X = x] = \begin{cases} \lambda e^{-\lambda x} & \text{if } \lambda \geq 0 \\ 0 & \text{if } \lambda < 0 \end{cases}.$$

Example 2.1.9 *Consider Example 2.1.7. The distance (also known as interarrival times) between two successive sensors obeys the exponential distribution.*

2.1.2.6 Normal (Central Or Gaussian) distribution

The *normal* (also called *central* or *Gaussian* distribution) is defined by

$$Pr[X = x] = \frac{1}{\sqrt{2\pi}\sigma} e^{-(x-\mu)^2/2\sigma^2}$$

(with parameters μ and σ^2). This is probably the most ubiquitous probability distribution because many inferences in probability theory depend on the normal "error" law of sampling.

Example 2.1.10 *Consider the two dimensional probability distribution, say $p(x, y)$, of the errors when measuring the position of a mobile from afar. The following is a natural assumption to make: knowledge of the error in the x (y) direction does not reveal anything about the error in the y direction (x). This implies probabilistic independence, that is, $p(x, y) = u(x) \cdot u(y)$, for some function u. Another natural assumption is that the error measurement should be independent of the angle ϕ of observation. When using polar coordinates $(x, y) = (r \cos \phi, \sin \phi)$, where $r = \sqrt{x^2 + y^2}$. Therefore*

$$u(x) \cdot u(y) = v(\sqrt{x^2 + y^2}), \tag{2.2}$$

for some function v. After setting $y = 0$, this reduces to $v(x) = u(x) \cdot u(0)$ and Equation 2.2 yields

$$\frac{u(x)}{u(0)} \cdot \frac{u(y)}{u(0)} = \frac{u(\sqrt{x^2 + y^2})}{u(0)}. \tag{2.3}$$

Solving the functional Equation 2.3, it can be shown that $p(x, y) = \frac{a}{\pi} \exp -a(x^2 + y^2)$, for some constant $a > 0$ (Exercise 2 for additional details).

2.1.3 Counting processes

A *counting* process is a stochastic process $\{N(t) : t \geq 0\}$ that counts the total number of events that have occurred up to time t. A counting process is said to have *independent increments* if the # of events that occur in disjoint time intervals are independent. The counting process possesses *stationary increments* if the distribution of the # of events depends only on the length of the interval.

The *Poisson process* of rate $\lambda > 0$ is defined as a counting process $\{N(t) : t \geq 0\}$ such that

1. $N(0) = 0$,
2. $N(t)$ has independent increments, and
3. the number of events in any interval of length t is Poisson distributed with mean λt, that is, for all $s, t \geq 0, n = 0, 1, 2, \ldots,$

$$Pr[N(t + s) - N(s) = n] = e^{-\lambda t} \frac{(\lambda t)^n}{n!}.$$

In proving a counting process is Poisson, conditions (1), (2) are easy to prove, but condition (3) is the hardest. For this reason we need a "criterion." This is supplied by the following theorem.

Theorem 2.1.11 (First Fundamental Characterization Theorem) *Consider a counting process* $\{N(t); t \geq 0\}$. *Then* $\{N(t); t \geq 0\}$ *is a Poisson with rate* λ *if and only if the following conditions are satisfied*

1. $N(0) = 0$
2. $\{N(t)\}$ *has independent and stationary increments*
3. $Pr[N(t) = 1] = \lambda t + o(t)$, *and*
4. $Pr[N(t) \geq 2] = o(t)$

Proof. Assume the counting process $N(t)$ satisfies the four conditions mentioned earlier. We show that the process is Poisson. To do this, define $p_n(t) := Pr[N(t) = n]$. We claim that

1. $\frac{d}{dt}(p_0(t)) = -\lambda p_0(t)$
2. $\frac{d}{dt}(e^{\lambda t} p_n(t)) = \lambda e^{\lambda t} p_{n-1}(t)$, for $n \geq 1$.

Indeed,

$$\begin{aligned} p_0(t+h) &= Pr[N(t+h) = 0] \\ &= Pr[N(t) = 0, N(t+h) - N(t) = 0] \\ &= Pr[N(t) = 0] \cdot Pr[N(t+h) - N(t) = 0] \\ &= p_0(t) \cdot Pr[N(h) = 0] \\ &= p_0(t) \cdot (1 - Pr[N(h) \geq 1]) \\ &= p_0(t) \cdot (1 - Pr[N(h) = 1] - Pr[N(h) \geq 2]) \\ &= p_0(t) \cdot (1 - \lambda h - o(h)) \end{aligned}$$

Hence,

$$\frac{dp_0(t)}{dt} = \lim_{h \to 0} \frac{p_0(t+h) - p_0(t)}{h} = -\lambda \cdot p_0(t) + \lim_{h \to 0} \frac{o(h)}{h} = -\lambda \cdot p_0(t)$$

The solution of this differential equation is $p_0(t) = e^{-\lambda t}$. The second part of the claim is proved in the same manner, and we omit the details here.

Further, we have

$$\begin{aligned} p_n(t+h) &= Pr[N(t+h) = n] \\ &= Pr[N(t) = n, N(t+h) - N(t) = 0] \\ &\quad + Pr[N(t) = n-1, N(t+h) - N(t) = 1] \\ &\quad + Pr[N(t) = n, N(t+h) - N(t) \geq 2] \\ &= p_n(t) \cdot p_0(h) + p_{n-1}(t) p_1(h) + o(h) \\ &= (1 - \lambda h) p_n(t) + \lambda h p_{n-1}(t) + o(h) \end{aligned}$$

Hence,

$$\frac{dp_n(t)}{dt} = \lim_{h \to 0} \frac{p_n(t+h) - p_n(t)}{h} = -\lambda \cdot p_n(t) + \lambda p_{n-1}(t),$$

and

$$\frac{d}{dt}(e^{\lambda t} p_n(t)) = e^{\lambda t} \frac{dp_n(t)}{dt} + \lambda e^{\lambda t} p_n(t) = \lambda e^{\lambda t} p_{n-1}(t).$$

So, using induction on $n \geq 1$ and the initial condition $N(0) = 0$, we can prove that $p_n(t) = e^{-\lambda t} \frac{(\lambda t)^n}{n!}$. ■

Let $N(t)$ be Poisson and X_n the r.v. that denotes the time between the $(n-1)$th and the nth events (known as *sequence* of *interarrival times*). Note that

$$Pr[X_1 > t] = Pr[N(t) = 0] = e^{-\lambda t}$$

and moreover by independence of increments

$$\begin{aligned} Pr[X_2 > t | X_1 = s] &= Pr[0 \text{ events in } (s, s+t] | X_1 = s] \\ &= Pr[0 \text{ events in } (s, s+t]] \\ &= e^{-\lambda t}. \end{aligned}$$

Similar arguments can be used to prove the following result.

Theorem 2.1.12 (Arrival and Service Times) *The random variables X_n defined above are mutually independent and identically distributed exponential with mean $1/\lambda$.* ■

The ith *service time* is a random variable Y_i that denotes the service time of the ith customer. Let $S_n = \sum_{i=1}^{n} X_i =$ arrival time of nth event and notice that

$$\begin{aligned} Pr[S_n \leq t] &= Pr[N(t) \geq n] \\ &= \sum_{j=n}^{\infty} e^{-\lambda t} \frac{(\lambda t)^j}{j!}. \end{aligned}$$

So, by differentiating the formula for $Pr[S_n \leq t]$, we obtain

$$Pr[S_n = t] = \frac{d}{dt} Pr[S_n \leq t] = \lambda e^{\lambda t} \frac{(\lambda t)^{n-1}}{(n-1)!},$$

which is the Gamma distribution. We also mention without proof the following result.

Theorem 2.1.13 (Second Fundamental Characterization Theorem) *Consider a sequence X_n, $n \geq 1$ of independent identically distributed exponential random variables with mean $1/\lambda$. Define a counting process $N(t)$ such that for each k, the kth event of N occurs at time $t = X_1 + X_2 + \cdots + X_k$. The resulting counting process $N(t)$ is Poisson with mean λ.*

Suppose each event of a Poisson $N(t)$ is of type i, where $i = 1, 2$. Moreover, if an event occurs at time t, then it is classified as being of type 1 with probability $P(t)$ and of type 2 with probability $1 - P(t)$.

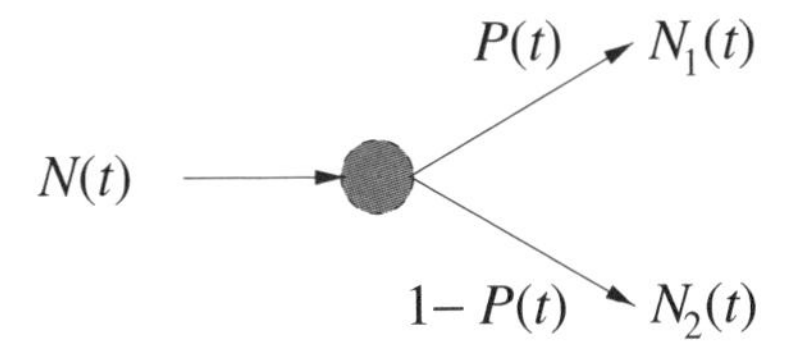

The following result, which is mentioned without proof, is known in the literature as the *Poisson random splitting theorem*.

Theorem 2.1.14 (Random Splitting of a Poisson) *Let $N_i(t) = \#$ of events of type i that occur by time t. Then $N_1(t)$, $N_2(t)$ are independent Poisson random variables with mean λtp and $\lambda t(1-p)$, respectively, where*

$$p = \frac{1}{t}\int_0^t P(s)\,ds.$$

2.2 Modeling traffic

Projections of network performance are vital to the development of all well-established network architectures and can be based either on experimental or analytic (theoretical) models. The experimental approach entails either making an after the fact analysis based on actual values or making a simple projection by scaling up from existing experience to an expected future environment. In the analytical approach after developing an analytic model (usually based on queuing theory), a simulation model is run, which is subsequently used to calibrate the network parameters that are under consideration.

2.2.1 Delay models

The development of efficient network algorithms is greatly influenced by transmission delays of packets from source to destination. The factors that are of interest are determining which network protocol gives the best delay-throughput characteristics under specified conditions, what size buffers must be employed by a network's users in order to keep the probability of buffer overflow below a particular value, what is the maximum number of voice calls that can be accepted by a network in order to keep the voice packet transfer delay to a minimum, and how many users a network link can support and still maintain a reasonable response time. Delay may be due to *processing* (time between when a packet is correctly received and when it is correctly assigned to an outgoing link), *queuing* (time between when a packet is assigned to a queue for transmission and when it starts being transmitted), *transmission* (time between when the first and the last bits of the packet are transmitted) and *propagation* (time between when the last bit is transmitted and when the bit is received).

The most basic result for modeling delay is Little's theorem, which gives a fundamental law that describes the input and output characteristics of a networked system in steady state. Little's theorem is concerned with time averages in the limit. In steady state, let N be the average number of customers in the system (in buffer and in service), λ the average packet arrival rate, and T the average time spent by each customer in the system (in queue and in service). With these definitions in mind the following theorem can be proved.

Theorem 2.2.1 (Little's Theorem) *Assuming the rate of arrival and departure are the same (in the limit), we have that in the steady state* $N = \lambda T$.

2.2.2 Queuing models

Queuing is concerned with predicting the overall effects of some change in load or design, for example, concern for system response time and/or throughput. A typical network consists of a set of processors located at the nodes of a graph such that each node has a queue. Data (typically packets) arrive at the nodes, are serviced and exit. Arriving and departing data usually follow either a random or a deterministic process. Queuing systems generally consist of a queuing station with a single server or many servers (which in the context of communications means a single or infinite transmission line). Following Kendall's notation (see Kendall (1953)), queuing systems are generally described using the notation $A/S/m$ or $A/S/m/K$, whereby the parameters A, S, m, K are explained in Table 2.1. In modeling, the number of processors may be 1 or any number m or even ∞.

Table 2.1 Parameters used in Kendall's notation $A/S/m$ and $A/S/m/K$ for queueing systems.

A	Description of the arrival process
S	Description of the service process
m	Number of processes involved
K	Limit of the number of packets in the system

The process may be Random (Memoryless Poisson, General) or Deterministic. M denotes negative exponential distribution, G denotes general independent arrivals or service times, And D denotes deterministic arrivals or fixed-length service. Poisson processes are represented as M (since Poisson processes have exponentially distributed interarrival times, they are also called *memoryless*).

The following additional notation is being used in the sequel. N_Q is the average number of packets waiting in queue, and T_Q is the average waiting time in queue. We know that from Little's Theorem, $N = \lambda T$ and $N_Q = \lambda T_Q$.

2.2.3 Birth–death processes

A *Markov* process is a process whose next state is fully predicted from the current state regardless of the previous states assumed by the process. The random processes used in the queuing systems to be studied in the sequel are special types of Markov processes on populations (usually, packets, calls, etc.) that obey the principles of the so-called *birth–death processes*. In this instance, *birth* means arrival, while *death* means departure of an item from the given population. A state is determined by the number of items. Transitions from

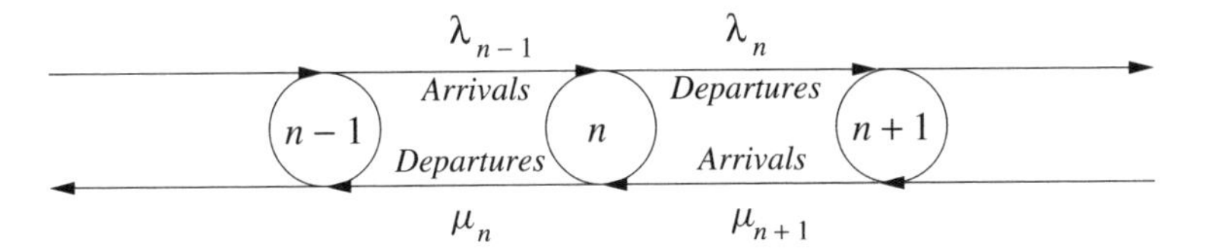

Figure 2.1 State transitions for a birth–death process.

any given state can be made only to one of neighboring states (Figure 2.1). It follows that if the system is currently in state n then it moves to state $n+1$ when a new item arrives and to state $n-1$ when an item departs.

Let p_n denote the steady-state probability in state n and λ_{n-1} (respectively, μ_n) the average arrival (respectively, departure) rate in state n. Consider state n (Figure 2.1). Then the total number of arrivals to state n are the $\lambda_{n-1}p_{n-1}$ from state $n-1$ plus the departures $\mu_{n+1}p_{n+1}$ from state $n+1$. The total number of departures from state n are the $\lambda_n p_n$ arrivals from state $n+1$ plus the departures $\mu_n p_n$ departures from state n. Therefore at equilibrium we have the following identity

$$\lambda_{n-1}p_{n-1} + \mu_{n+1}p_{n+1} = \lambda_n p_n + \mu_n p_n.$$

2.2.4 $M/M/1/\infty$ queuing system

$M/M/1/\infty$ is the simplest queuing system and is depicted in Figure 2.2. Customers compete for service on a first-in-first-out (FIFO) basis and service times are independent identically distributed random variables with common exponential distributions. Arriving customers get served if the server is free and queued otherwise. The arrival and departure processes are Poisson with arrival (respectively, departure) rates λ (respectively, μ). Let n be the number of customers in the system. The state of the system is fully described by the number n of customers present. From the state transition diagram depicted in Figure 2.3 we see that the state transitions are given by the equations

$$\begin{aligned} p_n &= \text{probability that the system is in state } n \\ &= \text{probability that there are } n \text{ packets in queue.} \end{aligned}$$

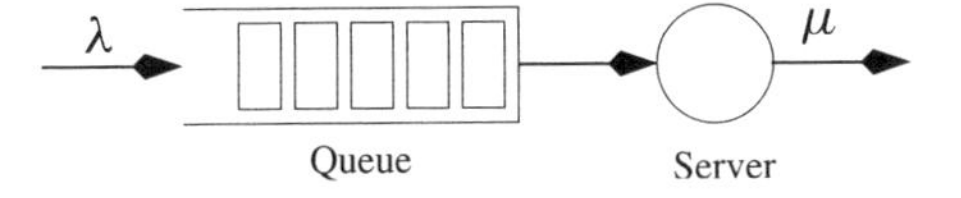

Figure 2.2 An $M/M/1/\infty$ queue.

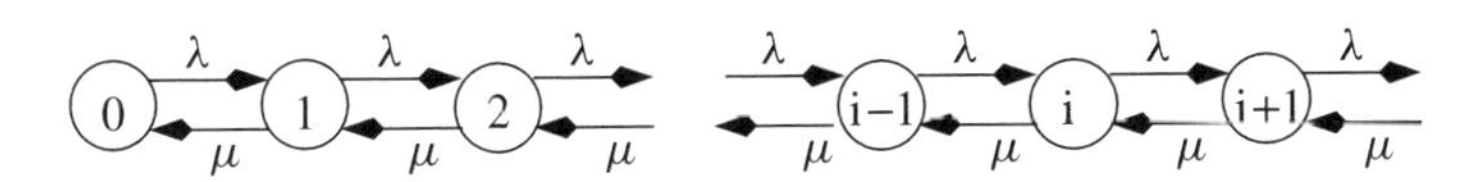

Figure 2.3 State transitions for an $M/M/1/\infty$ queue.

At equilibrium, rate process leaves = rate process enters, so we have the equations

$$p_{n+1}\mu = p_n\lambda$$
$$\sum_{n=0}^{\infty} p_n = 1,$$

which imply that for $\rho = \lambda/\mu$,

$$p_n = p_0\left(\frac{\lambda}{\mu}\right)^n$$
$$= p_0\rho^n,$$

where $n = 0, 1, \ldots$, and

$$1 = \sum_{n=0}^{\infty} p_n$$
$$= p_0 \sum_{n=0}^{\infty} \rho^n$$
$$= \frac{p_0}{1-\rho}.$$

So, $p_0 = 1 - \rho$ and $p_n = \rho^n(1-\rho)$. Let N be the expected number of packets in the system. We conclude that

$$N = \sum_{n=0}^{\infty} n p_n$$
$$= \rho(1-\rho)\sum_{n=0}^{\infty} n\rho^{n-1}$$
$$= \rho(1-\rho)\frac{d}{d\rho}\left(\sum_{n=0}^{\infty}\rho^n\right)$$
$$= \rho(1-\rho)\frac{d}{d\rho}\left(\frac{1}{1-\rho}\right)$$
$$= \rho(1-\rho)\frac{1}{(1-\rho)^2}$$
$$= \frac{\rho}{1-\rho}.$$

Note that ρ is the utilization factor of the system and $N = \frac{\rho}{1-\rho}$ is the average number of packets in the system. The average delay per packet (waiting time plus service time) is

$$T = \frac{N}{\lambda} = \frac{\rho}{\lambda(1-\rho)} = \frac{1}{\mu - \lambda}.$$

The average waiting time in queue is

$$T_Q = (\text{delay per packet}) - (\text{service time}) = \frac{1}{\mu - \lambda} - \frac{1}{\mu} = \frac{\rho}{\mu - \lambda}.$$

The average number of packets in queue is

$$N_Q = \lambda T_Q = \lambda \frac{\rho}{\mu - \lambda} = \frac{\rho^2}{1 - \rho}.$$

2.2.5 $M/M/m/\infty$ queue: *m* servers

An $M/M/m/\infty$ queuing system is depicted in Figure 2.4. Customers compete for service on a FIFO basis and service times are independent identically distributed random variables with common exponential distributions. The arrival and departure processes are Poisson with arrival (respectively, departure) rates λ (respectively, μ). Let n be the number of customers in the system. The state of the system is fully described by the number n of customers. From the state transition diagram depicted in Figure 2.5 we see that the state transitions can be derived from the equilibrium equations

$$\text{rate process leaves} = \text{rate process enters},$$

so we have the equations

$$\begin{aligned} \lambda p_{n-1} &= n\mu p_n \text{ for } n \leq m \\ \lambda p_{n-1} &= m\mu p_n \text{ for } n > m \end{aligned}$$

from which we conclude that for $\rho = \frac{\lambda}{m\mu} < 1$

$$p_n = \begin{cases} \frac{p_0 m^n \rho^n}{n!} & n \leq m \\ \frac{p_0 m^m \rho^n}{m!} & n > m. \end{cases}$$

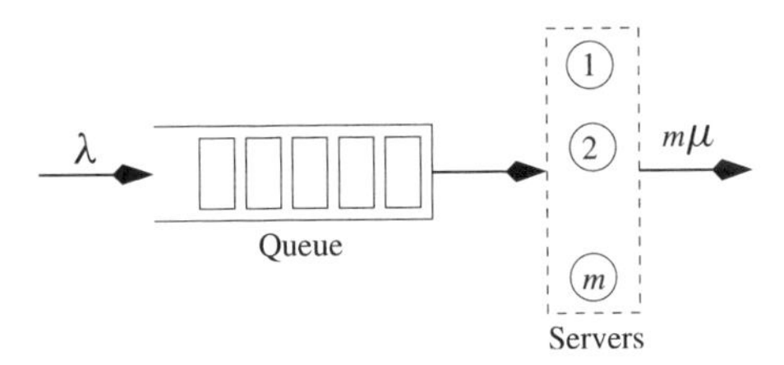

Figure 2.4 An $M/M/m/\infty$ queue.

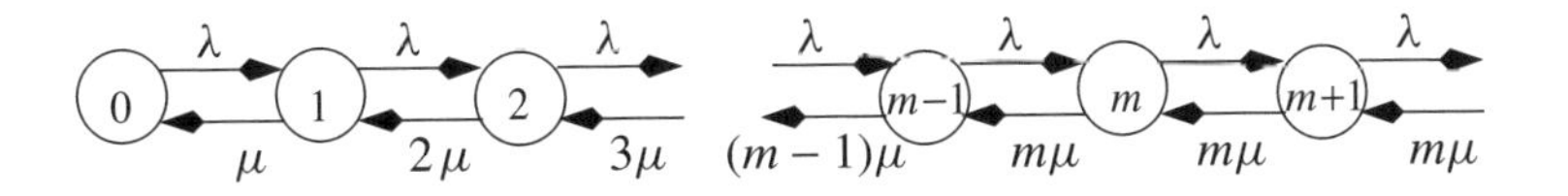

Figure 2.5 State transitions for an $M/M/m/\infty$ queue.

Using the condition

$$\sum_{n=0}^{\infty} p_n = 1$$

we obtain

$$\begin{aligned}
p_0 &= \left[1 + \sum_{n=1}^{m-1} \frac{(m\rho)^n}{n!} + \sum_{n=m}^{\infty} \frac{(m\rho)^n}{m!} \frac{1}{m^{n-m}}\right]^{-1} \\
&= \left[1 + \sum_{n=1}^{m-1} \frac{(m\rho)^n}{n!} + \frac{(m\rho)^m}{m!(1-\rho)}\right]^{-1} \\
&= \left[\sum_{n=0}^{m-1} \frac{(m\rho)^n}{n!} + \frac{(m\rho)^m}{m!(1-\rho)}\right]^{-1}
\end{aligned}$$

The probability P_Q of queuing is equal to

$$\begin{aligned}
P_Q &= \sum_{n \geq m} p_n \\
&= \frac{p_0 (m\rho)^m}{m!} \sum_{n \geq m} \rho^{n-m} \\
&= \frac{p_0 (m\rho)^m}{m!(1-\rho)} \\
&= \left[\sum_{n=0}^{m-1} \frac{(m\rho)^n}{n!} + \frac{(m\rho)^m}{m!(1-\rho)}\right]^{-1} \cdot \frac{(m\rho)^m}{m!(1-\rho)}.
\end{aligned}$$

This last identity is also known as *Erlang's loss formula.* The expected number N_Q of packets in queue is

$$\begin{aligned}
N_Q &= \sum_{n=0} n p_{m+n} \\
&= P_Q \frac{\rho}{1-\rho}.
\end{aligned}$$

The average delay per packet in the queue is

$$\begin{aligned}
T_Q &= \frac{N_Q}{\lambda} \\
&= \frac{P_Q \rho}{\lambda(1-\rho)}.
\end{aligned}$$

Since, $\rho = \frac{\lambda}{m\mu}$, the average delay per packet is

$$T = T_Q + \frac{1}{\mu}$$

$$= \frac{P_Q \rho}{\lambda(1-\rho)} + \frac{1}{\mu}$$
$$= \frac{P_Q}{m\mu - \lambda} + \frac{1}{\mu}.$$

The average number N of packets in the system is

$$N = \lambda T$$
$$= \frac{\lambda P_Q}{m\mu - \lambda} + \frac{\lambda}{\mu}$$
$$= \frac{\rho P_Q}{1-\rho} + m\rho.$$

2.2.6 Queues for channel allocation

In contrast to wireline networks where delay is given priority, the two most critical issues affecting QoS in wireless networks are the probability that an originating call is blocked and forced termination. In the sequel, we consider mobiles uniformly distributed and moving at random speed and direction.

Assume that m channels are allocated to each cell. We distinguish two types of calls, *originating* and *handoff*, and we denote the average arrival rates by λ_O and λ_H, respectively. Let B_O and B_H be the blocking probabilities of originating and handoff calls, respectively. For $i = 0, 2, \ldots, m$, let p_i be the probability that i channels are busy. Originating and handoff calls are given equal priority. All arrival processes are Poisson, the service rate is μ, the service time is exponential and all assumptions are equally applicable to all the cells in the system. A wireless system with originating and handoff calls is depicted in Figure 2.6. Originating and handoff calls arrive at the rate λ_O and λ_H, respectively, and are serviced by any of the channels $1, 2, \ldots, m$ at the rate μ. The Markov chain transition diagram for this system is depicted in Figure 2.7. The total request rate is $\lambda_O + \lambda_H$. The equilibrium equations are

$$i\mu p_i = (\lambda_O + \lambda_H) p_{i-1} \tag{2.4}$$

for $i = 0, \ldots, m$. Moreover

$$\sum_{i=0}^{m} p_i = 1. \tag{2.5}$$

Using Equation 2.4 recursively, as well as Equation 2.5 we obtain

$$p_i = \frac{(\lambda_O + \lambda_H)^i}{i!\mu^i} p_0$$

$$p_0 = \left(\sum_{i=0}^{m} \frac{(\lambda_O + \lambda_H)^i}{i!\mu^i} \right)^{-1}$$

$$B_O = B_H = p_m = \frac{(\lambda_O + \lambda_H)^m}{m!\mu^m} \cdot \left(\sum_{i=0}^{m} \frac{(\lambda_O + \lambda_H)^i}{i!\mu^i} \right)^{-1}.$$

The last formula is known as *Erlang's B formula.*

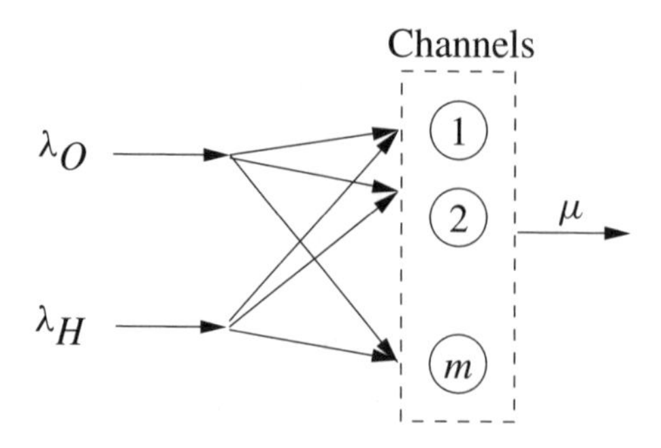

Figure 2.6 A wireless system with originating and handoff calls.

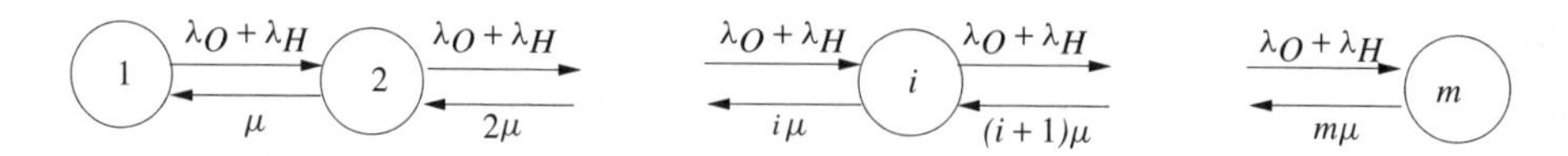

Figure 2.7 State transitions for a system with handoffs.

2.2.7 Queues with reserved channels for handoffs

Dropping an originating call is not as disastrous as dropping a handoff call. It is therefore reasonable to assume that handoff calls should be given priority over new originating calls in a wireless system. Figure 2.8 depicts a wireless system with m_R channels reserved for handoff while the remaining $m_c = m - m_R$ are shared by both originating and handoff calls. Notice that in this system a handoff call may be serviced by any of the m channels while an originating call may be serviced only by channels $1, 2, \ldots, m_c$. The system can be represented by a Markov chain transition diagram (Figure 2.9). The total request rate is $\lambda_O + \lambda_H$, of which λ_H is allocated for handoffs. The equilibrium equations are

$$i\mu p_i = (\lambda_O + \lambda_H)p_{i-1}, \text{ for } i = 0, \ldots, m_c, \tag{2.6}$$

$$i\mu p_i = \lambda_H p_{i-1}, \text{ for } m_c < i \leq m. \tag{2.7}$$

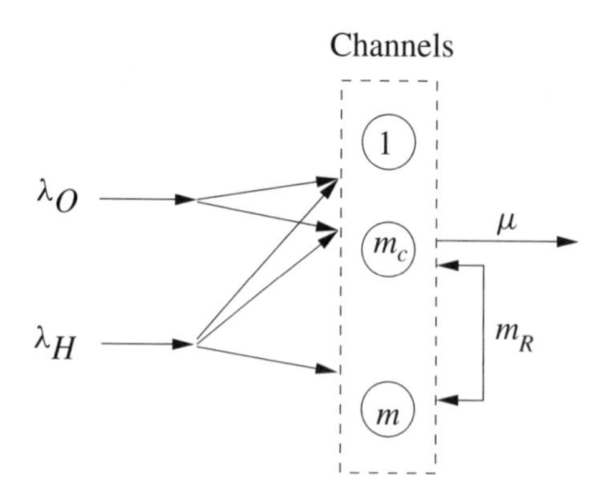

Figure 2.8 A wireless system with reserved channels for handoffs.

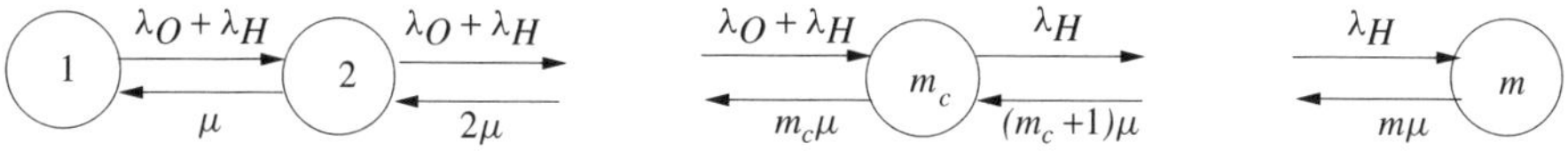

Figure 2.9 State transitions for a system with reserved channels for handoffs.

Moreover, Equation 2.5 is satisfied. Using Equations 2.6 recursively, as well as Equation 2.5, we obtain that

$$p_i = \frac{(\lambda_O + \lambda_H)^i}{i!\mu^i} p_0, \text{ for } i = 0, \ldots, m_c,$$

$$p_i = \frac{(\lambda_O + \lambda_H)^{m_c} \lambda_H^{i-m_c}}{i!\mu^i} p_0, \text{ for } m_c < i \leq m.$$

$$p_0 = \left(\sum_{i=0}^{m_c} \frac{(\lambda_O + \lambda_H)^i}{i!\mu^i} + \sum_{i=m_c+1}^{m} \frac{(\lambda_O + \lambda_H)^{m_c} \lambda_H^{i-m_c}}{i!\mu^i} \right)^{-1}$$

$$B_O = \sum_{i=m_c}^{m} p_i$$

$$B_H = p_m = \frac{(\lambda_O + \lambda_H)^{m_c} \lambda_H^{i-m_c}}{m!\mu^m} p_0.$$

2.3 Multiple access

In *point-to-point* networks, the received signal is a function of a single transmitted signal. On the contrary, in *broadcast* networks, a single transmission medium is shared and the received signal is a function of possibly more than one transmitted signal (Figure 2.10.) It is therefore important to understand how to mediate access to a shared channel. There are two possibilities. In a *centralized* system, a distinguished node (*master*) makes access decisions for the remaining nodes (*slaves*). In a *distributed* system, all nodes are equivalent and the access decision is derived together in a distributed fashion. Centralized schemes are less fault-tolerant (due to master failure) and generally less efficient. There are several methods to share a medium. In *static partitioning schemes*, the transmission medium is

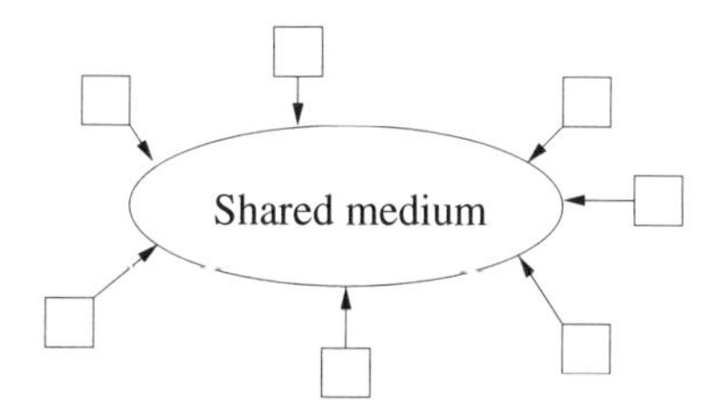

Figure 2.10 Hosts sharing a medium.

partitioned into separate dedicated channels. *MAC schemes* are dynamic, distributed and on-demand (Figure 2.11). In all cases, it is important that collisions are minimized. In the sequel we consider several such protocols and look at their performance.

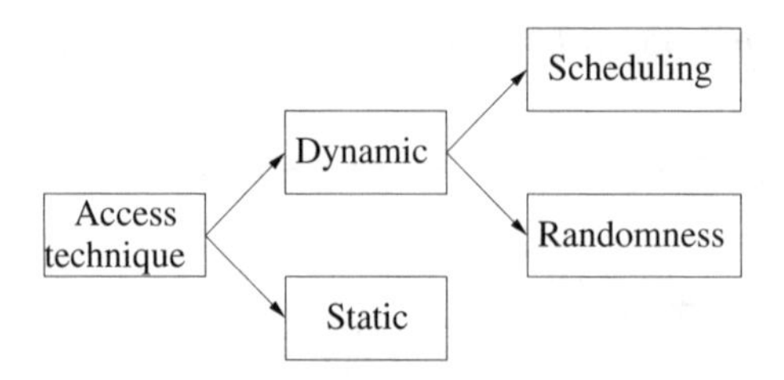

Figure 2.11 Types of MAC schemes.

2.3.1 Uncoordinated access

Assume we have n hosts in a wireless broadcast system whereby each host is within range of all other hosts and all the hosts share the same channel. If two or more stations talk at the same time, the message is garbled.

Given that the hosts are in no way coordinated but rather attempt to talk at random (respectively, stay silent) with the same probability p (respectively, $1 - p$), how should they select the value of p so as to maximize throughput? Assuming the system is synchronous, the following analysis is possible. The probability that a given node succeeds is

$$p(1-p)^{n-1},$$

while the probability that one of the n nodes succeeds is

$$np(1-p)^{n-1}.$$

As a function of p this last formula is maximized when

$$\frac{d(np(1-p)^{n-1})}{dp} = 0.$$

Since $\frac{d(p(1-p)^{n-1})}{dp} = (1-p)^{n-1} - (n-1)p(1-p)^{n-2}$ it is easy to see that this maximum value is attained when $p = 1/n$. It follows from the earlier observations that the probability that a given node succeeds is

$$\frac{1}{n}\left(1 - \frac{1}{n}\right)^{n-1} \approx \frac{1}{en},$$

asymptotically in n, where e is Euler's number, while the probability that one of the nodes succeeds is

$$n\frac{1}{n}\left(1 - \frac{1}{n}\right)^{n-1} \approx \frac{1}{e},$$

asymptotically in n. Given that a node that fails to access the channel attempts to retransmit independently of the other nodes, it is easy to see that the expected number of steps until a host succeeds to talk is equal to e, while the system throughput is $1/e$.

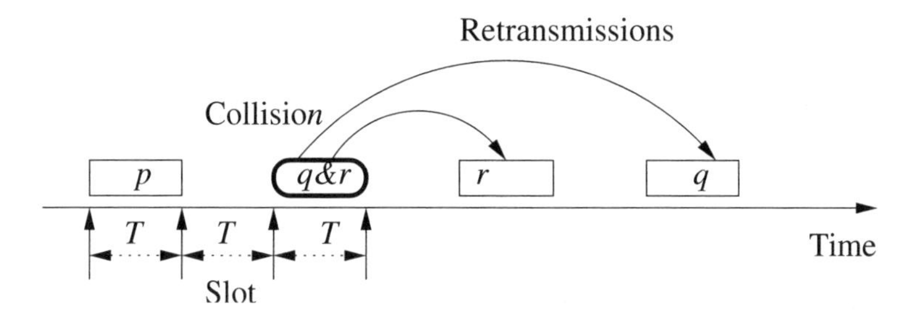

Figure 2.12 In the slotted Aloha model, packets are transmitted synchronously. When two packets q and r collide, retransmission is necessary.

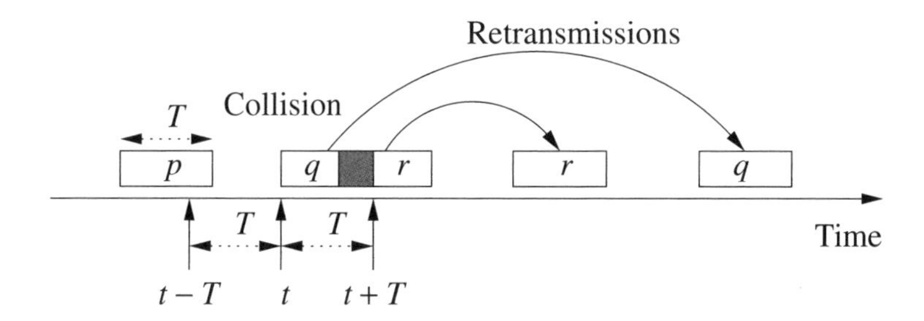

Figure 2.13 In the pure Aloha model packets are transmitted asynchronously. When two packets q and r collide retransmission is necessary.

2.3.2 Contention-based access

Aloha Ethernet comes in two varieties, *slotted* and *unslotted* (or *pure*) Aloha. In *slotted* Aloha, time is divided into slots that handle fixed-length packets with transmissions only at slot boundaries (Figure 2.12), while *unslotted* Aloha has no restrictions on packet size or time of transmission (Figure 2.13).

2.3.2.1 Slotted aloha

In slotted Aloha, the hosts are synchronized, packets have same length, and each packet requires one time unit (slot). Packet arrivals are Poisson: we have m hosts with total arrival rate of λ and therefore each station has an arrival rate of λ/m. Reception is perfect if no collision occurs. Otherwise at end of the slot each node obtains feedback. All colliding packets must be retransmitted (hosts with colliding packet are called *backlogged*). In addition, there is *no buffering* at the nodes. An alternative (equivalent) assumption is that there is an infinite # of processors.

We can analyze slotted Aloha as follows. A new or previously transmitted packet scheduled for transmission at a given time can be transmitted only if no other packet arrives in the *vulnerable period*. The probability of exactly n arrivals at a given node during a time slot is equal to $e^{-\lambda/m}\frac{(\lambda/m)^n}{n!}$. Therefore, the probability of 0 arrivals is $e^{-\lambda/m}$. Hence, probability of a new arrival of a packet during a slot is $a = 1 - e^{-\lambda/m}$. Assume n

of these hosts are blocked with backlogged packets that are retransmitted[1] with probability r, that is, retransmissions are geometrically distributed with probability of retransmission in the ith slot $r(1-r)^{i-1}$.

The arrival rate of new packets is $a = 1 - e^{-\lambda/m}$ and there are $m - n$ nonbacklogged nodes. Therefore, the expected # of attempted transmissions is

$$G(n) = (m-n)a + nr.$$

Success occurs if either one new packet arrives and no backlogged packets attempt to transmit or no new packet arrives and one backlogged packet attempts to transmit. Interpreting this statement mathematically, we see that

$$Pr[\text{Success}] = (m-n)(1-a)^{m-n-1}a(1-r)^n + (1-a)^{m-n}n(1-r)^{n-1}r.$$

Using the following approximations

$$\begin{aligned}(1-x) &\approx e^{-x},\\ (1-x)^y &\approx e^{-xy},\\ \frac{x}{1-x} &\approx x\end{aligned}$$

it is easy to derive

$$\begin{aligned}Pr[\text{Success}] &= \frac{(m-n)(1-a)^{m-n}a(1-r)^n}{1-a} + \frac{(1-a)^{m-n}n(1-r)^n r}{1-r}\\ &= (m-n)(1-a)^{m-n}a(1-r)^n + (1-a)^{m-n}n(1-r)^n r\\ &= e^{-(m-n)a-nr}((m-n)a+nr)\\ &= G(n)e^{-G(n)}.\end{aligned}$$

It follows from Theorem 2.1.11 that we can think of new plus backlogged packets as following a Poisson process with mean arrival rate $G(n)$. Therefore, the random variable N_n that counts the number of new packets plus those backlogged is Poisson with mean $G(n)$, that is,

$$Pr[N_n = k] = \frac{G(n)^k}{k!}e^{-G(n)}.$$

In particular,

$$\begin{aligned}Pr[N_n = 0] &= Pr[\text{idle}] = e^{-G(n)}\\ Pr[N_n = 1] &= G(n)e^{-G(n)},\end{aligned}$$

that is,

$$Pr[\text{successful transmission}] = G(n)e^{-G(n)}.$$

From this, we obtain the departure rate as a function of attempted transmission rate $G(n)$. At equilibrium, the arrival rate equals the departure rate $\lambda = G(n)e^{-G(n)}$. However the function

[1]Of course the retransmission probability r is unknown at the moment. However, we will determine in the sequel which choice of r will optimize throughput.

xe^{-x} is maximized when $x = 1$. Therefore, the probability of successful transmission is maximized when $G(n) = (m-n)a + nr = 1$, and the retransmission probability r thus derived is

$$r = \frac{m-n}{n}e^{-\lambda/m} - \frac{m-n-1}{n}.$$

The probability that a given test packet will avoid collision is the same as the probability that no other packet is transmitted, that is, $Pr[N_n = 0] = e^{-G(n)}$. Since the probability of a collision of the test packet with another packet is $1 - e^{-G(n)}$, the probability that it requires exactly k attempts is equal to

$$e^{-G(n)}(1 - e^{-G(n)})^{k-1}.$$

Hence, the expected number E of steps until success occurs is

$$E = \sum_{k=0}^{\infty} ke^{-G(n)}(1 - e^{-G(n)})^{k-1} = e^{G(n)}$$

and therefore the expected number of unsuccessful attempts is $e^{G(n)} - 1$.

2.3.2.2 Unslotted aloha

In pure Aloha, there are no time slots and therefore there is no need for synchronization or for limiting packet size. When a packet arrives, it immediately attempts transmission. If a collision occurs, the node waits a random time and retries. Two transmissions collide if the interdeparture interval is less than the packet transmission time. Assume that a new or previously transmitted packet is scheduled for transmission at time t (Figure 2.13). In unslotted Aloha a packet can be transmitted only if no other packet arrives in the *vulnerable period*, that is, the time interval $[t - T, t + T]$ (an interval of length $2T$). Assuming that retries are Poisson distributed with parameter x and that there are n backlogged packets, the total number of attempts follows the Poisson distribution with rate $G(n) = \lambda + nx$. An analysis similar to slotted Aloha is possible for unslotted Aloha. It can be shown that

$$Pr[\text{Success}] = e^{-2G(n)}$$

and the throughput is

$$G(n)e^{-2G(n)}.$$

2.3.2.3 Instability

For some values of the total arrival rate λ, a point is reached where the number n of backlogged packets increases without bound, that is, the protocol is unstable. A number of methods have been suggested for varying the retransmission probability r that a backlogged node will retransmit (Section 2.3.2.1).

The first type of algorithms use *backoff*. Backoff protocols rely on acknowledgements only. A function $p(x)$ is defined in advance. Each station i has a variable $backoff_i$ and attempts transmission if queue is not empty with probability $p(backoff_i)$; if it fails to transmit owing to a collision it increments the variable $backoff_i$; else it assigns 0 to the variable $backoff_i$. If the queue is empty and no transmission is made, the variable $backoff_i$

remains unchanged. For example, in *binary exponential backoff*, if a packet has collided k times, the backoff time is set to 2^{-k}. It has been shown that binary exponential backoff is unstable for any $\lambda > 0$ (if there are infinitely many potential stations). When there are only finite stations, binary exponential backoff becomes unstable with $\lambda > 0.568$. In *polynomial backoff*, a backlogged packet attempts with probability $r = c/(i + 1)^k$ where c and k are fixed and i is the number of unsuccessful attempts. Thus, when $k = 0$ we have *constant backoff*, when $k = 1$ we have *linear backoff*, and when $k = 2$ we have *quadratic backoff*. Polynomial backoff remains stable for any $\lambda < 1$ and $k > 1$.

In *pseudo Bayesian* systems , it is assumed that arriving packets are immediately backlogged so that attempts are made at a rate of $G(n) = nr$. Nodes maintain an estimate $\hat{n}$ for n and attempt to send with probability $\hat{r} = \min\{1, 1/\hat{n}\}$ and adjust $\hat{n}$ dynamically. The idea is to increase $\hat{n}$, if it is seen that no node is sending, and decrease $\hat{n}$ if it is seen that more than one node is sending. The updating rules used are

- $\hat{n} \leftarrow \hat{n} - 1 + \lambda$, for idle or success, and
- $\hat{n} \leftarrow \hat{n} + \lambda + \frac{1}{e-2}$, for collision.

If λ is either unknown or changing, then $\lambda = 1/e$ is being used. The value of $\hat{n}$ is sent with each transmission.

2.4 Demand assigned multiple access

Since channel efficiency is only $1/e$ for slotted Aloha and even worse for either pure Aloha or backoff protocols, practical systems must use reservation whenever possible. In reservation systems, senders reserve future time slots so as to be able to transmit without collision within such a reserved time slot. In polling systems, nodes take turns in accessing the network. If the hosts share a line, then a host is acting as a central controller issuing control messages to coordinate transmissions. Such a central controller may use radio transmissions in a certain frequency band to transmit outbound messages, and nodes may share a different frequency band to transmit inbound messages. If a central controller is absent, the hosts themselves may use a polling order list, in which case we have some kind of distributed control. In many instances, some kind of round-robin control is established between the hosts.

2.4.1 Bit-map

Assume that there are n hosts labeled 0 to $n - 1$. In the *bit-map* protocol, contention periods are exactly n slots long (Figure 2.14). The reservation algorithm is as follows.

- If the ith host has a packet to send, then it inserts a 1 bit in the ith slot and no other station is allowed to transmit during this slot.
- After n slots have passed by, each host has complete knowledge of which stations wish to transmit, in which case hosts may now begin transmitting in the numerical order assigned.

0	1		$i-1$	i	$i+1$		$n-1$
0	1		0	1	1		0

Figure 2.14 Array of n reservation slots with appropriate reservations by the hosts.

2.4.2 Binary countdown

In binary countdowns, all hosts wanting to use the channel broadcast their address as a binary string (with all strings assumed to be of the same length), say of length n. The arbitration rule for two hosts A and B is as follows. Starting from the highest to the lowest order (i.e. left to right in Figure 2.15), they compare their bits. The first time they differ, the host with the smaller bit drops out.

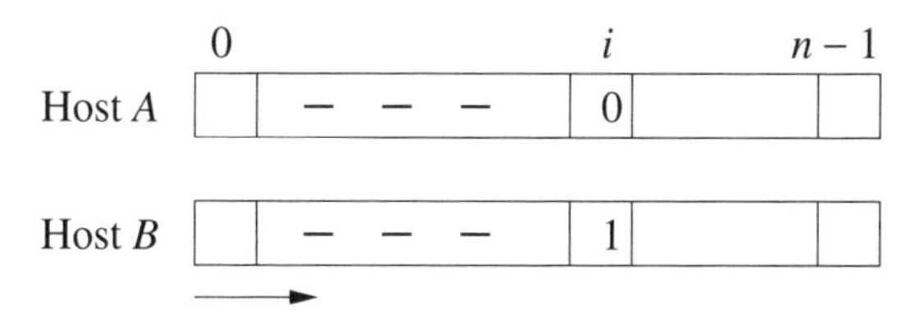

Figure 2.15 Arbitration between hosts A and B.

2.4.3 Splitting algorithms

The idea is to partition the set of hosts into *groups*. Each group is assigned to a slot. Thus a collision can only occur within a group.

Example 2.4.1 *Assume the nodes have unique (fixed-length) identifiers (e.g.* $0, 1, \ldots, n - 1$*). Assume n is even and divide nodes into pairs*

$$\{0, 1\}, \{2, 3\}, \ldots, \{n - 2, n - 1\}.$$

Host $2i$ *or host* $2i + 1$ *(possibly both) go in slot* i. *If a collision occurs, then they go in order.*

A *tree splitting* algorithm generalizes the above idea of host splitting. Let a collision occur in the kth slot (Figure 2.16). Divide the nodes into two parts: first, the set C_k of those involved in the collision, and second, the set N_k of those not involved in this collision. Nodes in N_k go into waiting mode. Now split C_k into two subsets C_k^0 and C_k^1 (e.g. by flipping a coin, or by looking at the identities of the nodes). The first subset C_k^0 transmits in slot $k + 1$ and the second subset C_k^1 transmits in slot $k + 2$.

This procedure forms a tree. During first slot, any node of the tree rooted at 1 may contend. If a collision occurs, in the next slot only nodes in the left subtree (rooted at 2) may contend and so on. Therefore in the algorithm, after a collision, all new arrivals wait and all nodes involved in a collision divide into subsets. Each successive collision causes that subset to split again.

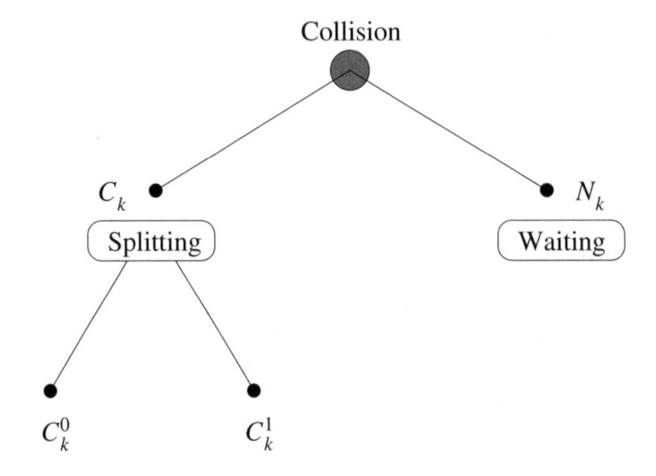

Figure 2.16 kth iteration of the splitting algorithm.

2.5 Carrier sense multiple access in IEEE 802.11

The IEEE 802.11 series of standards defines an ad hoc mode of operation, also termed the *peer-to-peer mode* by certain drivers. The link layer of IEEE 802.11 can behave as the distributed coordination function (DCF) or point coordination function (PCF). In the ad hoc mode, only the DCF is available and described hereafter. A number of nodes can independently, that is, without the concourse of an access point, form a multipoint point wireless link. Such a network is termed an *Independent Basic Service Set* (*IBSS*) in the IEEE 802.11 standards. Control of access to the medium is achieved using a scheme called *Carrier Sense Multiple Access with Collision Avoidance* (*CSMA/CA*).

The IEEE 802.11 protocol architecture consists of two layers, namely, the physical layer and the MAC layer. CSMA/CA is placed in the MAC layer. It uses the frame transmit/receive and carrier sensing services of the physical layer.

CSMA/CA is based on the listen-before-talk and contention principles. A node ready for frame transmission first listens for the presence of a signal on the medium. If the medium is free, then the node can transmit at once. If the medium is busy, then the node waits until the medium is free before transmitting. A scenario is pictured in Figure 2.17. At time t_1,

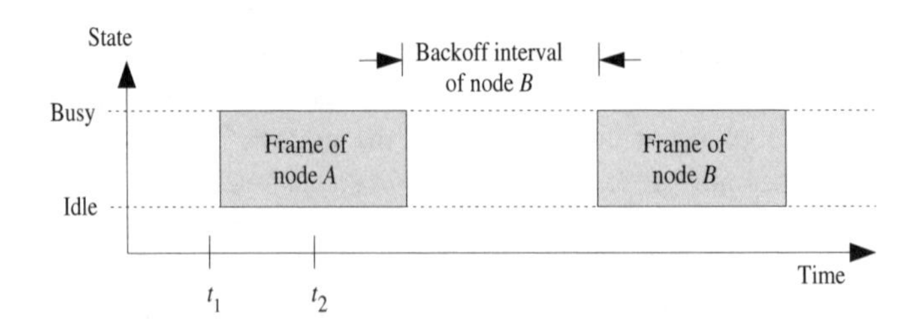

Figure 2.17 CSMA/CA scenario.

node A senses the medium. The medium must remain idle for a required time period. If this requirement is met, node A can transmit a frame. At time t_2, node B senses the medium. The medium is busy and node B must wait until node A's transmission ends. When the medium becomes idle again, node B does not transmit right away. It selects the duration of a wait period at random. This period of time is called the *backoff interval*, and begins when the medium becomes idle. If the medium remains idle for the whole duration of the backoff interval, then node B can transmit a frame.

Determination of the length of the backoff interval deserves more attention (Figure 2.18). After a successful frame transmission, the medium must be idle for a period of time corresponding to the value of a parameter called *DCF inter frame spacing* (*DIFS*). Then time is divided into slots of equal length. The number of slots is denoted as *CW* (contention window) and the duration of one slot is denoted as *ST* (slot time). The total duration of the *contention window* is then $CW \times ST$. The backoff time is determined as the start time of one slot selected at random, uniformly. In other words, the slot can be selected as a random integer uniformly selected in the interval $[0, CW]$.

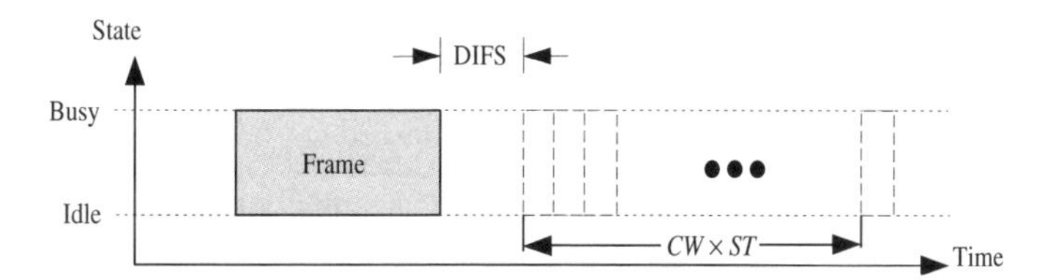

Figure 2.18 Determination of the backoff interval.

2.5.1 Persistence

There are variations of CSMA. The persistence scheme is a variation point (Figure 2.19). The IEEE 802.11 adopts a nonpersistent scheme. Other alternatives are the 1-persistent scheme and the p-persistent scheme. When sensing a busy medium, a persistent node *persists* sensing the medium until it becomes idle. A 1-persistent node transmits right after the medium becomes idle. The fact that transmission is attempted that early minimizes the medium idle time after the transmission of one frame. If, however, a number of nodes have

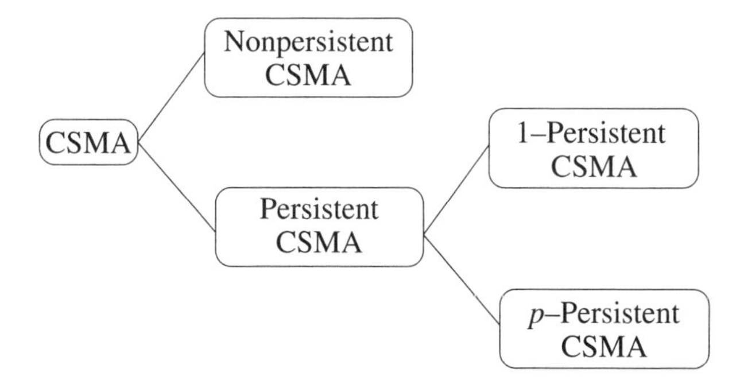

Figure 2.19 Various persistence schemes. They also come in slotted and unslotted varieties.

been simultaneously waiting for the medium to become idle, then they all try to transmit at the same time. A collision is guaranteed to occur.

A p-persistent node defers its transmission attempt and runs a probabilistic algorithm. Time is divided into slots of equal length. At the beginning of each slot, each node independently transmits with probability p or defers with probability $1 - p$. The probability p ($0 \leq p \leq 1$) is a parameter of the network. Probabilistic transmissions spread in time the transmission attempts after the medium becomes idle. The goal is to minimize the risks of collisions; however, this can be at the expense of lost transmission time. A network performs best when all the member modes operate according to the same persistence scheme and same probability p, when applicable. The p-persistent scheme performs better than the nonpersistent scheme or the 1-persistent scheme on loaded networks. The various types of persistent protocols are summarized in Table 2.2.

Table 2.2 Different types of persistence and the actions taken by the participating hosts.

Persistence	When medium busy	Action when idle
Nonpersistent	Waits random time	Immediate transmission
1-Persistent	Keeps listening	Immediate transmission
p-Persistent	Waits until next slot; Listens again	Transmits with probability p Defers with probability $1 - p$

The 1-persistent scheme is employed by Ethernet. The p-persistent scheme is employed by AX.25 (see Beech et al. (1998)).

2.5.2 Collision avoidance

CSCMA/CA needs to deal with collisions. At the same instant, several nodes may sense a free medium and decide to transmit. There is no central coordination. The transmissions interfere, resulting in a collision. Collisions are detected by the absence of acknowledgment. When a node experiences a collision, the affected frame is rescheduled for retransmission. The retransmission delay is selected at random and independently of the other nodes. This has the effect of spreading in time the retransmission attempts of the nodes involved in the collision and hence reducing the risk of repeated collisions.

CSMA/CA has to face the hidden terminal problem, which is defined as follows. At least three nodes are involved, namely, A, B and C (Figure 2.20). Nodes A and B are within communication range and so are nodes B and C. However, nodes A and C are not and cannot receive each other's transmissions. Let us suppose that both nodes A and C need to transmit a frame to node B. Node A listens for the presence of a signal on the medium, senses a free channel and starts transmission. While this transmission is in progress, node C may also start transmission. Node C senses a free medium because it is not within communication range of node A. Both the transmitted frames collide at node B.

In CSMA/CA, the hidden terminal problem is addressed using a request-to-send/clear-to-send (RTS/CTS) mechanism. Before transmission, every node sends an RTS signal to

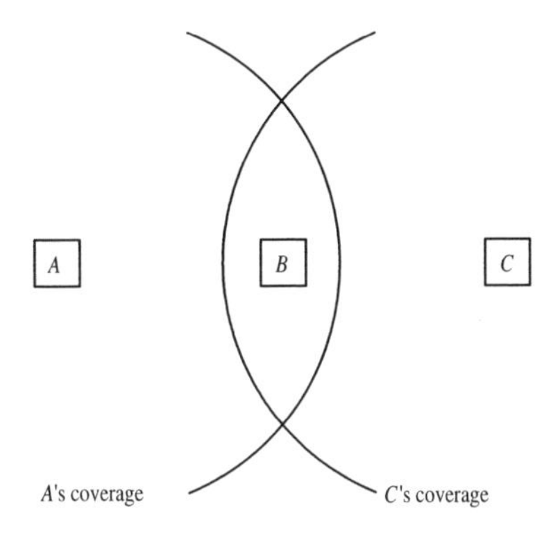

Figure 2.20 The hidden terminal problem.

the destination. If the destination is not busy, then it responds with a CTS signal. If the RTS/CTS scheme is applied to the aforementioned example, then before transmitting the data frame, node A sends an RTS signal to node B. Node B responds with a CTS, which is received by both nodes A and C. With respect to A, it means that the transmission of the data frame can start. With respect to C, it means that node B is busy and transmission should be put on hold. To complete the process, when node B has successfully received the frame, it confirms with an acknowledgment signal. The acknowledgment is received by both nodes A and C. With respect to A, it is interpreted as a successful transmission. With respect to C, it means that node B becomes available for communication.

The RTS and CTS frames include a duration field in their header. The value in this field is interpreted as the time required to transmit a data frame and the corresponding acknowledgment. Each node maintains a state variable called the *network allocation vector* (*NAV*) whose value reflects the current reservation, as a time duration, of the medium. This value is updated when RTS and CTS frames are overheard.

The RTS/CTS mechanism is optional. It can be disabled, always used or used for frames above a certain length.

2.6 Medium access control in ad hoc networks

MAC protocols in ad hoc networks must take into account the *hidden* and *exposed* sensor problems. To be more precise, the *hidden* sensor problem occurs when two sensors that lie outside the range of each other transmit at the same time to another sensor, while the *exposed* sensor problem refers to the inability of a sensor that is blocked, owing to transmission by a nearby transmitting sensor, from transmitting to another sensor. In this section we elaborate on MAC in ad hoc networks.

The channel access problem can be viewed as a graph coloring problem. Here, coloring may refer to either edges or vertices of the graph under consideration, and therefore edge-coloring is related to activating links while node-coloring is related to activating hosts, respectively, of the ad hoc network. However, as shown by Ephremides and Truong (1990), the well-known NP-complete problem of finding the maximum independent set in a graph reduces to the broadcast scheduling problem, and therefore the latter problem is NP-complete in ad hoc networks.

2.6.1 Neighbor aware contention resolution

Neighbor aware contention resolution (NCR) is a special leader election problem whereby a host decides its leadership locally among a set of contenders. It is required that each contention context be identifiable, for example, the time slot number in networks based on a time-division multiple access scheme. Knowledge of the contenders can be acquired by some means, for example, all neighbors of a host at distance at most k hops, for some $k \geq 1$. For each host u, let C_u be the set of contenders of u. To decide leadership and in order to limit communication overhead, a priority is assigned to hosts in C_u that depends on identifiers of the entities as well as the contention context. In general, conflicts can be avoided since it can be assumed that contenders have mutual knowledge and the order of contenders is based on priority numbers.

2.6.2 Multiple access protocols

In this section, we outline four access control protocols for ad hoc networks. For additional details, we refer the reader to Bao and Garcia-Luna-Aceves (2002, 2003).

In all the protocols described in the following text, it is assumed that the physical layer is capable of assigning codes via the direct sequence spread spectrum (DSSS) transmission technique. Such code assignments are based on schemes that can be transmitter oriented, receiver oriented, or per-link oriented. During each time slot, each node is assigned a new spreading code that is derived from the priority of the node, as well as the codes assigned to contending nodes. Priority is assigned to each entity e that depends on the identifier of the entity and the current contention context t using the identity

$$Priority_e = Hash(e \oplus t) \oplus e,$$

where $H(\cdot)$ is a hash function. If C_u is the set of contenders of entity u and t is the contention context, then u is elected leader if $Priority_u > Priority_v$, for all $v \in C_u$.

The previous technique is applicable when resources are equally divided among all contenders. When demands from different entities vary, then different portions of the shared resource are allocated. In any such approach, entities specify demands by an integer value chosen from a given integer set. Demands are then propagated to the contenders before the contention resolution process.

2.6.2.1 Node activation multiple access (NAMA)

The node activation multiple access (NAMA) protocol requires that the transmission from a node is received by the one-hop neighbors of the node without collisions. In other words, when a node is activated for channel access, the neighbors within two hops of the node should not transmit. Therefore, if $N_1(x)$ denotes the set of distance one neighbors of a node x, then the contender set of node u is defined by

$$C_u := N_1(u) \cup \bigcup_{v \in N_1(u)} N_1(v) \setminus \{u\}.$$

2.6.2.2 Link activation multiple access(LAMA)

The purpose of the link activation multiple access (LAMA) protocol is to determine which node is eligible to transmit, and find out which outgoing link from the node can be activated in the current time slot. A given node u, prior to determining its eligibility to transmit, initializes the priorities and code assignments of nodes within distance two hops. An eligible node u examines each reception code assigned to its distance one neighbors, and decides whether it can activate links to its distance one neighbor set $N_1(u)$. Hence, the set of contenders to node u are selected among the distance two neighbors with a given code.

2.6.2.3 Pairwise link activation multiple access (PAMA)

The pairwise link activation multiple access (PAMA) protocol differs from both NAMA and LAMA in that instead of node priorities link priorities are used for contention resolution. The protocol chooses the link with the highest priority for reception or transmission at the node, based on the priorities of the incident links to a node using the identity

$$Priority_{(u,v)} = Hash(u \oplus v \oplus t) \oplus u \oplus v.$$

In this case, the set of contenders of a link includes all other links incident to the endpoints of the link.

2.6.2.4 Hybrid activation multiple access (HAMA)

The hybrid activation multiple access (HAMA) protocol is a node activation protocol that is also capable of broadcast transmissions, while at the same time also maximizing the chance of link activations for unicast transmissions. The code assignment is based on a transmission oriented scheme whereby a node derives its own priority by comparing with the priorities of its neighbors. Participating nodes are assigned the states Receiver, Drain, Broadcast Transmitter, Unicast Transmitter, Drain Transmitter, and Yield, depending on roles assigned, and these in turn are being used to derive priorities.

2.6.3 Throughput analysis of NAMA

Following Bao and Garcia-Luna-Aceves (2002, 2003), in this section we analyze the NAMA protocol. Analysis of the other protocols can be found in the same citation.

Suppose that a random set of n points are selected randomly and independently with the uniform distribution in a unit square. Clearly, n is the *density* of the pointset in the unit square. The points represent the hosts of a wireless network and have identical radius r. The probability that a region R of area $|R|$ has exactly k nodes from the random pointset obeys the Poisson distribution and is equal to

$$\frac{(n|R|)^k}{k!} e^{-n|R|}. \tag{2.8}$$

The expected number, say n_1, of distance one neighbors of a given node is equal to $n\pi r^2$, since πr^2 is the area of a disk of radius r and n is the density of the set points. Let us now compute the expected number, say n_2, of distance two neighbors of a given node

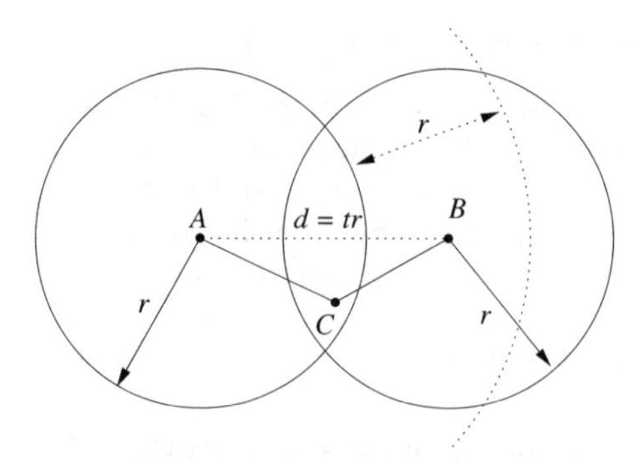

Figure 2.21 Distance two neighbors of a given node.

(Figure 2.21). Nodes A and B are at most at distance two hops of each other if there is a node C in the lune L, which is determined by the intersection of the disks of radius r centered at A and B, respectively, and at distance $d = tr$ from each other, for some $0 \leq t \leq 1$. The expected number of nodes in this lune is equal to $n|L(t)|$, where $|L(t)|$ is the area of $L := L(t)$. It is easy to determine the area of $L(t)$ as a function of t, namely

$$|L(t)| = 2r^2 \left(\arccos\left(\frac{t}{2}\right) - \frac{t}{2}\sqrt{1 - \left(\frac{t}{2}\right)^2} \right)$$

In particular, the probability that there is at least one node inside the lune $L(t)$ is equal to $1 - e^{-n|L(t)|}$. If we add up all the points in the annulus $(r, 2r)$ centered at node A multiplied by the probability of becoming two hop neighbors, then we obtain that the expected number of distance two neighbors of A is equal to

$$n_2 = n\pi r^2 \int_1^2 2t \left(1 - e^{-n|L(t)|}\right) dt.$$

It follows that the expected number of neighbors of A of distance at most two is equal to

$$n_1 + n_2 = n\pi r^2 \left(1 + \int_1^2 2t \left(1 - e^{-n|L(t)|}\right) dt\right).$$

Now consider protocol NAMA described earlier. Just like the Aloha CSMA/CD protocol described in Section 2.3.2, the expected number of contenders follows the Poisson distribution with mean $N = n_1 + n_2$. Since all nodes in a contention set have equal chance of succeeding, the probability that a node will succeed among its contenders is equal to the average among all possible numbers of the contenders, that is,

$$\sum_{k=1}^{\infty} \frac{1}{k+1} \frac{N^k}{k!} e^{-N} = \frac{1}{N} - \frac{1+N}{e^N}.$$

2.7 Bibliographic comments

Ross (2002) provides a nice introduction to probabilistic thinking with a computer science perspective. The seminal work by Jaynes (2003) is an outstanding reference for probability theory as the logic of science. Example 2.1.10 is based on a beautiful derivation of John Herschel (1850) and James Clark Maxwell (1860) that is mentioned in the book of Jaynes (2003), Chapter 7).

Delay models have been influenced by the work of Little (1961). For a data networks approach to delay modeling and queuing, the reader is referred to Bertsekas and Gallager (1992, 2nd edition). Also, Tanenbaum (1996, 4th edition) provides a good introduction to the foundations to computer networks, where the reader can also find additional discussion on demand assigned multiple access (bit-map and binary countdown) in Section 2.4. The analysis in Sections 2.2.6 and 2.2.7 on queuing for wireless channels is based on Zeng and Agrawal (2002). There are numerous reference works on queuing systems and networks but the two volumes Kleinrock (1975) (theory) and Kleinrock (1976) (applications) are the seminal references. Most studies on packet switching and network performance have been influenced by the research in Kleinrock and Tobagi (1975), Tobagi and Kleinrock (1976), and Tobagi and Kleinrock (1977).

There are several studies on MAC in ad hoc networks. Discussions in the present chapter are influenced by Bao and Garcia-Luna-Aceves (2002, 2003). Additional studies on WiMAX/802.16 and Bluetooth access control node discovery are given in Chapter 3.

2.8 Exercises

1. A memoryless token moves on the nodes of a n-node complete network. If the token is at a given node, it moves to any of the other $n - 1$ nodes with the same probability. The next time it moves, it does not remember its previous move or position. How many steps are needed until all nodes are visited? **Hint:** Work along the steps in proof of Example 2.1.6.

2. ($\star$) Prove that the solution of the functional Equation 2.3 is $u(x) = \sqrt{a/\pi}\exp(-ax^2)$, for some constant $a > 0$. Use this last identity to derive the distribution of errors $p(x, y) = \frac{a}{\pi}\exp -a(x^2 + y^2)$. **Hint:** Take logarithms on both sides of Equation 2.3 to derive a functional equation $w(x) + w(y) = w(\sqrt{x^2 + y^2})$, where $w(x) = \ln \frac{u(x)}{u(0)}$. Equation $w(x) + w(y) = w(\sqrt{x^2 + y^2})$ implies that the sum of a function of x plus the sum of the function of y is equal to a function of $x^2 + y^2$. Now take the second partial derivative $\frac{\partial^2}{\partial x \partial y}$ on both sides of the last equation. Since the left-hand side is equal to 0, so is the right-hand side. Conclude from this that $w(x) = cx^2$, for some constant $c > 0$.

3. Consider a system whereby packets arrive every t seconds at a regular rate with the first packet arriving at time 0. Assume all packets have equal length and require at time units for transmission ($a \leq 1$). Further, suppose that the delay and propagation time is D seconds. Argue that

   (a) the arrival rate is $1/t$, and

   (b) the time spent in system is $at + D$.

   Apply Theorem 2.2.1 to compute the average number of packets in the system. **Hint:** $N = a + D/t$.

4. Suppose that in a network of n nodes $1, 2, \ldots, n$ packets arrive at a given node i at a rate λ_i. Let n_i be the average number of packets in the system arriving at the node i, and assume t_i is the average delay of packets at node i.

   (a) Apply Little's theorem to the isolated node i.

   (b) Apply Little's theorem to the entire system given that N is the average number of packets in the system and T is average time a packet is in the system.

5. This exercise outlines the proof of Theorem 2.2.1. Suppose a sample history of a system is observed from time $t = 0$, and let the quantities measured be $N(t) = \#$ of packets in the system at time t, $a(t) = \#$ of packets that arrived in the interval $[0, t]$, $b(t) = \#$ of packets that departed in the interval $[0, t]$, and $T(i) =$ time spent in the system by ith customer. Consider Theorem 2.2.1. Prove the following:

   (a) The average arrival (respectively, departure) rate over interval $[0, t]$ is $a(t)/t$ (respectively, $b(t)/t$).

   (b) The average of the customer delay up to time t is

$$T_t = \frac{1}{a(t)} \sum_{i=1}^{a(t)} T(i).$$

   (c) The average of the number of customers up to time t is

$$N_t = \frac{1}{t} \sum_{i=1}^{t} N(i).$$

   (d) In many systems of interest, these quantities tend to a steady state, namely,

$$\lim_{t\to\infty} a(t)/t = b(t)/t, T := \lim_{t\to\infty} T_t, N := \lim_{t\to\infty} N_t.$$

   Use this to conclude that $N = \lambda T$.

6. This exercise refers to Sections 2.2.3 and 2.2.1. Consider a counting process $X(t)$ and let us define (assuming the limit exists) $p_n = \lim_{t\to\infty} Pr[X(t) = n]$. Intuitively, p_n is the proportion of time there are n packets in the system. We also define by a_n the proportion of packets that find n in the system when they arrive, and d_n the proportion of packets that leave behind n in system when they depart. Show that in a system in which packets arrive one at a time and are served one at a time, $a_n = d_n$.

   (a) Show that an arrival (respectively, departure) sees (respectively, leaves) n in system exactly when number goes from n to $n + 1$ (respectively, $n + 1$ to n).

   (b) Show that in any interval of time t,

$$|\# \text{ of transitions } (n \to n+1) - \# \text{ of transitions } (n+1 \to n)| \leq 1$$

   (c) Show that in the limit

$$\text{rate of transitions } (n \to n+1) = \text{rate of transitions } (n+1 \to n).$$

Conclude that if $X(t)$ is Poisson then $p_n = a_n$. (See Ross (2002) for additional details and discussion.)

7. This exercise refers to the birth–death process in Section 2.2.3.

   (a) Argue that $\lambda_0 p_0 = \mu_1 p_1$.

   (b) Prove that

$$p_n = \frac{\lambda_0 \lambda_1 \cdots \lambda_{n-1}}{\mu_1 \mu_2 \cdots \mu_n} p_0.$$

8. Consider a base station that is capable of handling an individual request per time unit on the average. When a request comes in, the base station handles it immediately. Just as it completes the current request, a new request arrives. Explain why, as the average arrival rate gets closer to 1 per unit time, the average departure rate gets worse. **Hint:** Use $M/M/1$ queues.

9. Give the proofs of the claims concerning random variables made in Section 2.1.1.

10. Compute the expectation and variance for the random variables defined in Section 2.1.2.

Random Variable	Expectation	Variance
Bernoulli	p	$p(1-p)$
Binomial	np	$np(1-p)$
Geometric	$1/p$	$1/p^2$
Poisson	λ	λ
Exponential	$1/\lambda$	$1/\lambda^2$

11. This exercise refers to Section 2.3.1.

    (a) Why is being synchronous important in the analysis of Section 2.3.1?

    (b) Consider a round-robin algorithm whereby the ith host sends after host $i-1 \bmod n$. How well would a round-robin algorithm perform?

    (c) What happens to round-robin when a new host joins in?

    (d) Why do all deterministic channel allocation algorithms have similar problems?

12. Explain in detail why the retransmission probability in slotted Aloha must be chosen so as to satisfy $r = \frac{m-n}{n} e^{-\lambda/m} - \frac{m-n-1}{n}$.

13. Eight hosts $a, b, \ldots, h$ contend for a shared channel using the tree splitting algorithm.

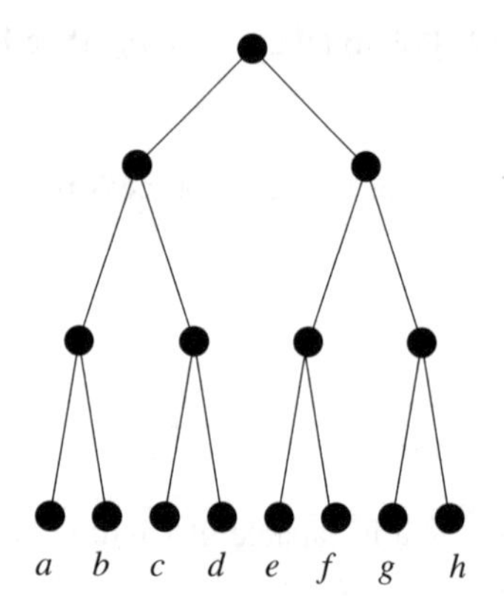

Suddenly a, d, g, h become ready to transmit at once. In how many slots (steps) is contention fully resolved?

14. In the context of the 802.11 protocol, what does the concept of fairness mean? How does this protocol address fairness?

3

AD HOC WIRELESS ACCESS

> To a large extent I agree with you about 'thinking in analogies', but I do not think of the brain as 'searching for analogies' so much as having analogies forced upon it by its own limitations...
> Turing (1948)

The above quotation is from a letter by Alan Turing, then working at Bletcley Park, to Jack Good, a statistician from Cambridge (cited by Hodges (1983), page 388). Thinking in analogies as well as searching for them will be a vital component of our presentation of access control in the present chapter.

Ad hoc networks are self-organized communication systems where the infrastructure (the participants, the routers or the network configuration) is dynamically created and maintained. The building components are the very same nodes interested in exchanging messages. In general, self-organization requires complex protocols that are only now starting to be studied in detail. This chapter is devoted exclusively to access control in Bluetooth networks as well as WiMAX/802.16, although similar techniques may be applicable to more general ad hoc networks.

Bluetooth is an industrial specification, also known as IEEE 802.15.1, that provides a way to connect and exchange information between "personal devices." It is available in various new products like digital assistants, mobile phones, laptops, PCs, printers, digital cameras and video game consoles via short range Radio Frequency (RF). It is used in situations when two or more devices are in close proximity to each other, but it is less effective for setting up networks and remote access, in which case WiFi is better suited.

Section 3.1 describes management issues in Bluetooth: architecture, network formation rules, and delay. A critical review of its efficiency for node discovery (in both 2-node and multiple node systems) is provided in Section 3.2 whereby a model is discussed and detailed performance analysis is provided. Section 3.3 discusses Bluetooth topology formation algorithms including bluetree, tree scatternet, bluenet, bluestar and loop scatternet. Finally Section 3.4 describes the WiMAX/802.16 mesh mode.

Principles of Ad hoc Networking Michel Barbeau and Evangelos Kranakis

3.1 Management of Bluetooth networks

Bluetooth was originally conceived as cable replacement technology. It is the first de facto standard for ad hoc networking brought about by several companies. Its particular design is less suited for other applications, but the industry has already been supplying Bluetooth modules.

3.1.1 Architecture

The basic topological units of Bluetooth are the *Piconets*. They are *star* networks that are being managed by a single *master* that implements centralized control over channel access. All other participants are called *slaves*. Communication is strictly

- master → slave, and
- slave → master,

while direct slave-to-slave communication is impossible. Moreover, a master has at least one and at most seven slaves. Two such piconets are depicted in Figure 3.1.

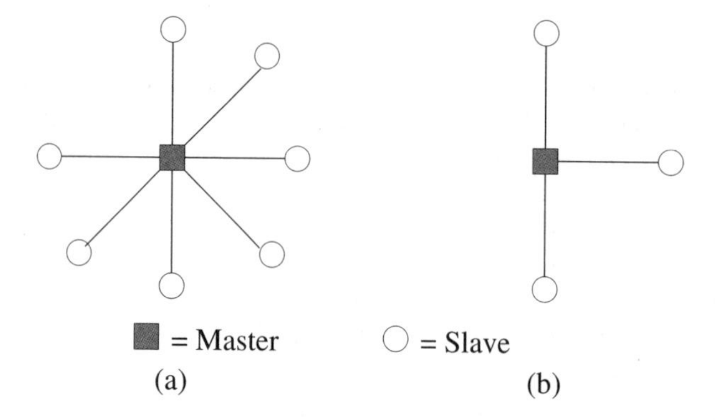

Figure 3.1 In the piconet (a) the master has seven slaves, in piconet (b) it has three.

Masters and slaves play double roles. If a slave may want to set up a new piconet or take over an existing piconet, during the existence of a piconet, the role of master and slave can be switched. Piconets are joined to form *scatternets*. A node can be the slave in two piconets, or become master in a new piconet and slave in the old piconet.

If you want to communicate with more than eight nodes at the same time, multiplexing is required. Moreover, nodes would need to alternate between their respective piconets. An example of a 14-node scatternet is depicted in Figure 3.2. Bluetooth does not provide for slave-to-slave communication. To solve this problem, one has

1. to either channel traffic through a master (this increases communication and power consumption)
2. or one of the two slaves could setup its own piconet or even switch roles with a master.

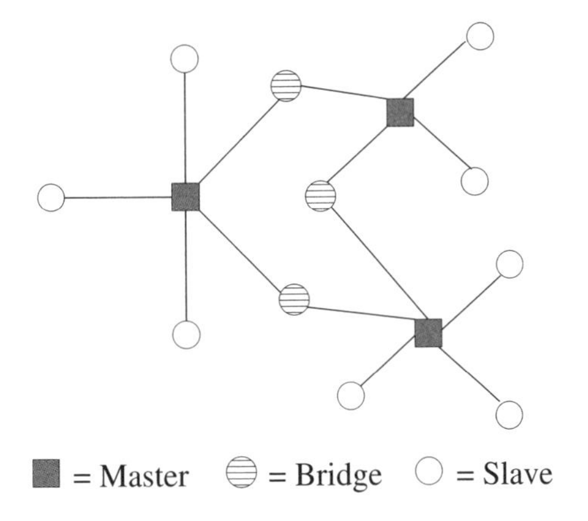

Figure 3.2 Example of a 14-node scatternet.

3.1.1.1 Bluetooth network formation rules

Bluetooth networks were conceived as a technology that would replace cables and wires at the home and office. As such it has always been envisioned that they would be operational over a small geographic area. Salonidis et al. (2001) proposed a set of rules that imply "smallness." Scatternets are collections of piconets satisfying the following rules:

1. The scatternet is a connected network formed from piconets.
2. It has masters and slaves. Slaves are of two types: "pure" slaves (i.e. slaves belonging to a single piconet) and "bridge" slaves (i.e. slaves that belong to multiple piconets).
3. Two masters must always share a bridge.
4. Two masters can share only a single slave.
5. A bridge may connect only two piconets.
6. A piconet can have at most seven slaves.

We can prove the following theorem.

Theorem 3.1.1 (Salonidis et al. (2001)) *If the nodes of a scatternet network satisfy requirements 1–6 mentioned earlier then its maximum size is at most* 36.

Proof. Consider a scatternet with n nodes. It is required to estimate the maximum possible value of n. Let p be the number of piconets in this scatternet (this is also the number of masters). For each $i = 1, 2, \ldots, p$ let n_i be the number of slaves of the i-th piconet. Let s_i be the number of pure slaves and b_i the number of bridge slaves of the i-th piconet. It is true that

$$n_i = s_i + b_i,$$

for $i = 1, 2, \ldots, p$. Note that always $n_i \leq 7$.

The ith master must have $b_i = p - 1$ bridges, because it must be connected to all other masters via a bridge and a bridge can only have two masters. At the same time it must also have s_i slaves, where

$$s_i = n_i - b_i = n_i - (p-1) \leq 7 - (p-1). \tag{3.1}$$

The number of masters is p and since two masters can have at most one bridge, the total number of bridges is at most

$$\binom{p}{2} = \frac{p(p-1)}{2}. \tag{3.2}$$

As a consequence,

$$n \leq \#\text{ masters } + \#\text{ pure slaves } + \#\text{ bridges} \tag{3.3}$$

Using Inequalities 3.1, 3.2, and 3.3 we obtain

$$\begin{aligned} n &\leq p + \sum_{i=1}^{p} s_i + \frac{p(p-1)}{2} \\ &\leq p + \sum_{i=1}^{p} (7 - (p-1) + \frac{p(p-1)}{2} \\ &= p + 8p - p^2 + \frac{p^2}{2} - \frac{p}{2} \\ &= \frac{17p}{2} - \frac{p^2}{2}. \end{aligned}$$

In particular, the number p of masters of a scatternet with given number n of nodes must satisfy

$$p^2 - 17p + 2n \leq 0. \tag{3.4}$$

So the maximum value of n is bounded above by the maximum value of $\frac{17p - p^2}{2}$. This maximum value is obtained when $p = 17/2$. Therefore

$$\begin{aligned} n_{\max} &\leq \left\lfloor \frac{17(17/2) - (17/2)^2}{2} \right\rfloor \\ &= \left\lfloor \frac{17^2}{8} \right\rfloor \\ &= \left\lfloor \frac{289}{8} \right\rfloor \\ &= 36. \end{aligned}$$

This completes the proof of the theorem. ■

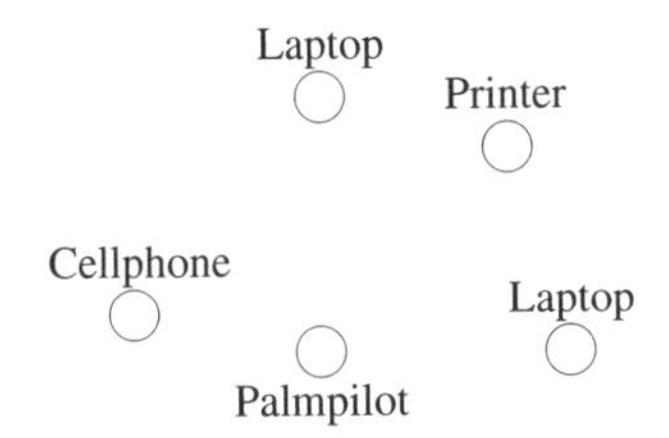

Figure 3.3 A typical set of Bluetooth nodes.

3.1.2 The Bluetooth asymmetric protocol

Next, the problem of link establishment in Bluetooth is addressed. For example, consider the system consisting of five Bluetooth devices (two laptops, a printer, a palm-pilot and a cellphone) depicted in Figure 3.3. How is the Bluetooth network formed, which node becomes master and which nodes slaves and what is its architecture? To form a network one way or the other, the nodes will have to follow the steps enlisted below:

1. Start
2. Synchronization
3. Discovery
4. Paging
5. Connection established.

Implementation of these steps depends on the network formation delay (which includes synchronization delay, discovery delay and paging) as examined in the sequel.

3.1.2.1 Network formation delay

The Bluetooth asymmetric protocol has three main phases (see Salonidis et al. (2001)).

- Nodes can use the same frequency, but may be out of phase due to internal clock differences. To overcome this problem, sender and receiver hop at frequencies of different speeds. This causes some frequency synchronization delay.
- Next follows the discovery protocol, in which nodes enter either the Inquiry or the InquiryScan state.
- Finally, we have the paging procedure in which the sender uses the receiver's page hopping sequence to initiate a device access code packet that can be heard only by the receiver device.

Figure 3.4 depicts the asymmetric Bluetooth protocol while Figure 3.5 describes the time requirement necessary for discovery. According to Salonidis et al. (2001), the link formation delay, abbreviated by LF, is determined by three random variables.

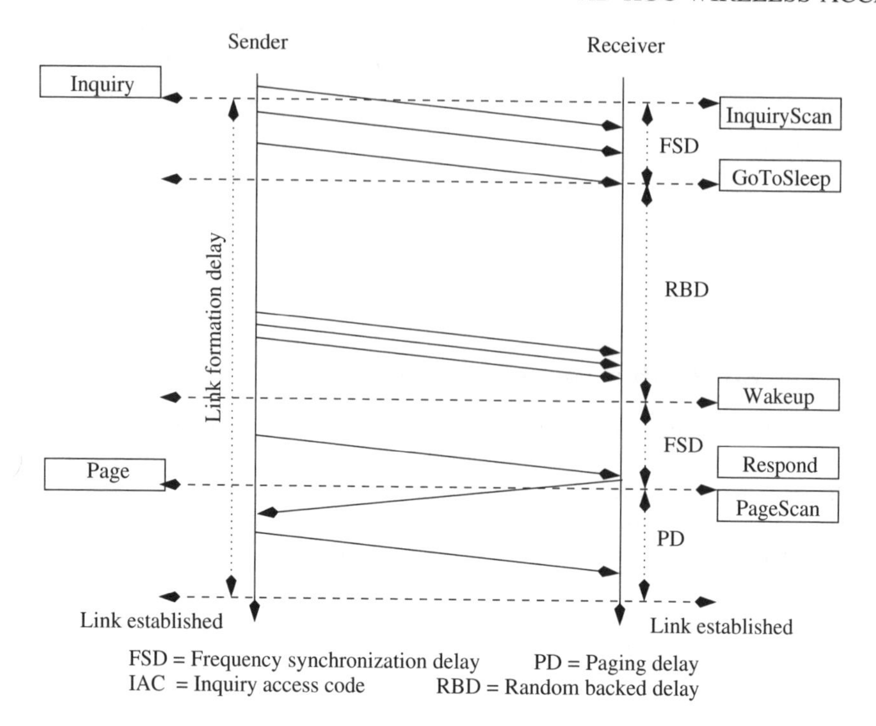

Figure 3.4 The asymmetric Bluetooth protocol.

1. FS (frequency synchronization) delay: about 10 ms.
2. RB (random backed) delay: about 639.75 ms.
3. P (paging) delay: about 625 μs.

Paging is considered negligible since it follows the inquiry state. The Bluetooth asymmetric protocol implies the following delay equation

$$\mathrm{LF} = 2 \cdot \mathrm{FS} + \mathrm{RB} + \mathrm{P} \approx 659.75 \text{ ms}.$$

It is important to note that item 2 (also called *discovery delay*) is also the main cause of delay in the Bluetooth asymmetric protocol. The condition for success of the protocol is depicted in Figure 3.5 for two participating nodes (represented as random variables X_i, $i = 1, 2$) that may talk (T) or listen (L), whereby a sufficiently long period must be established with only one node talking.

To understand synchronization delay, note that if LF is the r.v. measuring the link formation delay and X is the r.v. measuring activity of the merged schedule (time only process talks) then for the connection to be established, the merged schedule X must exceed the link formation delay LF. In particular, if we let $P = Pr[LF \leq X]$ then tt can be proved (using reliability theory) that

$$E[T_c] = \frac{E[X]}{2} + \frac{(E[X|LF > X] + E[X])(1 - P)}{P} + E[LF].$$

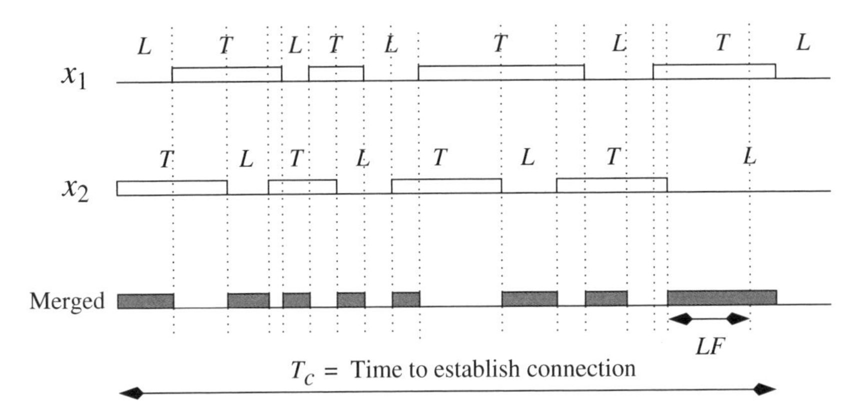

Figure 3.5 Condition for success of the asymmetric Bluetooth protocol.

Concerning the discovery delay procedure, Bluetooth supports the paradigm of *spontaneous connectivity*. The procedure used for node discovery is called *Inquiry* and connections are established on the basis of information exchange.

1. Bluetooth node is set into Inquiry mode by the application.
2. Sends Inquiry messages to probe for other nodes.
3. Other Bluetooth nodes (within the range) only listen.
4. They reply to Inquiry messages only when they have been set explicitly to *InquiryScan* mode.

To prevent collisions and since Inquiry needs to be initiated periodically, some type of randomness must be employed in order to determine the time interval between two Inquiries. This technique is called *Collision Avoidance*.

Once a unit has discovered another unit, connection establishment is very fast. In an ideal scenario, the expected delay for link formation (discovery plus connection) is about 1 s when both nodes follow the uniform distribution between the Inquiry and InquiryScan. In practice this takes several seconds.

The Bluetooth scenario is geared towards settings where a dedicated unit (e.g. a laptop) is responsible for discovering other units. Lengthy connection establishment prevents the use of Bluetooth in fast settings. In a mobile setting, there are other problems like two Bluetooth devices traveling at a relative speed of 36 km/h would never be able to setup connection.

3.1.2.2 Delay management

In an ideal scenario, the expected delay for link formation (Discovery plus Connection) is about 1 s when both nodes follow the uniform distribution between the Inquiry and InquiryScan. In practice, this takes several seconds. To save power, units do not continuously listen to inquiry messages. Instead, they listen for a very short period of time

Table 3.1 Bluetooth power consumption example at 3.3 V.

CPU	Bluetooth mode	Consumption (mW)
Powered down	Detached	<11
Running	Detached	29
Running	Standby	50
Running	InquiryScan	100
Running	Inquiry	200
Running	Send	94
Running	Receive	94

(11.25 ms by default), which, under regular conditions, suffices for the inquiry message to get through with sufficiently high probability. Then the unit enters idle mode for a much longer interval (typically 1.28 s). However, the inquiring unit needs to send inquiry messages (and alternately listen for potential replies) during the entire interval, since it cannot know when the target unit is actually listening. Table 3.1 is based on Kasten and Langheinrich (2001) and depicts energy consumption of Bluetooth devices at various modes.

Default setting of Bluetooth is *Standby* (everything but the internal clock shut-off). No connections can be opened. In *Connect* state, we distinguish four power modes of operation.

- *Active:* Unit actively participates in the channel.
- *Sniff:* Only listen to channel at specified times.
- *Hold:* Unit can take time off for sleeping.
- *Park:* Special mode for slaves that do need to participate in a piconet.

3.1.3 Bluetooth protocol architecture (IEEE 802.15)

The Bluetooth protocol architecture does not follow any of the Open Systems Interconnection (OSI), Transmission Control Protocol/Internet Protocol (TCP/IP) or 802 models and is depicted in Figure 3.6. The protocols can be divided into four categories: Baseband, Link Manager, Logical Link Control and Adaptation Layer (L2CAP), and Service Discovery Protocol (SDP) (see Rathi (2000) for an introductory essay). The baseband and the Link control layers enable the physical RF link to form a piconet among Bluetooth devices. Two types of physical links are provided: Synchronous connection oriented (SCO) and asynchronous connectionless (ACL). ACL packets are used for data only, while the SCO packets may contain audio only or a combination of audio and data. The Link Manager Protocol is responsible for the link setup between Bluetooth units. This layer is concerned with security issues, like authentication and confidentiality. It also deals with control and negotiation of baseband packet sizes. The Bluetooth logical link control and adaptation layer supports higher level multiplexing, segmentation and reassembly of packets, as well as Quality of Service (QoS) communication and groups. It does not provide reliability, but uses the baseband ARQ to ensure reliability. Finally, the SDP is the basis for discovery of services on all Bluetooth devices. Using the SDP device information, services and the

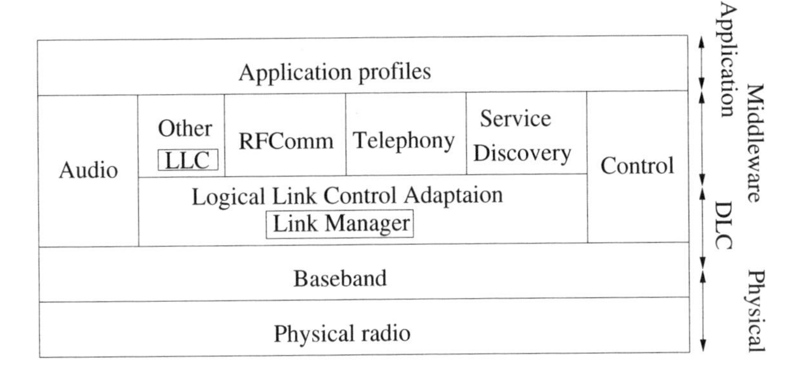

Figure 3.6 Summary of the Bluetooth protocol.

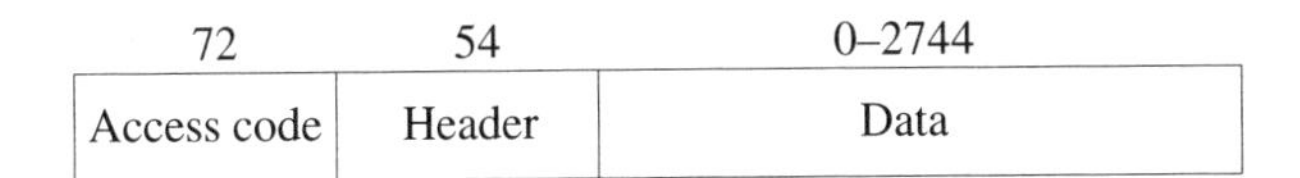

Figure 3.7 Bluetooth frames.

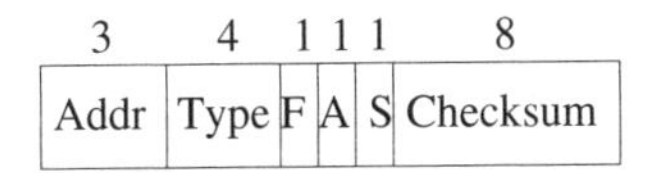

Figure 3.8 Addr identifier in Bluetooth protocol.

characteristics of the services can be queried and then a connection between two or more Bluetooth devices may be established. Other SDPs such as Jini and UpnP may be used in conjunction with the Bluetooth SDP protocol.

The Bluetooth frame depicted in Figure 3.7 includes an access code identifying the master, so that the slaves can tell which traffic belongs to them.

In the Header, (Figure 3.8) Addr identifies which of the active devices frame is intended for. Type identifies frame type.

3.2 Model for node discovery in Bluetooth

During node discovery, a node tries to find out which other nodes are within its reach. The node broadcasts a message and waits to receive a response. Upon receiving such a response from another node, the former node becomes aware of the identity of the latter that knows its presence. The Bluetooth standard resorts to role pre-assignment by using so called *asymmetric* protocols (see for example Bluetooth, SIG, Inc., (2001b)), whereby nodes have preassigned roles that establish beforehand which node will be sending (discovering other nodes) and which nodes will be receiving (waiting to be discovered). Unfortunately, this is unrealistic since there are no centralized entities in an ad hoc network that could assign the roles (see Salonidis et al. (2001)). An alternative is to use *symmetric* protocols whereby a node can be both discovering nodes and waiting to be discovered. However, in symmetric protocols nodes must compete for access to the communication channel without being aware of the presence of other nodes which, in turn, reduces their chance to coordinate their transmissions in order to minimize collisions.

3.2.1 Avoiding collisions

Collisions in packet radio networks are caused by either *direct* or *secondary* interference Garcés and Garcia-Luna-Aceves (2000, March 26–30). Direct interference occurs when two nodes transmit to each other at the same time while secondary interference occurs when nodes unaware of each other's presence attempt to transmit at the same time. Currently, there are two solutions to this problem in ad hoc networks. One is frequency hopping as it is used in Bluetooth (see Bluetooth, SIG, Inc., (2001b)). Frequency hopping minimizes the probability of collisions by having nodes transmitting in multiple channels. The other solution, used when a single broadcast channel is shared among all nodes, are randomized backoff protocols (see Tanenbaum (1996, 4th edition), among others), such as implemented in wireline networks. Collision avoidance in a shared broadcast channel has been widely studied by Garcés and Garcia-Luna-Aceves (2000, March 26–30). Existing collision avoidance protocols are based on the exchange of control messages among the nodes in order to dynamically establish a transmission schedule with the highest possible throughput. However, all these protocols assume that nodes are already aware of the presence of other nodes. In other words, they assume that node discovery has already taken place.

Existing protocols take advantage of the frequency hopping characteristics of Bluetooth and cannot be applied in single broadcast channel systems. In both cases, collisions are minimized by randomly assigning roles to each node. With a shared broadcast channel, however, once there are more than two nodes sending, collisions will occur. The goal here is to provide a formal characterization of node discovery protocols and suggest a simplified model for analysis and comparison in shared channel networks. Our optimization criteria is the time it takes for two nodes to discover each other in the presence of other nodes also engaged in node discovery. The protocols we propose have the same probabilistic flavor as in Law et al. (2001), but we consider additional techniques to minimize collisions. In addition, we also consider the *exponential backoff* approach where, to avoid collisions, a node backs off for a period of time before answering a message. In frequency hopping systems, exponential back off is also used but only as an internal step of the protocol (e.g. in Bluetooth, before a node sends an *inquiry response* message Bluetooth, SIG, Inc., (2001b)). This comparison is important since backoff protocols are a standard way to deal with collisions in shared channel systems. Our analysis shows that probabilistic approaches might be an efficient alternative to backoff protocols. This result is relevant not only for shared channel systems but also for frequency hopping systems that could use a similar approach to the one we describe in all situations where exponential backoff is used.

3.2.2 Details of the node discovery model

In the sequel, we concentrate on systems with a shared broadcast channel such as those found in the Ricochet[1] or the Near-Term Digital Radio[2] networks by Ramanathan and Hain

[1] http://www.metricom.com
[2] http://www.gordon.army.mil/tsmtr/ntdr.htm

(2000). First, we propose a model that can be used to analytically compute lower bounds for node discovery protocols. Second, we describe several protocols that shed light on the problem of node discovery and perform better than existing solutions.

We assume a system with k nodes. The nodes do not have an ID and the nodes do not know k. The nodes communicate by broadcasting messages. A node x_i, $1 \leq i \leq k$, can be in one of the two states: *talking* (T) or *listening* (L). The state of each node is determined by the node discovery protocol run at that node. We assume that the nodes are synchronized, that is, they change states at the same point in time and remain in a given state for a period of time that is identical for all nodes. We also disregard the initial phase when not all nodes are operative and assume that all nodes start working at the same time.

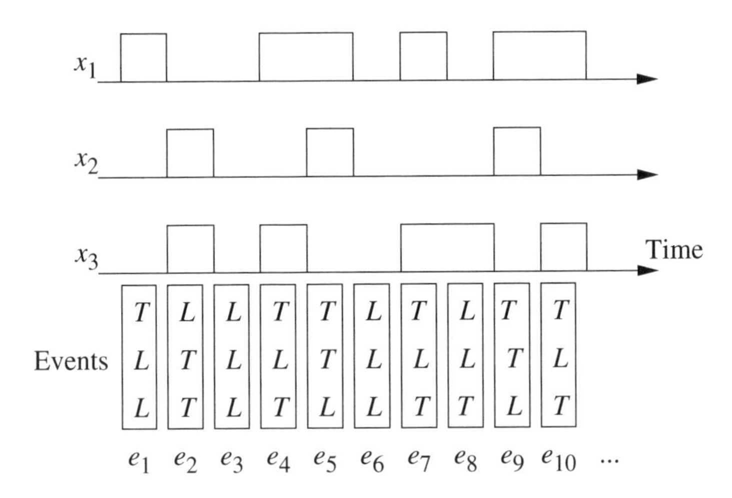

Figure 3.9 Schematic representation of the model for $k = 3$.

This model is shown in Figure 3.9. Each node is depicted as a series of transitions between states (high being T and low being L). A *run*, R, of the protocol is a totally ordered sequence of *events*, e_j, $(1 \leq j \leq m)$, where each event contains the state of the k nodes in the system at that point in time. For simplicity, we will assume that e_i precedes e_j in a run if $i < j$. In addition, two events e_i, e_j are consecutive in a run with e_i preceding e_j if $j = i + 1$. Thus, for $k = 2$, the possible events are

$$e_i \in \left\{ \begin{matrix} L \\ L \end{matrix}, \begin{matrix} L \\ T \end{matrix}, \begin{matrix} T \\ L \end{matrix}, \begin{matrix} T \\ T \end{matrix} \right\}.$$

We will represent the state of a node x_i during event e_t as $x_i^{e_t}$. Next, we define when messages can be received.

Definition 3.2.1 (Receiving a Message) *A node x_i receives a message from a node x_j, $i \neq j$, during event e_t iff $x_i^{e_t} = L$, $x_j^{e_t} = T$, and $\forall x_k$, $k \neq j$, $x_k^{e_t} = L$.*

Note that this definition implies several things. First, if more than one node talks during an event, no message will be received during that event. Second, if a node receives a message during an event, all other nodes that are listening during that event will also receive the message. That is, if a node x_i receives a message from x_j in an event, all other nodes in the system will also receive the same message. This is a simplified model that, nevertheless, is a faithful representation of shared broadcast channels.

The node discovery protocol *terminates* when two nodes have found each other. This happens when a node successfully sends a message to another node and the second node manages to successfully send a reply message to the first node immediately afterwards. More formally we have the following definition.

Definition 3.2.2 (Protocol Termination) *A run R terminates when the following two consecutive events occur:*

- $e_o \mid (\exists x_i^{e_o} = T) \wedge (\forall j \neq i,\ x_j^{e_o} = L)$, *and*
- $e_p \mid (\exists x_k^{e_p} = T,\ k \neq i) \wedge (\forall l \neq k,\ x_l^{e_p} = L)$

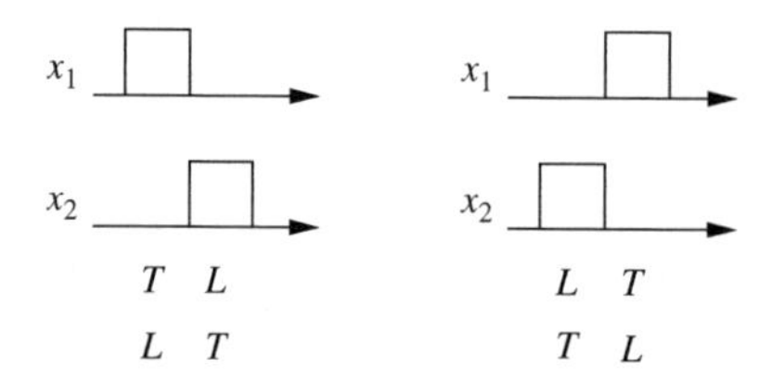

Figure 3.10 The two possible terminating event sequences ($k = 2$).

The termination of the protocol represents the point when two nodes have discovered each other. Figure 3.10 shows the two possible terminating sequences for $k = 2$:

$$\begin{matrix} L\,T \\ T\,L \end{matrix} \ \text{or}\ \begin{matrix} T\,L \\ L\,T \end{matrix}.$$

This definition also has important implications for the model. The most relevant one is that the two events must be consecutive. This is because the second phase of node discovery usually involves a message that is explicitly addressed to one node (to the sender in the previous event). If this message is lost because of interference or because the intended recipient is not listening, then node discovery will not occur. The reason is that there are no acknowledgements in the protocol and a node has no way of knowing whether a message has been received or not. Although here we do not differentiate between types of messages, this behavior is captured by the constraint that the two events must be consecutive.

We define the *length* of a run to be the number of events from its start until its termination. Since the termination of the protocol requires two events, it follows that the minimum length of a run is 2. In the rest of this chapter, we will focus on the design and analysis of node discovery protocols that minimize the expected length of a run.

3.2.3 Protocols for node discovery

We now propose several node discovery protocols for $k = 2$. These protocols will be later generalized to an arbitrary number of nodes. However, the discussion in the simpler case of

$k = 2$ is very helpful to illustrate the problem of node discovery. In the sequel we consider five protocols: $\mathcal{SP}$ (Sleep Protocol), $\mathcal{RP}$ (Random Protocol), $\mathcal{AP}$ (Answering Protocol), $\mathcal{LP}$ (Listen after Talking Protocol), $\mathcal{CP}$ (Conditional Protocol). In brief, these protocols operate as follows:

- $\mathcal{SP}$: After reception, the node that received the message suspends activity for a period of time chosen uniformly and independently in the time interval 1.30 time units.
- $\mathcal{RP}$: Nodes decide at random whether they should talk or listen.
- $\mathcal{AP}$: A node will immediately answer if it receives a message.
- $\mathcal{LP}$: A node talking will immediately switch to listening.
- $\mathcal{CP}$: A node always listens after having talked and answers if it receives a message.

Following is a more detailed discussion of these five protocols:

3.2.3.1 Random protocol

In the *Random Protocol* ($\mathcal{RP}$), each node decides at random whether to talk or to listen. This protocol is similar in spirit to that proposed by Law et al. (2001) except that in our protocol a node can change states from event to event. In the device discovery protocol of Law et al. (2001), nodes are permanently assigned to a state and remain in that state for the duration of the protocol. Obviously, this does not work in shared single broadcast channels. The link establishment protocol of Salonidis et al. (2000) is also similar to $\mathcal{RP}$ in that the nodes change their state at random. However, unlike Salonidis et al. (2000) in our protocol nodes remain in a given state for the entire duration of an event before considering switching state. Except for these differences, $\mathcal{RP}$ is equivalent to the protocols of Law et al. (2001) and Salonidis et al. (2000) but on a different network model.

3.2.3.2 Answering protocol

$\mathcal{RP}$ tries to minimize collisions by forcing nodes to switch from talking to listening with certain probability. However, this ignores the pattern of events needed for the protocol to terminate. Since this pattern is known, it is possible to make the nodes behave in such a way so as to favor the occurrence of such patterns. The first example of this adaptation is the *Answering Protocol*, $\mathcal{AP}$. In $\mathcal{AP}$, a node will immediately answer if it receives a message from another node. In this way, at least one of the nodes is doing the correct thing after a message has been successfully sent from one node to the other. For comparison, in Bluetooth, when a node receives an *inquiry* message, it backs off for a random period of time, listens again on the same frequency and when it receives a second *inquiry* message from the same node then it responds immediately Salonidis et al. (2001). In $\mathcal{AP}$, we do not introduce a delay and we do not wait for a second message. As soon as it receives a message, a node answers immediately. Such an approach will clearly not work for $k > 2$. However, understanding $\mathcal{AP}$ helps to illustrate the behavior of the backoff mechanism that will be used in the $\mathcal{SP}$ protocol.

3.2.3.3 Listening protocol

An alternative way to modify the $\mathcal{RP}$ protocol is based on the fact that after sending a message (i.e. talking), a node needs to listen in order to discover the other node. This idea leads to the *listening after talking* protocol, LP. For comparison, in frequency hopping systems like Bluetooth, nodes actively trying to discover other nodes alternatively send messages and listen for answers according to a pre-established pattern of frequencies. This is the same mechanism exploited by the $\mathcal{LP}$ protocol. However, in single channel systems, there is no pattern of frequencies to follow, all messages are sent on the same channel. Thus, a node needs to listen more frequently because, when it talks, it might cause both direct interferences (with nodes that are trying to respond) and hidden terminal problems (for nodes that are either sending or responding to other nodes). The $\mathcal{LP}$ protocol takes this into account by forcing nodes to listen after they have sent.

3.2.3.4 Conditional protocol

The protocols $\mathcal{LP}$ and $\mathcal{AP}$ are complementary in the sense that $\mathcal{AP}$ describes what to do after listening while $\mathcal{LP}$ describes what to do after talking. Both can be combined into a *conditional protocol* where a node always listens after having talked and answers if it receives a message. The interesting aspect of this protocol is that, for $k = 2$, it is enough to reach the events $\frac{L}{T}$ or $\frac{T}{L}$ to establish a link. After such event occurs, the system is deterministic and the next event will complete the node discovery procedure. Thus, the probability of node discovery is given by $2pq$.

The protocols discussed in the preceding text introduce a probabilistic component that is common in frequency hopping systems but which, to our knowledge, is not used in single shared channel systems. In these systems, the most widespread approach to avoid collisions is to introduce delays to avoid simultaneous transmissions from several nodes. In conventional networks, collisions can be detected using CSMA-like systems Kleinrock and Tobagi (1975). When a collision is detected, the node waits for a random period of time before re-transmitting in the hope of avoiding a new collision. In our model, we are not concerned with standard collision detection (which takes place at the lower layers of the communication stack) but with the simultaneous response to a message from a node trying to discover other nodes. Of course, for $k = 2$, this does not make much sense since there cannot be simultaneous responses. However, it is worthwhile to analyze this type of protocol in the simpler case to get a precise idea of its behavior.

3.2.3.5 Sleep protocol

To consider the effects of exponential backoff on our model, we introduce the *sleep protocol*, $\mathcal{SP}$. In $\mathcal{SP}$, after reaching either $\frac{L}{T}$ or $\frac{T}{L}$, the node that received the message suspends its activity for a period of time. This period can be modeled as a uniform random variable with values between 1 and n (For an estimate of the value of n, we can look at the case of Bluetooth protocols. In Bluetooth, a node talking can scan all frequency ranges in 10 μs or 20 μs depending on where (which country) the system operates. Upon receiving an *inquiry* message, a node suspends its activity for of 639.375 μs. Assuming that the duration of an event in our model is the time it takes for a node to scan all frequencies, then the inactivity period varies between roughly 30 and 60 time intervals.) After this period, the node wakes up and sends a message back to the original sender.

3.2.3.6 Analysis of protocol $\mathcal{AP}$

Next we carefully analyze only the performance of the AP and give the expected running time of the other protocols in Table 3.2. Additional details can be found in the exercises and in Alonso et al. (2003b).

Table 3.2 Expected running time of Bluetooth protocols.

Bluetooth protocol	E[# Steps]
$\mathcal{SP}$: Sleep (Bluetooth)	$\frac{1}{2pq^2} + \frac{30}{q}$
$\mathcal{RP}$: Random	$\frac{2-p^2-q}{2p^2q^2}$
$\mathcal{AP}$: Answering	$\frac{2-p^2-q^2}{2pq^2}$
$\mathcal{LP}$: Talk/Listen	$\frac{2-q^2}{2p^2q}$
$\mathcal{CP}$: Conditional	$\frac{2-q^2}{2pq}$

Theorem 3.2.3 (Alonso et al. (2003b)) *The expected length of a run of the $\mathcal{AP}$ protocol, $R_{\mathcal{AP}}$ is given by the following expression:*

$$E(R_{\mathcal{AP}}) = \frac{2-p^2-q^2}{2pq^2}.$$

Proof. We will consider different sub-runs, $S_{R_{\mathcal{AP}}}$, of the protocol $R_{\mathcal{AP}}$. A subrun has two phases. In phase 1, the protocol acts as an RP until either $\genfrac{}{}{0pt}{}{L}{T}$ or $\genfrac{}{}{0pt}{}{T}{L}$ is produced. The expected length of this phase is $\frac{1}{2pq}$, since the probability of the events of interest is $2pq$ and it is a sequence of independent events. Phase 2 consists of one event. The possible events are restricted in phase 2. If phase 1 finished with $\genfrac{}{}{0pt}{}{T}{L}$, phase 2 can be either $\genfrac{}{}{0pt}{}{L}{T}$ or $\genfrac{}{}{0pt}{}{T}{T}$. If phase 1 finished with $\genfrac{}{}{0pt}{}{L}{T}$, phase 2 can be either $\genfrac{}{}{0pt}{}{T}{L}$ or $\genfrac{}{}{0pt}{}{T}{T}$. If phase 2 is the event $\genfrac{}{}{0pt}{}{T}{T}$, the subrun fails and a new subrun starts. Otherwise the protocol terminates.

More precisely, the protocol acts randomly until either of the following two states occurs

$$\genfrac{}{}{0pt}{}{L}{T} \text{ or } \genfrac{}{}{0pt}{}{T}{L}.$$

Then the protocol has the following state transitions:

$$\begin{pmatrix} L & \rightarrow & T \\ T & & Z \end{pmatrix} \text{ or } \begin{pmatrix} T & \rightarrow & Z \\ L & & T \end{pmatrix},$$

where $Z = L$ or T. When $Z = T$ the outcome is

$$\begin{matrix} T \\ T \end{matrix},$$

the subrun fails and protocol repeats. Essentially you are flipping two random variables X, Y,

$$\begin{matrix} X \\ Y \end{matrix}$$

and a subrun occurs when $X \neq Y$. Hence,

$$Pr[\text{Subrun occurs}] = Pr[X \neq Y] = 2pq$$

and expected number of steps until a subrun occurs is

$$\frac{1}{2pq}.$$

Recall that when finally a subrun occurs, another experiment is performed and the probability of this last experiment succeeding is q. Expected length of a run until a subrun of the protocol occurs is $1 + \frac{1}{2pq}$. The probability that a subrun fails is p. Hence,

$$\begin{aligned} E[\#\text{ steps}] &= \sum_{i=1}^{\infty} i\left(1 + \frac{1}{2pq}\right) p^{i-1} q \\ &= \left(1 + \frac{1}{2pq}\right) q \sum_{i=1}^{\infty} i p^{i-1} \\ &= \left(1 + \frac{1}{2pq}\right) q \frac{1}{(1-p)^2} \\ &= \left(1 + \frac{1}{2pq}\right) \frac{1}{q} \\ &= \frac{2 - p^2 - q^2}{2pq^2}. \end{aligned}$$

This completes the proof of the theorem. ■

3.2.3.7 Protocol comparison

The table summarizes the running times of the various protocols considered. In general, the objective of these protocols is to minimize the delay involved in discovering other nodes.

It is easy to compare the performance of the five protocols proposed as a function of the probability of talking p. The symmetric characteristics of the protocols $\mathcal{RP}$, $\mathcal{AP}$, $\mathcal{LP}$, and $\mathcal{CP}$ is clearly recognizable. $\mathcal{RP}$ has the narrowest range of values of p where it is feasible, with a minimum for $p = 0.5$. $\mathcal{AP}$ benefits from lower values of p (minimum at $p = 0.3$) since this diminishes the probability that the first sender talks at the send time the response is sent. In contrast, $\mathcal{LP}$ benefits from higher values of p (minimum at $p = 0.65$) since it compensates these higher values by forcing nodes to listen after they talk. Finally, $\mathcal{CP}$ benefits from these two properties and offers an acceptable response for the widest range of values of p (minimum at $p = 0.4$).

For $\mathcal{SP}$ we have used $n = 30$. $\mathcal{SP}$ follows the same pattern as $\mathcal{AP}$. This is not surprising as $\mathcal{SP}$ is, indeed, an $\mathcal{AP}$ protocol with a delay between the two terminating events. The delay caused by the inactive period before sending the response is what causes the curve to be shifted upwards and to the left. The faster deterioration in behavior as p increases that can be observed in $\mathcal{SP}$ is due to the hidden penalty in case the subrun does not lead to node discovery: the inactive period is wasted time that increases the overall delay. This wasted time is proportional to p since the probability that the first sender is listening when the respond is sent decreases with higher p.

This comparison helps to understand the nature of the protocols but it is not entirely practical. The probability of sending, p, is a parameter of the protocol but does not need to be the same for all of them. Thus, to decide which protocol is the best, we need to compare the best values for each one of them. This is done in Figure 3.11, where it is clear that any of the proposed protocols performs better than $\mathcal{SP}$ unless $n = 1$, in which case $\mathcal{SP}$ reverts to $\mathcal{AP}$.

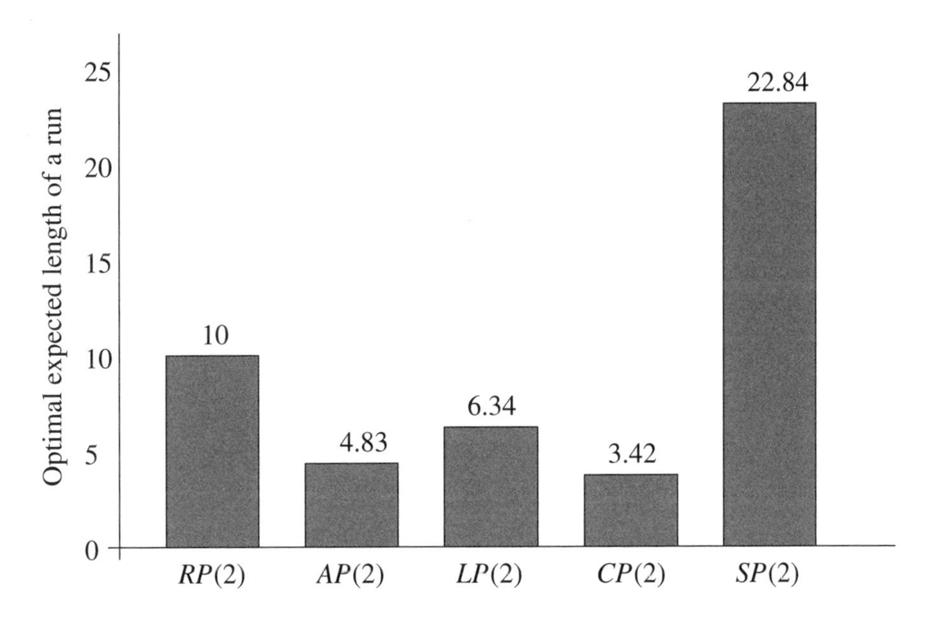

Figure 3.11 Optimal values for each protocol.

3.2.4 Multiple nodes competing for air-time

With $k > 2$, an event is now a vector of k elements, s_i, $1 \leq i \leq k$, with values in the set $\{L, T\}$. For instance, for $k = 4$, a run of the protocol terminates with a sequence of the form:

$$\begin{matrix} T & L \\ L & T \\ L & L \\ L & L \end{matrix}$$

(by Definition 2). For $k = 4$, there are 12 such sequences.

We will refer to the $\mathcal{AP}$, $\mathcal{LP}$, and $\mathcal{CP}$ protocols in a k-node environment as $\mathcal{AP}(k)$, $\mathcal{LP}(k)$, and $\mathcal{CP}(k)$, respectively. However, of these three protocols, only $\mathcal{LP}(k)$ makes sense in practice. Both $\mathcal{AP}(k)$ and $\mathcal{CP}(k)$ force a node to answer immediately when it receives a message. When there are k nodes, the fact that a given node receives a message implies that there are $k - 1$ nodes that have also received the same message. If all of them reply immediately , no message will get across because of the resulting interference. Thus, these nodes must either wait for a period of time (as in the $\mathcal{SP}$ protocol) or respond only with certain probability. If this probability is p, as it would be reasonable to assume, then $\mathcal{AP}(k)$ reverts to $\mathcal{RP}(k)$ and $\mathcal{CP}(k)$ becomes $\mathcal{LP}(k)$.

3.2.4.1 Analysis of protocol $\mathcal{RP}(k)$

Theorem 3.2.4 (Alonso et al. (2003b)) *The expected length of a run of the $\mathcal{RP}(k)$ protocol, $R_{\mathcal{RP}(k)}$ is given by the following expression:*

$$E(R_{\mathcal{RP}(k)}) = \frac{1}{k} \cdot \left(\frac{1}{pq^{k-1}} + \frac{1}{(k-1)p^2q^{2k-2}} \right).$$

Proof. Protocol $\mathcal{RP}(k)$ can be analyzed using the technique of Li (1980). For k users we have k independent and identically distributed random variables $X_1, X_2, \ldots, X_k$, one for each user. For any subset $S \subseteq \{1, 2, \ldots, k\}$ let $A_S := b_1 b_2 \cdots b_k$, where

$$b_i = \begin{cases} T & \text{if } i \in S \\ L & \text{if } i \notin S. \end{cases}$$

It is clear that the probability that A_S occurs is equal to $p^{|S|}q^{k-|S|}$. Consider an alphabet with the 2^k letters A_S, where $S \subseteq \{1, 2, \ldots, k\}$. Let us abbreviate $A_i := A_{\{i\}}$. Consider the patterns $P_{i,j} := A_i A_j, i \neq j$. Clearly, the algorithm succeeds when a string pattern $P_{i,j}$ occurs, for some $i \neq j$. The probability that a given pattern $P_{i,j}$ occurs is exactly $\frac{1}{p^2q^{2k-2}}$. Consider the following random variable

$$N = \text{ waiting time until one of the patterns } P_{i,j}, i \neq j \text{ occurs.}$$

We want to compute $E(N)$, the expected number of steps until the protocol succeeds, i.e. the expected length of a run $E(R_{\mathcal{RP}(k)}) = E(N)$.

According to the method of Li (1980) (see also Blom and Thoburn (1982)), the value $E(N)$ will be derived from the solution of a linear system which we now set up. Let the *stepping probability* $\pi_{i,j}$ be the probability that pattern $P_{i,j}$ occurs before any other pattern. It is, of course, clear that

$$\sum_{(i,j),i\neq j} \pi_{i,j} = 1. \tag{3.5}$$

We are interested in measuring the *overlap* between any two different patterns $P_{i,j}$, $P_{i',j'}$, $i \neq j, i' \neq j'$; if the last letter of $P_{i,j}$ is equal to the first letter of $P_{i',j'}$ then we say that we have an *overlap* between $P_{i,j}$ and $P_{i',j'}$ Thus an overlap between $P_{i,j}$ and $P_{i',j'}$ is possible only if $j = i'$. Consider the quantity

$$e_{(i,j),(i',j')} = \begin{cases} 1/(p^2q^{2k-2}) & \text{if } i = i', j = j' \\ 1/(pq^{k-1}) & \text{if } j = i' \\ 0 & \text{if } j \neq i' \end{cases}. \tag{3.6}$$

Consider the following Equations:

$$\sum_{(i,j),i\neq j} e_{(i,j),(i',j')} \cdot \pi_{i,j} = E(N), \text{ for all } (i', j'), i' \neq j'. \tag{3.7}$$

According to the main Theorem of Li (1980), the expected value $E(N)$ and the stepping probabilities are obtained as solutions of the system of Equations 3.5 and 3.7.

We now proceed to solve this system. Fix a pair (i, j) and look at coefficients $e_{(i,j),(i',j')}$ of $\pi_{i,j}$ in the system (3.7). We add all the equations in system 3.7. If we use Identities (3.6) and make elementary calculations, we see that the coefficient of $\pi_{i,j}$ in the resulting equation is equal to

$$\sum_{(i',j')\neq(i,j),i'\neq j'} e_{(i,j),(i',j')} = \frac{1}{p^2q^{2k-2}} + \frac{k-1}{pq^{k-1}}.$$

Using Identity (3.5), it follows that

$$\left(\frac{1}{p^2q^{2k-2}} + \frac{k-1}{pq^{k-1}}\right) \cdot \left(\sum_{(i,j),i\neq j} \pi_{i,j}\right) = \frac{1}{p^2q^{2k-2}} + \frac{k-1}{pq^{k-1}}$$
$$= k(k-1)E(N).$$

This completes the proof of the theorem. ■

3.2.4.2 Analysis of protocol $\mathcal{LP}(k)$

In protocol $\mathcal{LP}(k)$ nodes are forced to listen after they have sent. The behavior of this protocol can be represented with a finite automaton (Figure 3.12). The automaton has four states. S is the initial state representing the state of the system when the protocol is initiated and all events where there is no node in state L or there are j nodes in state T with $1 < j < k$. T_1 represents all events where exactly one node is in state T. T_k represents the event when all nodes are in state T. E is the final state reached, once node discovery has been completed. Figure 3.12 also reflects the probability of the transitions among the states.

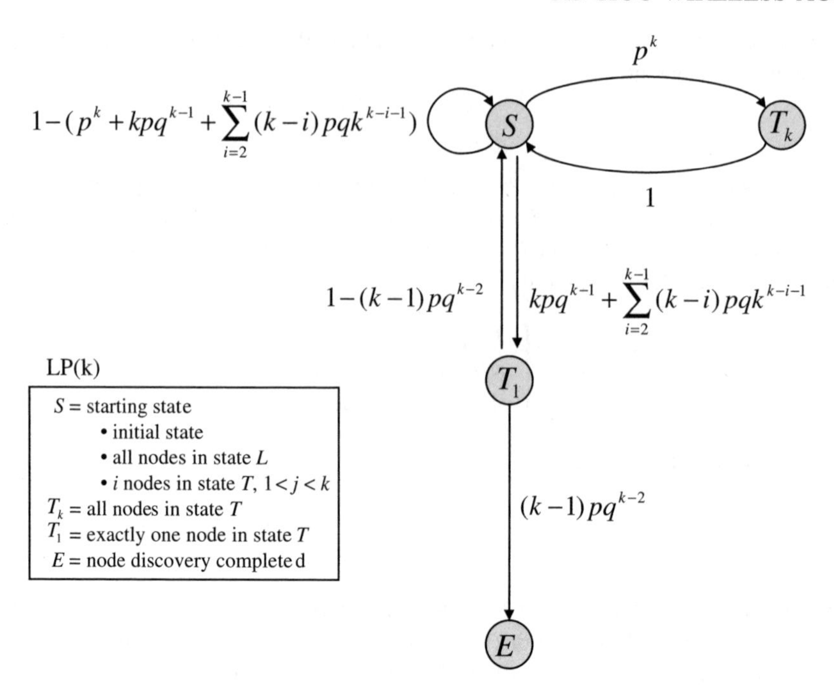

Figure 3.12 The $\mathcal{LP}(k)$ protocol as finite automaton.

The analysis of the protocol can be done on the basis of this automaton. For simplicity, we use the following abbreviations:

$$A = kpq^{k-1}$$
$$B = \sum_{i=2}^{k-1}(k-i)pq^{k-i-1}$$
$$C = (k-1)pq^{k-2}$$
$$D = p^k$$

We can prove the following theorem.

Theorem 3.2.5 (Alonso et al. (2003b)) *The expected length of a run of the $\mathcal{LP}(k)$ protocol, $R_{\mathcal{LP}(k)}$ is given by the following expression:*

$$E(R_{\mathcal{LP}(k)}) = \frac{1+A+B+D}{C(A+B)}.$$

Proof. Using the finite automaton as the basis, the expected number of steps needed by the protocol can be calculated by solving the following set of equations:

$$E = 0$$
$$T_1 = C(1+E) + (1-C)(1+S)$$

$$T_k = 1 + S$$
$$S = (A + B)(1 + T_1) + (1 - A - B - D)(1 + S) + D(1 + T_k)$$

where each equation represents the expected number of steps required to go from each of the four states to state E. Solving for S, the result follows. This completes the proof of the theorem. ■

3.3 Bluetooth formation algorithms

A Bluetooth formation algorithm is necessary not only to create a connected network topology but also to assign roles of master, slave or bridge to a given a set of Bluetooth devices. Some of the important issues that need to be addressed include:

1. Design of network architecture, (e.g. number of piconets per scatternet, bridges per piconet, piconets attached to a bridge, slaves belonging to a master), and how these may influence packet collisions, synchrony, timing, switching and overall network performance.

2. Network management and role assignment of master, slave, bridge, respectively, to various participating devices so as to optimally distribute resources and enhance network efficiency.

3. Mobility and topology changes support, including, addition of nodes, and deletion of nodes and how to efficiently configure and update the new network parameters and maintain stability.

4. Modes of operation (e.g. active, sniffing, parking, hold) and how they affect resource allocation and energy consumption.

Taking these four important issues into account helps our understanding of Bluetooth formation algorithms.

A standard approach to building Bluetooth networks requires learning the Bluetooth device interconnections first and using these for the construction of a connected system. In a sense, one has to create a structure of logical connections that respect physical device limitations (e.g. bounded degree requirement). Invariably, one is led to constructing spanning connected structures. There are several possibilities for doing this including breadth first search and depth first search trees as well as various types of minimum weight spanning trees. However, constructing bounded degree spanning trees is an NP-complete problem (see Garey and Jhonson (1979)).

In the sequel, we discuss the details of several scatternet formation algorithms and their performance.

3.3.1 Bluetooth topology construction protocol

Salonidis et al. (2000, 2001) propose and analyze a symmetric protocol for *2-node link formation* (i.e. node discovery with 2 nodes). The protocol is based on random schedules where each node switches between sending and listening at random and stays in each state

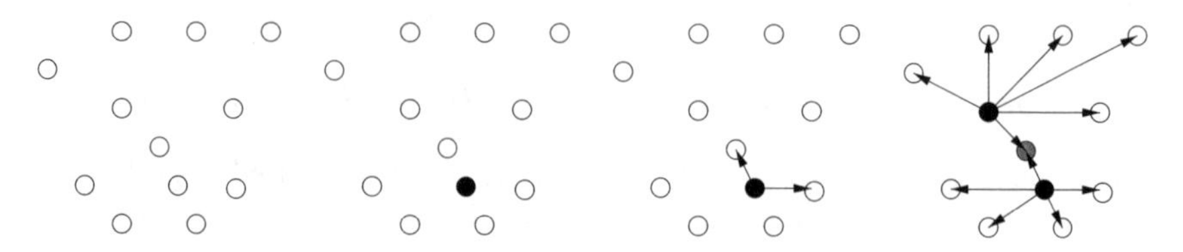

Figure 3.13 BCTP scatternet formation algorithm: initially all nodes are alternating status until one is elected master. The resulting network has two masters and one bridge.

for a random period of time. A random schedule is needed because the mean connection time (the time needed for the two nodes to discover each other) is infinite if the nodes switch state deterministically Salonidis et al. (2000).

This *alternating states* protocol has been used as the basic building block for a scatternet formation protocol Salonidis et al. (2001). The first algorithm, called *Bluetooth Topology Construction Protocol* (*BTCP*), is due to Salonidis et al. (2001) and the resulting Bluetooth network is according the rules specified in Subsection 3.1.1. The algorithm (Figure 3.13) assumes that all hosts are within one hop from each other and is as follows.

1. Initially nodes collect information about each other by alternating between Inquiry and InquiryScan states.

2. Node with highest Bluetooth device address is elected leader among all participating nodes.

3. The designated masters are connected by paging from the leader.

4. The masters page their slaves and bridges wait to be paged by their two masters.

Connection establishment delay is caused if all hosts start at the same time and if this delay becomes longer, connectivity cannot be guaranteed.

3.3.2 Bluetree

The following algorithm due to Zaruba et al. (2001) essentially constructs a breadth first search tree, called *Bluetree* (Figure 3.14).

1. Bluetooth devices acquire Bluetooth addresses of their neighbors.

2. An arbitrary is selected to become the root, also called *blueroot*, of the tree and acts as the master while all intermediate neighboring nodes become slaves.

3. The resulting slaves page their neighbors that are not direct neighbors of the blueroot and these devices become bridges.

4. The procedure is iterated until the leaves of the tree are reached.

An alternative to the above algorithm is also presented in Zaruba et al. (2001) whereby multiple hosts are initially becoming bluetrees. These in turn form a forest of bluetrees

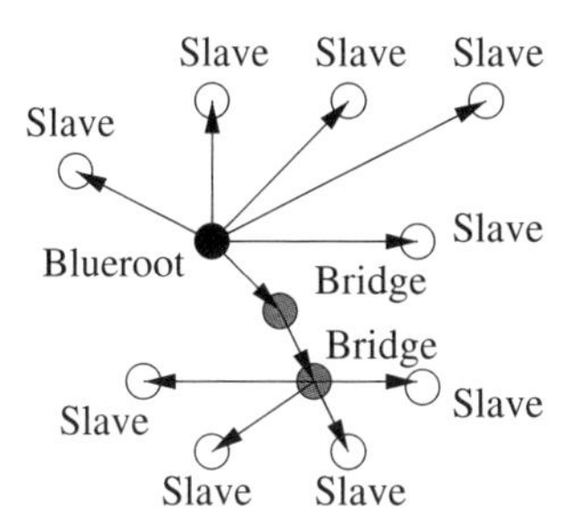

Figure 3.14 Bluetree scattermet formation algorithm.

that are subsequently merged to form a single bluetree. Of course, this algorithm has the advantage that the setup time is reduced since more nodes are initiating the Bluetooth network formation procedure.

3.3.3 Tree scatternet

The tree scatternet algorithm of Miu et al. (2002) resembles the Bluetree algorithm in that it also creates a scatternet with a distinct tree structure.

1. The topology created is a collection of rooted trees (this includes isolated nodes) and the basic operation of the algorithm is to merge the trees into a single tree.

2. These trees have coordinating nodes whose purpose is to detect the presence of other trees. Once detected, they inform their corresponding roots. The participating roots page each other and eventually merge their trees.

It is also possible to detect node insertions and deletions and update the resulting scatternet.

3.3.4 Bluenet

One of the main limitations of the bluetree algorithm is that the participating host configurations may form bottlenecks which are imposed by the hierarchical nature of the formation algorithm. The Bluenet algorithm of Wang et al. (2002) tries to overcome this limitation and it has two phases.

1. First, is the discovery phase whereby each node learns of its neighbors.

2. Master nodes are then selected randomly and they invite up to d (d is some fixed parameter) of their neighbors to form a piconet by becoming slaves.

3. Nodes that become slaves stop exploring in the next phase.

4. In the second phase, the remaining isolated nodes are connected to piconets and become bridges, in case they get connected to more than one piconet.

5. In the third and final phase, the piconets get connected together and form a scatternet.

The final phase is very critical since it must take into account the sizes of the resulting piconets formed, which may have to split again in order to maintain not only the required piconet sizes of the Bluetooth specification but also attain connectivity. As a consequence, the algorithm may have convergence problems that are not addressed clearly by the authors.

3.3.5 Scatternet formation algorithm

Law et al. (2001) propose a probabilistic protocol for node discovery. In this protocol, a node decides with probability p to start discovering other nodes or, with probability $1 - p$, to listen until it is discovered by other node. If a node does not manage to discover any other node in a period of time, it gives up. Similarly, if a node does not hear from any other node in a period of time, it also gives up. The protocols aim at establishing only one-to-one connections. The number of connections established in each round of the protocol is the smaller of the number of nodes in discovering mode and the number of nodes waiting to be discovered.

The algorithm assumes that all nodes are within the range of each other and consists of the following steps:

1. The set of Bluetooth devices is partitioned into sets of interconnected devices (called *components*) which may be either isolated nodes, piconets or scatternets themselves.

2. Components have leaders and when they get interconnected, only one leader survives while the others back down.

3.3.6 Loop scatternet

The loop scatternet algorithm assumes that all hosts are within the range of each other (Figure 3.15).

1. In the first phase, the algorithm forms a ring of p piconets.

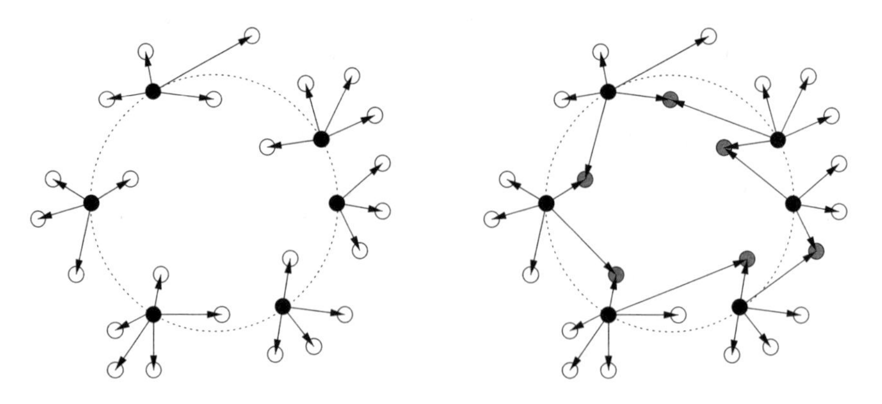

Figure 3.15 The two phases of the loop scattermet formation algorithm.

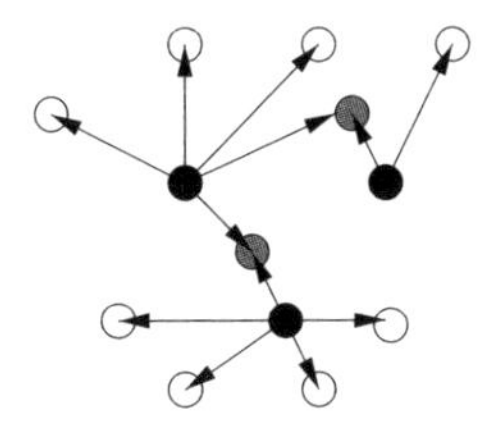

Figure 3.16 Bluestar scattermet formation algorithm.

2. In the second phase, more bridges are created in order to reduce the diameter of the Bluetooth network. This is done by attaching a bridge between the i-th and the $i \pm \sqrt{p}$ piconets, where p is the number of piconets in the first phase.

3.3.7 Bluestar

The bluestar algorithm, due to Petrioli et al. (2003), has three phases. (Figure 3.16).

1. The first phase consists of a (symmetric, i.e. if node u discovers v then node v also discovers u, and vice versa) node discovery of neighboring nodes.
2. The second phase is the bluestar phase whereby piconets are formed. Bluetooth device roles are determined by weights and roles assigned to masters and slaves in increasing order of weights.
3. The last phase connects piconets in order to form the scatternet.

3.4 Mesh mode of WiMAX/802.16

WiMAX/802.16 is a broadband wireless access technology (IEEE et al. (2004)). It is illustrative of a type of infrastructure network which has ad hoc features. A WiMAX/802.16 wireless access network consists of a base station (BS) and mobile stations (MSs). The BS provides network attachment to the MSs. There are two kinds of topology supported: the point-to-multipoint topology and mesh topology. In a point-to-point multipoint topology, the BS communicates directly with each MS and vice versa. In contrast, in a mesh topology, the BS may communicate with a MS through another MS called a *sponsor node*. The coverage of the BS can then be extended to MSs that are not within physical-layer communication range. In this manner, the connectivity is augmented. MSs can also exchange traffic directly together. In a mesh topology, links are established between nodes according to the ad hoc model. Example nodes and links forming a mesh topology is pictured in Figure 3.17. There is one mesh BS and four mesh MSs (indexed 1, 2, 3 and 4). The lines represent links established according to an ad hoc mode, discussed in more detail in the sequel. The nodes and links form a mesh. Each node has a permanent 48-bit MAC address. For the sake of efficiency, after an authorization phase, each node in a mesh network is given a 16-bit node ID. A neighbor is defined as a node with which a direct radio link is possible. When established, a link is identified by an ID. The node at the source of the link is the transmitter. The node at the sink of the link is the receiver.

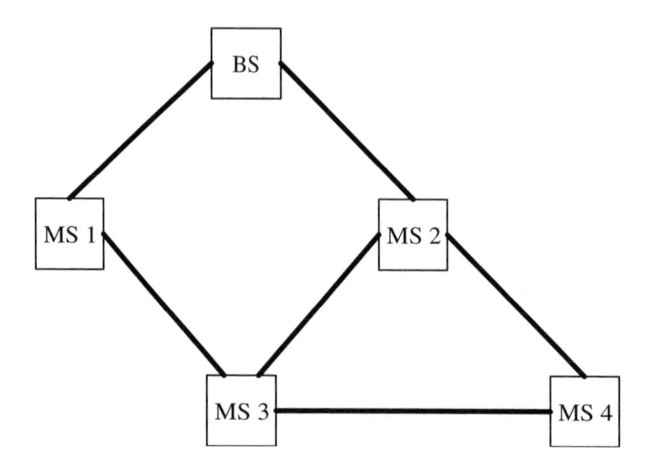

Figure 3.17 Links in a WiMAX/802.16 mesh network.

Table 3.3 Management connections.

Type	Usage
Basic	Short and urgent management messages
Primary	Delay tolerant management messages
Secondary	Encapsulated management messages

The service is link-layer connection oriented. There are two types of connections: management and transport. Management connections are created when an MS joins a network, as depicted in Table 3.3. There are three types of management connections: basic, primary and secondary. The basic connection is used to exchange, between a BS and an MS, short and urgent management messages. For example, messages that affect the profile of a connection. Delay tolerant management messages are transferred over the primary connection, for example, key management messages. The secondary connection is optional and is used to support IP related network management messages, such as IP address configuration (Dynamic Host Configuration Protocol (DHCP)) or network management (Simple Network Management Protocol (SNMP)) messages. Transport connections are used for up-link and down-link data flows. They are provisioned, that is, established according to a subscription profile, or established on demand. A transport connection can be either unicast (i.e. a one BS to one MS association) or multicast (i.e. a one BS to many MSs association).

Figure 3.18 depicts the connections established over the links of the mesh network represented in Figure 3.17. There are six connections: labeled a, b, c, d, e and f. Each connection has two end points. The BS is always one of them, while the other is one of the MSs. Connections a and c are one hop. All others are multihop and reflect the possibility for nodes to use other nodes, as sponsors, to get connected to a BS.

Link IDs enter in the composition of connection IDs. This composition establishes relationships between links and bandwidth (connections) assigned to the links. The mesh BS serves as a gateway to an infrastructure network (termed *backhaul services*). When

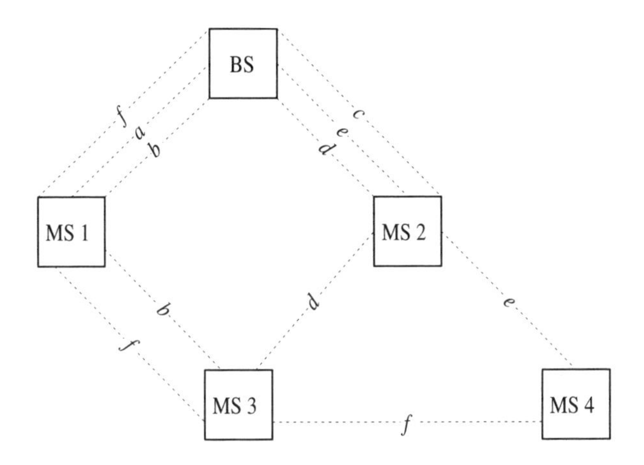

Figure 3.18 Connections in a WiMAX/802.16 mesh network.

an MS joins a mesh network, the management connections are established with the BS. Provisioned transport connections are also established. On-demand transport connections are established dynamically while an MS is attached to a BS. In a mesh topology, connections may be supported, from a BS to a mesh MS, by a path consisting of several links going through several sponsor MSs.

A number of security mechanisms, relying on an operator secret key, are defined to secure a mesh topology. There are, however, no key update mechanisms. Whenever the operator secret key is compromised, then the whole management of the network is compromised.

3.4.1 Scheduling

WiMAX/802.16 uses a time division multiple access (TDMA) scheme. A schedule is an allocation of time slots to nodes. Schedules are required to avoid traffic collisions. Scheduling refers to the process of establishing a schedule. The schedule relevance refers to a time interval in which the information in a schedule applies. The time relevance of a schedule is expressed as a number of frames. In the mesh mode, there are two forms of scheduling: centralized and distributed. In centralized scheduling, the BS of the mesh network gathers bandwidth requests from MSs, grants bandwidth and communicates grants to MSs. The schedule itself, although, is always computed individually by the MSs. Distributed scheduling can be either coordinated or uncoordinated. In coordinated scheduling, each node coordinates its schedule with its two-hop neighborhood. Uncoordinated scheduling is done on a node-to-node basis. Note that there are possibilities of collision if there is no coordination with two-hop neighbors.

The format of protocol data units (PDUs) in the mesh mode is pictured in Figure 3.19. In the mesh mode, the generic header (over 48 bits) and mesh subheader (over 16 bits) are always used. The generic header contains parameters not specific to the mesh mode, that is used also in a traditional point-to-multipoint topology. The mesh subheader contains solely the node ID of the sender of the PDU. The Cyclic Redundancy Code (CRC) checksum is optional.

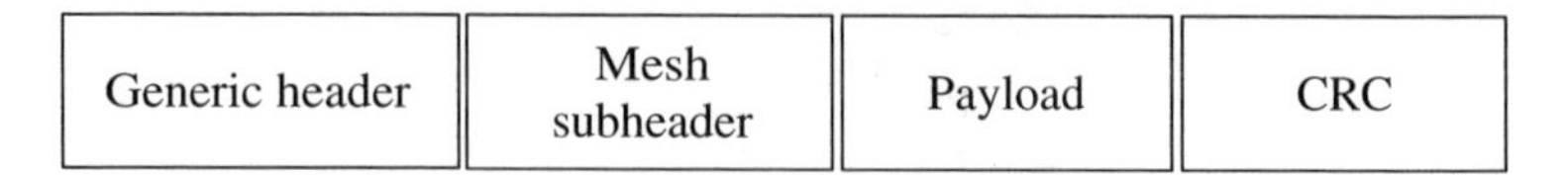

Figure 3.19 MAC PDU format of the WiMAX mesh mode.

3.4.2 Management messages

The payload of PDUs may consist of management or transport messages. There are five management messages defined specifically for the operation in the mesh mode: MSH-NCFG, MSH-NENT, MSH-DSCH, MSH-CSCH and MSH-CSCF. Every management message is sent as the payload of a single MAC PDU.

The MSH-NCFG message (mesh network configuration) is used for scanning, neighborhood discovery, obtaining network parameters and network entry. The MSH-NCFG message is sent using broadcast by every node. It contains a list of discovered neighbor MSs and a list of discovered mesh BSs (up to eight hops away). Lists do not have to be exhaustive. It also contains other control information used by receivers to construct their own neighbor list. The MSH-NCFG message is used by a new mesh MS to get the parameters of a network (Figure 3.20). While MSH-NCFG messages sent by neighbors are received, a new mesh MS builds its own list of neighbors (described in more detail in the sequel). The mesh MS waits until two conditions are satisfied. Firstly, a repeated MSH-NCFG message is received (i.e. from the same sender node). Repetition is interpreted as a completeness of neighborhood discovery. Secondly, a repeated MSH-NCFG message

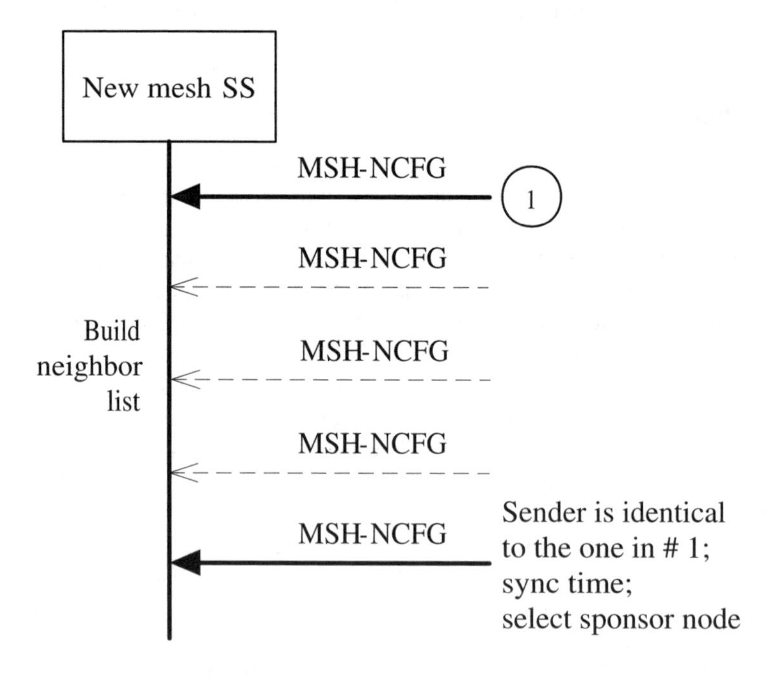

Figure 3.20 Acquisition of network parameters in mesh mode.

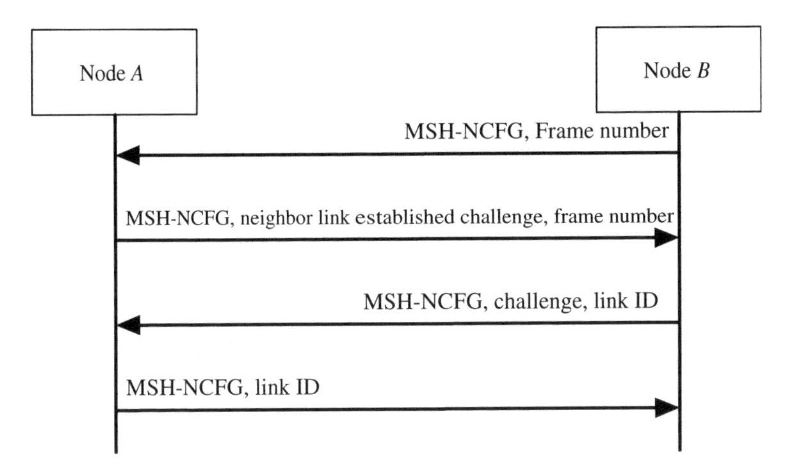

Figure 3.21 Establishment of a link in mesh mode.

is received with the ID of an operator to which the new mesh MS subscribes. A sponsor is a node whose assistance is used to join a network. The new mesh MS selects a sponsor, among the discovered neighbors, of an operator to which it subscribes. The new mesh MS synchronizes its time with the sponsor (assuming a null propagation delay) and begins network entry using the sponsor node.

After the completion of a successful authorization and a registration with a mesh BS (reviewed in the sequel), a new mesh MS has a node ID. It can establish links with several other nodes. The establishment of a link between a node *A* and a node *B* is accomplished using the MSH-NCFG message and is pictured in Figure 3.21. Node *A* saves the frame number from the last MSH-NCFG message received from node *B*. Node *A* sends an MSH-NCFG message of type neighbor link establishment containing a challenge to node *B*. The challenge consists of a hash message authentication code (HMAC) over the secret key of the operator, frame number from node *B*, *A*'s node ID and *B*'s node ID. Node *B* verifies the challenge from node *A* and replies with an MSH-NCFG message containing its own challenge. This challenge consists of an HMAC computed over the secret key of the operator, frame number from node *A*, *B*'s node ID and *A*'s node ID. Node *B* also sends the ID of the link from node *B* to node *A*. Node *A* verifies the challenge from node *B* and returns the ID of the link from node *A* to node *B*.

3.4.3 Mesh network

When it solicits entry in a mesh network, a node transmits The MSH-NENT (mesh network entry) message. Note that the node ID is null (in the mesh subheader) until the transmitter is authorized and assigned a node ID by a mesh BS. The message contains the ID of a sponsor node (selected when network parameters were obtained). The network entry process is pictured in Figure 3.22. The new mesh MS sends an MSH-NENT message of type Entry Request. The sponsor node opens a channel (a User Datagram Protocol (UDP) tunnel) with its BS and sends an MSH-NCFG message of type Entry Open as a response. The MAC address of the new node is contained in the MSH-NCFG message. Reception

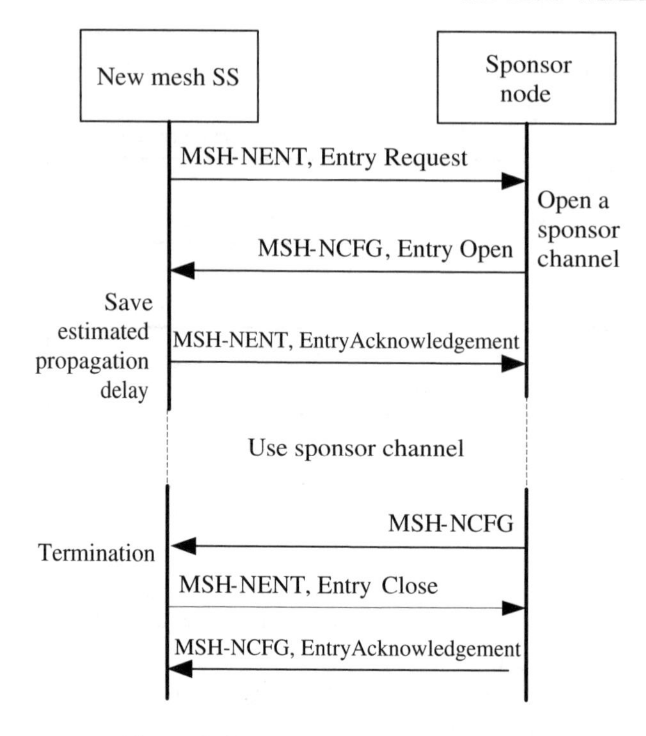

Figure 3.22 Network entry procedure.

is acknowledged by the new node using an MSH-NENT message of type Entry Acknowledgement. The new mesh MS is ready for the next step: authorization. Authorization is performed with a mesh BS, but through the sponsor node which tunnels the authorization information, request and response to the mesh BS. Registration follows. It is also performed with the mesh BS using the sponsor node as an intermediary. Successful completion of the registration process results in the assignment of a node ID to the new mesh MS.

The integrity of the MSH-NENT message is partially protected with an HMAC computed over the MAC address, serial number of the node and a network operator provided secret key. Whenever one node is compromised, so is the whole network.

Termination of membership in a mesh network is also pictured in Figure 3.22. Membership ends when the sponsor node stops including the MAC address of the mesh MS in the MSH-NCFG messages, which are sent regularly.

The MSH-DSCH message (mesh distributed scheduling) applies solely when distributed scheduling is used. Each node transmits this message on a periodic basis to inform its neighbor about its schedule, available bandwidth, bandwidths requests and bandwidth grants. The exact scheduling algorithm is left open.

The MSH-CSCH message (mesh centralized scheduling) is transmitted by either a mesh BS or a mesh MS. When the MSH-CSCH message is transmitted by a mesh BS, it carries grants of bandwidth to the mesh MSs. The message is repeated from mesh MS to mesh MS until a maximum number of hops is reached. When the MSH-CSCH message is transmitted by a mesh MS, it transports bandwidth requests to a mesh BS. The message is repeated from mesh MS to mesh MS until the mesh BS is reached. The exact scheduling algorithm is left open.

The MSH-CSCF message (mesh centralized configuration) is transmitted by a mesh BS. The message is repeated from mesh MS to mesh MS until a maximum number of hops is reached. It is used to inform the members of a network about its structure.

Each node maintains a list of neighbors. A node learns about its neighbors by tracking the arrival of MSH-NCFG messages. The activity of each neighbor is individually tracked with respect to the next time at which the arrival of an MSH-NCFG message from that neighbor is expected (Next Xmt Time) and the minimum time interval separating the arrivals of MSH-NCFG messages (Xmt Holdoff Time). Each entry in the list consists of the following information:

- MAC address of the neighbor.
- Hop count: neighbor's distance in hops.
- Node ID of the neighbor.
- Xmt Holdoff Time: expected inter arrival time for MSH-NCFG messages from that neighbor (derived from data contained in the MSH-NCFG message).
- Next Xmt Time: Time at which the reception of an MSH-NCFG message from the neighbor is expected (derived from data contained in the MSH-NCFG message). It can be interpreted as the lifetime of the entry for that neighbor (termed the *eligibility*).
- Reported flag: It is a flag set to true if the information about this neighbor was in the last MSH-NCFG message sent.
- Synchronization Hop Count: A hierarchy, in a mesh network, is organized for the purpose of clock synchronization. Nodes with external accurate clock synchronization capability have Synchronization Hop Count valued to zero. Otherwise, each node sets its Synchronization Hop Count using the lowest value among its neighbors' values plus one (values are contained in the MSH-NCFG message). A node without accurate clock synchronization capability always synchronizes its clock with the neighbor having the lowest Synchronization Hop Count. Conflicts are resolved by using the neighbor with the lowest Node ID.

In a mesh network, traffic flows according to a tree model. A mesh BS is at the root. With centralized scheduling, the assignment of the bandwidth is determined by the mesh BS and communicated to mesh MSs using the MSH-CSCH message. Each mesh MS determines its schedule individually according to the bandwidth assignments. In addition, the forwarding policies within each mesh MS are unspecified. With decentralized scheduling, the topology is a graph. Links are permitted between any two mesh MSs. End-to-end traffic can flow directly between mesh MSs without the involvement of a mesh BS. The forwarding policies within each mesh MS are unspecified.

3.4.4 Sleep mode

An MS may enter in the sleep mode to save batteries. The serving BS buffers the data destined to a sleeping MS. A sleep interval embeds periodic wakeful subintervals during which the MS monitors the eventual transmission of traffic indication messages.

Xiao (2005) studied the performance of the sleep mode protocol of WiMAX/802.16 under variations of parameter values defining the structure of a sleep cycled, that is, the minimum duration and maximum duration of a sleep interval during a cycle. Xiao examined the expected number of cycles in sequences of sleep mode cycles, expected duration of sequences, frame response time and energy consumption. Xiao gives an equation for the expected duration of sequences. This is central in his work as other results follow from this equation. The work of Xiao is reviewed and revisited in the following text.

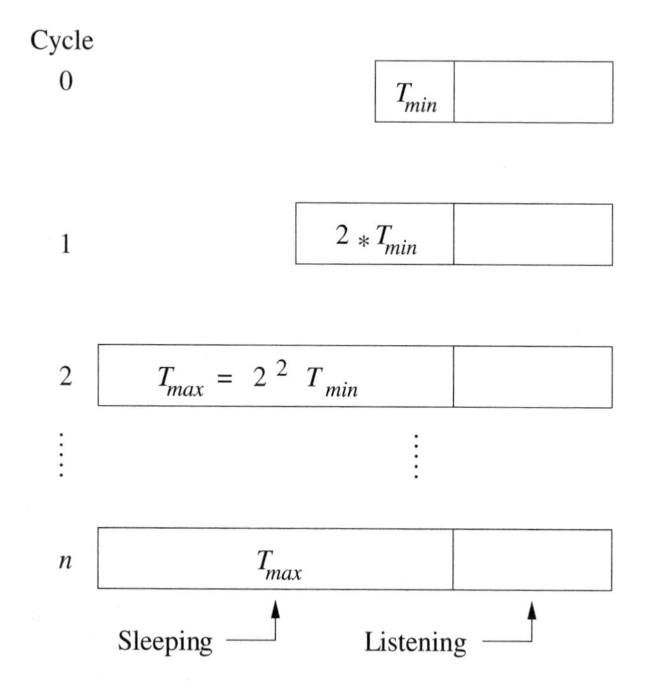

Figure 3.23 The sleep mode protocol of WiMAX/802.16.

The sleep mode consists of cycles, indexed from zero to n. (Figure 3.23). For $i = 0, 1, 2, \ldots, n$, a cycle is modeled as the sum

$$T_i + L \text{ time units}$$

The term T_i denotes the time period during which the MS sleeps. The sleep time is increased from one cycle to another. The term L represents the time period during which the MS is listening. The length of the listening interval is invariant.

The length of a sleep interval is bounded by a minimum value $T_{\min}$ and a maximum value $T_{\max}$. The first interval in a new sequence of sleep mode cycles starts with a sleep interval of length T_{min}, that is, $T_0 = T_{min}$. The length is doubled from one interval to the other, until the maximum value T_{max} is reached. For $i = 1, 2, \ldots, n$, we have $T_i = \min(2^i T_{min}, T_{max})$.

For a sequence of sleep cycles, the following quantities can be calculated: the expected number of cycles, expected duration, expected energy consumption as well as expected frame response time. We hereafter calculate the expected number of cycles. The calculation of the expected duration, expected energy consumption and expected frame response time is left in exercise.

It is assumed that the arrival of frames destined to an MS follows a Poisson distribution. The distribution has a parameter λ, which corresponds to the mean number of frames per unit of time. According to the Poisson distribution, the time between the arrival of two frames, that is, the inter arrival time, is distributed according to an exponential law with parameter $1/\lambda$, which is also the mean.

The Boolean random variable e_i is used to represent the arrival of frames during the cycle with index i. The variable is true if there is at least one frame arrival during the i-th interval, otherwise it is false. We have,

$$\begin{aligned} Pr[e_i = True] &= 1 - Pr[\text{zero events in } T_i + L] \\ &= 1 - e^{-\lambda(T_i+L)}. \end{aligned}$$

An important quantity required for the sequel is the calculation of the expected value of n, that is, $E[n]$. Since, the value of n ranges from 0 to ∞, we have

$$E[n] = \sum_{i=0}^{\infty} i\, Pr[n = i] \text{ cycles} \tag{3.8}$$

The term $Pr[n = i]$ represents the probability of success in exactly the i-th iteration, which is also the probability of failure in iterations zero to $i - 1$ and success in the i-th. The sleeping cycles are independent random variables. First, we have

$$\begin{aligned} Pr[n = 0] &= Pr[e_0 = True] \\ &= 1 - e^{-\lambda(T_{min}+L)}. \end{aligned}$$

For $n \geq 1$,

$$\begin{aligned} Pr[n = i] &= Pr[e_0 = False] \cdots Pr[e_{i-1} = False] Pr[e_i = True] && (3.9) \\ &= e^{-\lambda(T_0+L)} \cdots e^{-\lambda(T_{i-1}+L)} (1 - e^{-\lambda(T_i+L)}) && (3.10) \\ &= e^{\sum_{j=0}^{i-1} -\lambda(T_j+L)} (1 - e^{-\lambda(T_i+L)}). && (3.11) \end{aligned}$$

This result is put in Equation 3.8 to define $E[n]$.

The length of a sequence of sleep mode cycles is denoted by the letter D:

$$E[D] = \sum_{i=1}^{\infty} (\text{cycle } i \text{ duration})\, Pr[n = i] \text{ time units.} \tag{3.12}$$

The amount of energy consumed while sleeping is denoted as E_S units of energy per unit of time, while the amount of energy consumed while listening is denoted as E_L units of energy per unit of time The energy consumption for cycle with index i is

$$E_i = \sum_{j=0}^{i} (T_k E_S + L E_L) \text{ energy units.} \tag{3.13}$$

The model of energy consumption of a sequence of sleep mode cycles is modeled as

$$\mathcal{E} = \sum_{i=1}^{\infty} E_i \, Pr[n = i] \text{ energy units.} \tag{3.14}$$

The frame response time corresponds to the delay a frame destined to an MS has to wait before it is delivered. The frame response time is denoted by the letter R and defined as

$$E[R] = \sum_{i=1}^{\infty} \frac{(T_i + L)}{2} Pr[n = i] \text{ time units.}$$

It is assumed that if the cycle is of length i, the frame causing the escape from the sequence of sleep cycles will arrive at any moment during the last cycle with uniform probability.

Equation 3.11 can be simplified, that is, the summation of i terms can be replaced by a sum of two terms, thereby reducing the time complexity of this calculation from $\mathcal{O}(i)$ to $\mathcal{O}(1)$. Let m be equal to $\log_2 \frac{T_{\max}}{T_{\min}}$. For $i = 0, 1, 2, \ldots$, note that

$$T_i = \begin{cases} T_{\min} 2^i & \text{if } i \le m \\ T_{\min} 2^m = T_{\max} & \text{if } i > m \end{cases}$$

We use the fact that $2^0 + 2^1 + \ldots + 2^{i-1} = 2^i - 1$. For $i \le m$, we have

$$\begin{aligned} \sum_{j=0}^{i-1} -\lambda(T_j + L) &= \sum_{j=0}^{i-1} -\lambda(2^j T_{\min} + L) \\ &= -\lambda[T_{\min} \sum_{j=0}^{i-1} 2^j + \sum_{j=0}^{i-1} L)] \\ &= -\lambda[T_{\min}(2^i - 1) + iL] \end{aligned}$$

Hence, for $1 \le i \le m$, we have

$$Pr[n = i] = e^{-\lambda(T_{\min}(2^i - 1) + iL)} (1 - e^{-\lambda(T_{\min} 2^i + L)})$$

Hence, for $i > m$, we have

$$\begin{aligned} Pr[n = i] &= e^{-\lambda[T_{\min}(2^m - 1) + mL]} (e^{-\lambda[T_{\max} + L]})^{(i-1-(m-1))} (1 - e^{-\lambda(T_{\max} + L)}) \\ &= e^{-\lambda[T_{\min}(2^m - 1) + mL]} (e^{-\lambda[T_{\max} + L](i-m)})(1 - e^{-\lambda(T_{\max} + L)}) \\ &= \frac{e^{-\lambda[T_{\min}(2^m - 1) + mL]}}{e^{-\lambda[T_{\max} + L]m}} (1 - e^{-\lambda(T_{\max} + L)}) e^{-\lambda(T_{\max} + L)i} \\ &= e^{-\lambda[T_{\min}(2^m - 1) - mT_{\max}]} (1 - e^{-\lambda(T_{\max} + L)}) e^{-\lambda(T_{\max} + L)i} \end{aligned}$$

To summarize, we have

$$Pr[n = i] = \begin{cases} 1 - e^{-\lambda(T_{\min}+L)} & \text{if } i = 0 \\ e^{-\lambda[T_{\min}(2^i-1)+iL]}(1 - e^{-\lambda(T_{\min}2^i+L)}) & \text{if } 1 \leq i \leq m \\ cp^i & \text{if } i > m \end{cases}$$

with

$$c = e^{-\lambda[T_{\min}(2^m-1)-mT_{\max}]}(1 - e^{-\lambda(T_{\max}+L)})$$

and

$$p = e^{-\lambda(T_{\max}+L)}$$

To be effectively computable, the infinite sum must be resolved to an equation with a finite number of terms.

When all sleeping cycles T_i's are of equal length, say T, then the calculations can be simplified as follows. For $i = 0, 1, 2, \ldots$, we have

$$Pr[0 \text{ events in } T_i + L] = e^{-\lambda(T+L)}$$

Let p be equal to $e^{-\lambda(T+L)}$. Then, n is a geometric random variable for which we can compute

$$\begin{aligned} E[n] &= \sum_{i=0}^{\infty} i Pr[n = i] \\ &= \sum_{i=0}^{\infty} ip^i(1 - p) \\ &= \frac{p}{1 - p} \text{ cycles} \end{aligned}$$

Under the same assumption, the expected frame response time is

$$E[R] = \frac{(T + L)}{2} \text{ time units}$$

With sleep cycles starting with length $T_{\min}$, doubled from one cycle to the next until value $T_{\max}$ is reached then it stays the same, we can resolve the equation for the expected value of n to a finite sum. With m equal to $\log_2 \frac{T_{\max}}{T_{\min}}$ and p equal to $e^{-\lambda(T_{\max}+L)}$,

$$\begin{aligned} E[n] &= \sum_{i=1}^{m} i Pr[n = i] + \sum_{i=m+1}^{\infty} i Pr[n = i] \\ &= \sum_{i=1}^{m} i Pr[n = i] + \sum_{i=0}^{\infty} ip^i - \sum_{i=0}^{m} ip^i \\ &= \sum_{i=1}^{m} i Pr[n = i] + \frac{p}{(1 - p)^2} - \sum_{i=0}^{m} ip^i \text{ cycles.} \end{aligned} \tag{3.15}$$

A plot of the expected value of a sleep cycle sequence length, as of function of values of λ ranging from 0.02 to 0.2 frames per cycle, is pictured in Figure 3.24, together with results obtained by simulation.

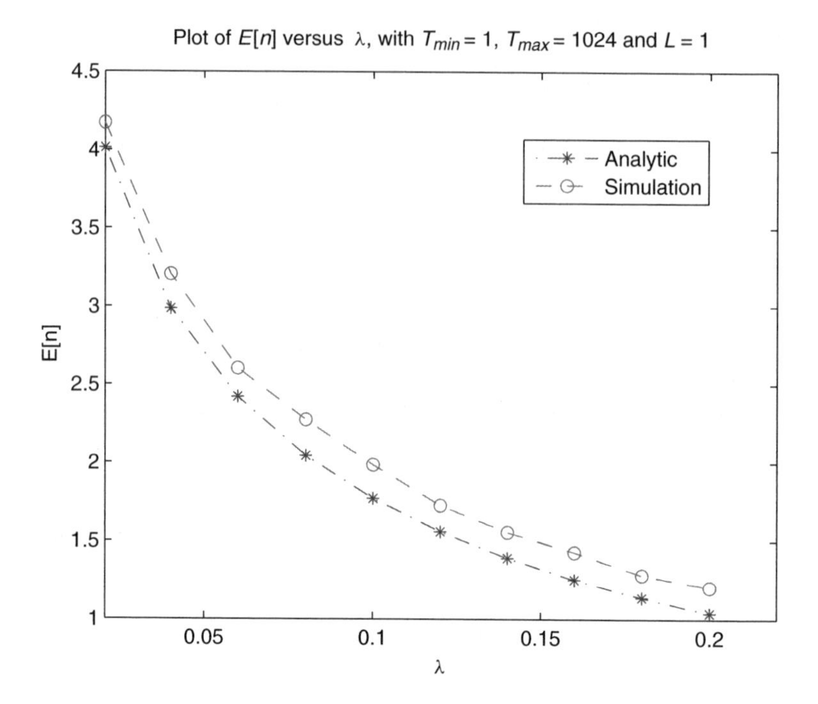

Figure 3.24 Expected value of a sleep cycle sequence length.

3.5 Bibliographic comments

The name Bluetooth is derived from the nickname of a 10th century king of Denmark, Harald Blatand who engaged in diplomacy leading different (usually warring) parties to negotiate with each other, thus bestowing a fitting name to this technology. The technology is available in various new products, like phones, keyboards and headsets for transferring files and is used in situations when two or more devices are in close proximity to each other. It is less effective for setting up networks and remote access, in which case WiFi is better suited. It uses the same frequency as WiFi but with less power consumption, thus resulting in weak connections.

The presentation of node discovery in Bluetooth given in Section 3.2 follows closely the publications of Alonso et al. (2003b) and Alonso et al. (2003a). For additional experimental comparisons and extensions to the multiple node protocols presented, we refer the reader to Alonso et al. (2003a). Gasieniec et al. (2001) recently studied the *wakeup problem*, which is related to the node discovery problem; however, their results are not directly applicable in our context.

Node discovery started to attract attention as the interest in ad hoc networks increased. To our knowledge, only two node discovery protocols have been proposed in a model that is comparable with ours, both in the context of Bluetooth. A scatternet formation protocol that uses design theory and projective spaces is due to Barrière et al. (2003a) and assumes that all hosts are within the range of each other. If n is the number of nodes of the scatternet, then the algorithm can update the scatternet for joins and leaves using only local rearrangements in time $O(\log^2 n \log\log^2 n)$ while the message complexity of these operations is $O(\log^4 n \log\log^4 n)$. The degree of the resulting network can be any arbitrary but fixed value d and the diameter is polylogarithmic in n. Additional information on Bluetooth formation algorithms can be found in the survey paper Wong et al. (2006).

3.6 Exercises

1. (See Salonidis et al. (2001)) If you also simplify Equation (3.4), then it is easy to see that the minimum number $p_{\min}(n)$ of masters of a scatternet with given number n of nodes must satisfy

$$p_{\min}(n) \geq \frac{17 - \sqrt{289 - 8n}}{2}$$

2. Is it possible to construct a scatternet of n nodes for each $n \leq 36$?

3. ($\star$) (See Alonso et al. (2003b)) Compute the expected running times of the protocols in Table 3.2.

4. ($\star$) (See Alonso et al. (2003b)) Consider the $\mathcal{SP}$. The expected length of a run for $\mathcal{SP}$ depends on the probability distribution of the time the node remains silent. The purpose of this exercise is to show that protocol $\mathcal{SP}$ is optimal for the uniform distribution. That is, let $J_1, J_2, \ldots, J_k$ be i.i.d. random variables assuming values in the range $1..n$. Prove that the probability that the random variables $J_1, J_2, \ldots, J_k$ do not collide is maximized by the uniform distribution. The proof is by induction on k. The base case is $k = 2$ and the result follows from the following assertions:

   (a) Let p_i be the probability that the random variable obtains the value i. Show that $Pr[J_1, J_2 \text{ collide}] = \sum_{i=1}^{n} p_i^2$.

   (b) It is enough to show that $\sum_{i=1}^{n} p_i^2 \geq \frac{1}{n}$. **Hint:** Consider the new distribution $\frac{p_1}{1-p_n}, \ldots, \frac{p_{n-1}}{1-p_n}$ and use induction on n.

   This completes the base case $k = 2$. Assume that the result is true for $k - 1$ i.i.d. random variables. To prove it for k, let $[n]^k$ denote the family of k-element subsets of the set $\{1, 2, \ldots, n\}$. Prove that

$$Pr[J_1, \ldots, J_k \text{ do not collide}] = \sum_{i=1}^{n} p_i (1 - p_i)^{k-1} \sum_{S \in [n \setminus \{i\}]^{k-1}} \prod_{j \in S} \frac{p_j}{1 - p_i}$$

   Now notice that for each i, the quantity

$$\sum_{S \in [n \setminus \{i\}]^{k-1}} \prod_{j \in S} \frac{p_j}{1 - p_i}$$

   represents the probability that $k - 1$ i.i.d. random variables with probability distribution given by

$$\frac{p_1}{1 - p_i}, \ldots, \frac{p_{i-1}}{1 - p_i}, \frac{p_{i+1}}{1 - p_i}, \ldots, \frac{p_n}{1 - p_i}$$

do not collide. Now use the induction hypothesis. Additional details of the proof can be found in Alonso et al. (2003b).

5. ($\star$) In this exercise, we state and prove a fundamental result in the theory of random processes. Let $\{W_i : i \geq 1\}$ be independent and identically distributed random variables with a finite mean (i.e. $E[W] < \infty$) and let N be a stopping time for $W_1, W_2, \ldots$ (i.e. the event $\{N = n\}$ is independent of $W_{n+1}, W_{n+2}, \ldots$, for all $n \geq 1$) such that $E[N] < \infty$. Then

$$E\left[\sum_{i=1}^{N} W_i\right] = E[W]E[N].$$

Hint: We outline an elementary proof of Wald's identity due to Kolmogorov and Prochorov (see Klimow (1979) page 216). Define the random variable $S_N := \sum_{k=1}^{N} W_k$ and consider the indicator random variables I_k defined by

$$I_k = \begin{cases} 0 & \text{if } k > N \\ 1 & \text{if } k \leq N. \end{cases}$$

Observe that $Pr[I_k = 1] = Pr[N \geq k]$ and I_k is independent of X_k. Moreover, we have that $S_N = \sum_{k=1}^{N} W_k = \sum_{k \geq 1} I_k W_k$. Now notice that since

$$|E[I_k W_k]| = |E[I_k]||E[W_k]| \leq Pr[N \geq k]E[W]$$

we can derive easily that

$$\sum_{k \geq 1} |E[I_k W_k]| = |E[W]| \sum_{k \geq 1} Pr[N \geq k] = |E[W]|E[N] < \infty.$$

Now conclude that $E[S_N] = E[W] \cdot E[N]$. For a martingale approach and additional information, see Ross (1996, 2nd edition).

6. Wald's identity (proved in Exercise 5) can also be used to provide a new analysis of the node discovery protocols. Analyze the expected running time of the protocols $\mathcal{RP}, \mathcal{SP}, \mathcal{AP}, \mathcal{LP}, \mathcal{CP}$ using only Wald's identity. For additional details, see Alonso et al. (2003a).

7. (See Alonso et al. (2003a)) We can extend the model to take into account communication frequencies. We assume a system with K nodes communicating by broadcasting messages. At any given time, a node can be at a given frequency $i = 1, 2, \ldots, f$ either T (Talking) or L (Listening). The state of a node is denoted by the pair (S, i) where $S = T$ or $S = L$ and $i = 1, 2, \ldots, f$. The nodes are synchronized: they change states at the same time and remain in a given state for a period of time that is identical for all nodes. An event E describes the state of the K nodes of the system:

$$E = \begin{pmatrix} S_1 & i_1 \\ S_2 & i_2 \\ \vdots & \vdots \\ S_K & i_K \end{pmatrix},$$

where (S_k, i_k) is the state of the k-th node. For each event E we denote by k^E the state of the kth node in event E. Define the conditions for a node k to receive a message from another node l. Also define what is a run of the protocol. For additional details, see Alonso et al. (2003a).

8. (See Alonso et al. (2003a)) We continue Exercise 7. A node is represented by a random variable X assuming the values (S, i), where S is either T or L and $i = 1, 2, \ldots, f$. Associated with this random variable is a probability distribution of the frequencies.

    (a) If F_i is the probability that a node is in frequency i, p is the probability that a node is talking, and q is the probability that a node is listening then show that $p_i = Pr[X = (T, i)] = pF_i, q_i = Pr[X = (L, i)] = qF_i$.

    (b) Conclude that $p + q = 1, \quad \sum_{i=1}^{f} F_i = 1, \quad \sum_{i=1}^{f} p_i + \sum_{i=1}^{f} q_i = 1$.

    For additional details, see Alonso et al. (2003a).

9. (See Alonso et al. (2003a)) We continue Exercise 7. Given the model above, we can analyze two types of node discovery protocols: *dynamic* and *static* frequency allocation. In either case, the protocol succeeds if two nodes discover each other. The two types of protocols differ in the way they allocate frequencies. In the first type, with *static* frequency allocation, the first node talks and the second listens in a given frequency and, in the next step, the second node talks and the first listens in the same frequency. In the second type, with *dynamic* frequency allocation, the first node talks and the second listens in a given frequency and, in the next step, the second node talks and the first listens in the same or in a different frequency. For additional details, see Alonso et al. (2003a).

10. ($\star$) (See Alonso et al. (2003a)) We continue Exercise 7. By Random Protocol (**RP**), we understand a protocol in which each node decides at random whether to talk or listen. Show that the expected waiting time for protocol **RP** with static frequency allocation to succeed is given by

$$\frac{1}{\sum_{i=1}^{f} \frac{2}{\left(\frac{1}{p_i q_i} + \frac{1}{p_i^2 q_i^2}\right)}}.$$

    For additional details, see Alonso et al. (2003a).

11. ($\star$) (See Alonso et al. (2003a)) We continue Exercise 10. Show that the expected waiting time for the protocol **RP** with dynamic frequency allocation to succeed is given by

$$\frac{1 + \sum_{j=1}^{f} p_j q_j}{2\left(\sum_{j=1}^{f} p_j q_j\right)^2}.$$

    For additional details, see Alonso et al. (2003a).

12. ($\star$) (See Alonso et al. (2003a)) Extend the results of Exercises 10 and 11 to K nodes.

(a) Show that the expected waiting time for protocol **RP** with K nodes and static frequency allocation to succeed is given by

$$\frac{1}{\left(2\binom{K}{2}\sum_{i=1}^{f}\frac{1}{\left(\frac{1}{p_i^2q_i^2(1-p_i)^{2K-4}}+\frac{1}{p_iq_i(1-p_i)^{K-2}}\right)}\right)}.$$

(b) Show that the expected waiting time for protocol **RP** with K nodes and dynamic frequency allocation to succeed is given by

$$\frac{1+\sum_{j=1}^{f}p_jq_j(1-p_j)^{K-2}}{2\binom{K}{2}\left(\sum_{j=1}^{f}p_jq_j(1-p_j)^{K-2}\right)^2}.$$

For additional details, see Alonso et al. (2003a).

13. For the WiMAX/802.16 sleep mode, derive equations expressing the expected duration, expected energy consumption and expected frame response time.

4

WIRELESS NETWORK PROGRAMMING

...one had always assumed that there would be no particular difficulty in getting programs right.
M. V. Wilkes, 1946

This is what M. V. Wilkes, computer pioneer and director of Cambridge's University Mathematical Laboratory had to say about the programming experience. He also recalls his experiences while visiting the Moore school in 1946 that "I can remember the exact instant of time at which it dawned on me that a great part of my future life would be spent in finding mistakes in my own programs" (see Bashe et al. (1986), page 321). The programming methodology is an important aspect in networking practice and this chapter describes how to use the Packet Socket Application Programming Interface (API) in order to access a WiFi/802.11 wireless interface on a Linux system.

4.1 Structure of information

The types of information dealing with a wireless interface are the address representation, configuration data and frame structure.

Wireless interface addresses follow the representation of Ethernet addresses. An address is a 6-byte string. There are three types of addresses: unicast, multicast and broadcast. An unicast address is the identity of one station on a wireless medium. A broadcast address designates all the stations on a wireless medium. A multicast address identifies a subgroup of stations on a wireless medium.

A wireless interface address is typically shown as six hexadecimal values separated by columns. Each value is between `00` and `FF` and represents the content of one byte. Here is an example:

```
00:60:1D:1E:31:18
```

Principles of Ad hoc Networking Michel Barbeau and Evangelos Kranakis

An address can be represented in the C++ language as follows:

```
#define WLAN_ADDR_LEN 6
struct WLANAddr
{
   unsigned char data[WLAN_ADDR_LEN];
   char * wlan2asc();
   Outcome str2wlan(char * s);
};
```

The field `data` stores the six-byte address value. The function `wlan2asc()` returns the address in a human readable form. The function `str2wlan()` defines the address from a human readable form.

The data about the configuration of a wireless interface on a Linux system can be obtained with two commands: `ifconfig` and `iwconfig`.[1] Here is sample data returned by the `ifconfig` command:

```
# ifconfig
eth0      Link encap:Ethernet  HWaddr 00:60:1D:1E:31:18
          inet addr:134.117.5.12  Bcast:134.117.5.255  Mask:255.255.255.0
          UP BROADCAST RUNNING MULTICAST  MTU:1500  Metric:1
          RX packets:5786 errors:0 dropped:0 overruns:0 frame:0
          TX packets:352 errors:0 dropped:0 overruns:0 carrier:0
          collisions:0 txqueuelen:100
          RX bytes:870844 (850.4 Kb)  TX bytes:52664 (51.4 Kb)
          Interrupt:10 Base address:0x100
```

In this example, the wireless interface name is `eth0`. It is interesting to observe that on Linux, a wireless interface is presented to processes as an *Ethernet like* interface. Wireless specific information is obtained with the `iwconfig` command. Here is an example:

```
# iwconfig
eth0      IEEE 802.11-DS  ESSID:"ENTRUSTW"  Nickname:"localhost"
          Mode:Managed  Frequency:2.422GHz  Access Point: 00:60:1D:F1:F0:27
          Bit Rate:11Mb/s   Tx-Power=15 dBm   Sensitivity:1/3
          Retry limit:4   RTS thr:off   Fragment thr:off
          Encryption key:off
          Power Management:off
          Link Quality:22/92  Signal level:-70 dBm  Noise level:-92 dBm
          Rx invalid nwid:0  Rx invalid crypt:0  Rx invalid frag:6
          Tx excessive retries:0  Invalid misc:0   Missed beacon:0
```

In this example, the wireless interface is configured in the *infrastructure* mode, also called the *managed mode*. The `iwconfig` command can be used to reconfigure a wireless interface with new parameters, specific to wireless operation. Here is an example:

```
iwconfig eth0 essid mynet
iwconfig eth0 mode Ad-hoc
iwconfig eth0 channel 3
iwconfig eth0 rate auto
```

The first line reconfigures the wireless network name (formally termed the *extended service set ID* [*ESSID*]). It is the name of a local network to which the wireless interface is participating. The second line sets the operating mode of the wireless interface. The ad hoc mode is selected, which is a single cell network without any access point. It means

[1] Root user privileges are required.

that only single hop communication is supported by the link layer. The third line sets the operating frequency or channel of the wireless interface. Values below 1000 are channel numbers. Values greater than 1000 are frequencies in Hertz. The suffixes k, M or G can be concatenated to the value (e.g. 2.422 G) to scale the frequency in kilo-, mega- or gigahertz. The *iwlist* command may be used to retrieve the available channel numbers and frequencies.[2] The fourth line puts the wireless interface in a mode in which it self-selects the data rate.

From the point of view of a process, a wireless interface frame consists of a header and a body. The format is pictured in Figure 4.1. The header consists of a destination address (6 bytes), a source address (6 bytes) and a type (2 bytes), for a total of 14 bytes. The type field is a network protocol selector. It designates the type of the packet in the body of the frame (e.g. Internet Protocol [IP]]). The body of a frame is of variable length, from 0 to 2312 bytes, and contains the data. In practice, the length is constrained by the maximum transmission unit (MTU). It is a configuration parameter of the interface which can be accessed with the *ifconfig* command and often sets to 1500 bytes.

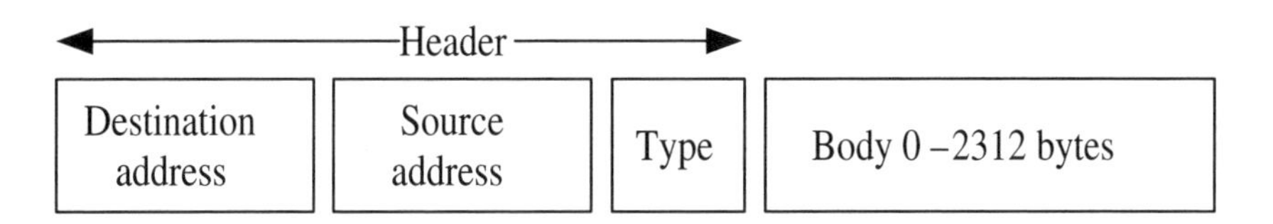

Figure 4.1 Format of WiFi/802.11 frames.

Note that the header fields *frame control*, *duration* and *sequence control information*, as specified in Reference IEEE (1999a), are not visible to processes. The *frame check sequence* is also invisible.

The format of the header can be represented in the C++ language as follows:

```
struct WLANHeader
{
   WLANAddr destAddr;
   WLANAddr srcAddr;
   unsigned short type;
};
#define WLAN_HEADER_LEN 14
#define IP_TYPE x0800
```

The type value for IP packets is `x0800`. An unreserved type such as `0x3900` can be used instead to send frames that bypass the TCP/IP stack at destination.

4.2 Socket

A socket is the software abstraction of the end-point of a communication channel provided to a process by the operating system, in general, and by Linux in particular. Usage of

[2]Depending on local regulations, some channels or frequencies may not be available on your system.

sockets for network or transport service access is well documented in books such as the one by Donahoo and Calvert (2001) or the one by Stevens (1998). The following is about data link service access using packet sockets, a topic that has not been the object of the same level of attention.

A process can access a wireless interface using a type of socket called *packet socket*. It is available on Linux, but not necessarily on all flavors of Unix.[3] A packet socket is the end-point of an Ethernet or wireless channel. In the following text, the focus is on wireless channels. Wireless frames can be received or sent through such a socket.

A process needs to maintain a number of pieces of information in order to use a wireless interface, namely, a configuration, a socket descriptor, an interface index, an address and an MTU. This information can be stored collectively in a C++ structure as follows:

```
struct Ifconfig
{
   int sockid;
   int ifindex;
   WLANAddr hwaddr;
   int mtu;
};
```

The `sockid` field is formally called the *socket descriptor*. It is a numerical reference shared by the process and operating system to designate a specific socket. The `ifindex` field is the index of the interface assigned by the operating system to which the socket is associated. A socket and a wireless interface are initially two separate entities. They need to be paired together explicitly. Afterwards, frames can be received or sent using the wireless interface and socket programming abstraction. The `hwaddr` field stores the address, which is extracted from the wireless interface. It is used to define the value of the source address field in constructed frames. The field `mtu` is the MTU associated to the wireless interface. Its value can be extracted or set programmatically. It can be used to control the length of frames and to determine an adequate size for buffers.

A variable of type `Ifconfig` needs to be declared:

```
Ifconfig ifconfig;
```

To begin communication with a wireless interface, a process issues a system call to create a socket abstraction. This system call consists of the `socket()` function, whose detailed signature is as follows:

```
packet_socket =
   socket(PF_PACKET, int socket_type, int protocol);
```

This call creates a link layer socket. It returns a socket descriptor.

The constant `PF_PACKET` selects the creation of a socket termed a *packet interface on device level* or *packet socket*, in the Linux terminology. The formal parameter `socket_type` may actually take either the value `SOCK_DGRAM` or `SOCK_RAW`. With `SOCK_DGRAM`, link layer header less frames must be sent or are received. With the value `SOCK_RAW`, the sent frames must include a link level header while the received frames

[3]Note that on-line documentation is available with the `man 7 packet` command.

include their link level header, as pictured in Figure 4.1. An example of the SOCK_RAW type is given in the sequel. The formal parameter protocol should be the protocol number of IEEE 802.3 (Ethernet), which in the network byte order form is defined as htons(ETH_P_ALL). In other words, Ethernet interfaces or wireless interfaces are presented the same way to processes. Note that all incoming frames are passed to the socket, in addition to being passed to the TCP/IP protocol stack implemented in the Linux kernel.

Here is an example packet socket creation:

```
ifconfig.sockid =
   socket(PF_PACKET, SOCK_RAW, htons(ETH_P_ALL))
```

A socket with the following characteristics is created: low level packet interface (PF_PACKET), raw packets including link level header (SOCK_RAW) and all frames are received (ETH_P_ALL). If successful, a socket id is returned which is used to refer to the created socket in the sequel. In case of error, the socket() system call returns the -1 value. Error handling is omitted.

When a process is finished with a socket, the resource is released using the close() system call, whose use is as follows:

```
close(ifconfig.sockid);
```

Note that the failure to close sockets creates memory leaks in processes. It is a frequent and hard to pinpoint bug in network programming.

4.3 Parameters and control

In the following text, we discuss configuration of a wireless interface, configuration of a socket and association of the former to the latter.

The parameters of a wireless interface can be retrieved or changed using the ioctl() system call. The parameters of a socket can be changed using the setsockopt() system call. When created, a socket is not bound to a wireless interface. This association has to be created explicitly using the bind() system call.

The ioctl() system call has the following signature:

```
int ioctl(int d, int request, ...);
```

The parameter d is a socket descriptor. A socket is required first and then the wireless interface can be configured. The socket can be seen as the channel through which the wireless interface is accessed.

The second parameter is the code of a request. The third parameter is a pointer to a memory area, which contains either an input parameter, an output parameter or an input/output parameter. The following call to ioctl() fetches the index of a wireless interface. A value of type struct ifreq must be provided to the invocation. It contains the name of the wireless interface (stored in the ifr_name field), which can be obtained with the ifconfig command and stored in a variable defined as follows (can also be read from a configuration file):

```
char device[] = "eth0";
```

The construction of the input parameter and the system call itself are as follows:

```
struct ifreq ifr;
strcpy(ifr.ifr_name, device);
ioctl(ifconfig.sockid, SIOGIFINDEX, &ifr);
ifconfig.ifindex = ifr.ifr_ifindex;
```

The index of the interface is returned in the `ifr_ifindex` field of the `ifr` variable.

The following call to `ioctl()` fetches the hardware address:

```
ioctl(ifconfig.sockid, SIOCGIFHWADDR, &ifr);
```

The hardware address of the interface is returned in a field named `ifr_hwaddr.sa_data` of the `ifr` variable.

The following call to `ioctl()` fetches the MTU:

```
ioctl(ifconfig.sockid, SIOCGIFMTU, &ifr);
ifconfig.mtu = ifr.ifr_mtu;
```

The MTU of the interface is returned in the `ifr_mtu` field of the `ifr` variable.

The `setsockopt()` system call can be used to configure a socket in the promiscuous mode.

Setting the promiscuous mode makes the socket receive all the frames transmitted on the channel, including its own frames. The is done using the `setsockopt()` system call and a parameter of type `struct packet_mreq` with the `mr_type` field set to value `PACKET_MR_PROMISC` as follows:

```
struct packet_mreq mr;
memset(&mr,0,sizeof(mr));
mr.mr_ifindex = ifconfig.ifindex;
mr.mr_type = PACKET_MR_PROMISC;
setsockopt(ifconfig.sockid, SOL_PACKET,
   PACKET_ADD_MEMBERSHIP, (char *)&mr, sizeof(mr));
```

Finally, binding a wireless interface to a socket is done using the `bind()` system call and a parameter of the `struct sockaddr_ll` type with the `sll_ifindex` field set to the value of the interface index, extracted before:

```
struct sockaddr_ll sll;
memset(&sll, 0, sizeof(sll));
sll.sll_family = AF_PACKET;
sll.sll_ifindex = ifconfig.ifindex;
sll.sll_protocol = htons(ETH_P_ALL);
bind(ifconfig.sockid, (struct sockaddr*)&sll, sizeof(sll));
```

4.4 Receiving frames

Frames can be received using the `recvfrom()` system call:

```
int recvfrom(int s, void *buf, size_t len, int flags,
   struct sockaddr *from, socklen_t *fromlen);
```

It waits for a frame to arrive. It is by default a blocking call. The parameter `s` is the socket descriptor, the value returned by the `socket()` system call. The parameter `flags` controls some options. If the parameter `from` is not null, that is, it points to a structure of type `struct sockaddr`, then when a frame is received its source address is returned in it. The parameter `fromlen` is an input/output parameter. In input, it is set to the size of the `struct sockaddr` structure. In output, it contains the actual size of the source address of the frame. The `recvfrom()` function normally returns the length of the received frame. Otherwise, it returns the value -1.

Receiving a frame requires memory for storing the incoming data (whose maximum size is MTU), a variable for storing its effective length, a variable for storing the source address and a variable for storing the length of the source address. These variables are defined as follows:

```
unsigned char * buff = new unsigned char[ifconfig.mtu];
unsigned int i;
struct sockaddr_ll from;
socklen_t fromlen = sizeof(struct sockaddr_ll);
```

The call to `recvfrom()` can be embedded in the following `while` loop:

```
while (true)
{
   i = recvfrom(ifconfig.sockid, buff, ifconfig.mtu, 0,
      (struct sockaddr *) &from, &fromlen);
   if (i == -1)
   {
      printf("cannot receive data: %s\n", strerror(errno));
      usleep(10000);
   } else {
      break;
   }
}
```

It loops until a non-empty frame has been received on the wireless channel. The call to `recvfrom()` waits until a frame is received or an error is reported. If an error is reported, then the returned result (stored in variable `i`) is equal to -1, the process reports the error and sleeps for 10 ms before retrying. Else, the process exits the loop using the `break` statement. Afterwards, the content of the frame can be processed.

4.5 Sending frames

Frames can be sent using the `sendto()` system call. The detailed signature of the `sendto()` system call is as follows:

```
int sendto(int s, const void *msg, size_t len, int flags,
   const struct sockaddr *to, socklen_t tolen);
```

The parameter `s` is the socket descriptor, the value returned by the `socket()` system call. A pointer to the frame being sent is provided as the parameter `msg`. The length of the frame, in bytes, is given by the parameter `len`. Some options can be selected with the parameter `flags`. The address of the destination is provided in the parameter *to*, together with the parameter *tolen* specifying its size. If the submitted frame is too long for the interface, then the frame is not transmitted. Errors are indicated by a return value of −1.

The following send buffer variable is defined to store a frame:

```
# define BUFFSIZE 256
unsigned char buff[BUFFSIZE];
```

With a socket of type `SOCK_RAW`, frames are sent with an *Ethernet like* header. The supplied frame should contain the *Ethernet like* header. A variable of type `WLANHeader` may be declared to facilitate the construction of the header:

```
WLANHeader hdr;
```

For the sake of simplicity, let us define the following destination address in the ASCII form:

```
char * rp = "00:60:1D:F1:F0:27";
```

The address needs to be converted to the binary form. The following variable is used for that purpose:

```
WLANAddr daddr;
```

The header is constructed as follows:

```
daddr.str2wlan(rp);
memmove(&hdr.destAddr, daddr.data, WLAN_ADDR_LEN);
memmove(&hdr.srcAddr, ifconfig.hwaddr.data, WLAN_ADDR_LEN);
hdr.type = htons(IP_TYPE);
memmove(buff, &hdr, WLAN_HEADER_LEN);
```

First, the destination address is converted from ASCII to binary and loaded in the destination address field of the header. The source address, already in binary form, is loaded in the source address field of the header. Then, the type field is set to the IP type (assuming an IP payload). Finally, the header is stored into the frame buffer.

Let us define the body of the sent frame as the following data:

```
char * dp = "This is a short message!";
```

The data is loaded in the frame buffer as follows:

```
memmove(buff+WLAN_HEADER_LEN, dp, strlen(dp));
```

A *to* address structure needs to be constructed before sending the frame. It is a parameter required by the `sendto()` system call.

```
struct sockaddr_ll to;
int tolen = sizeof(to);
to.sll_family = AF_PACKET;
to.sll_ifindex = ifconfig.ifindex;
```

The *to* address consists of the family *AF_PACKET* and interface index.

Finally the frame is sent as follows:

```
int sentlen = sendto(
   ifconfig.sockid, buff, WLAN_HEADER_LEN+strlen(dp), 0,
   (sockaddr *) &to, tolen);

if (sentlen == -1 )
{
   printf("sendto failed\n");
   return NOK;
}
return OK;
```

4.6 Exercises

1. Modify the example of this chapter such that one thread is created for each incoming message, but there is no dynamic memory allocation, except at initialization time. The receiver has a pool of pre-allocated messages, that is, a list of messages. Each entry in the list has a flag saying whether the message is unprocessed or free. When a frame is received, a free message is located and used to store that frame. The message is returned to the pool after it has been processed.

2. Read Section 5.4.1 about the Neighbor Discovery Protocol (NDP). Implement the NDP over the WiFi/802.11 wireless interface on Linux.

5

AD HOC NETWORK PROTOCOLS

...Everybody was there and they were all ready to point the finger, right? If it didn't work. Fortunately...it worked beautifully.
Leonard Kleinrock, 1969

This is how Leonard Kleinrock, the pioneer in packet switching technology recalls the excitement during the historic ARPAnet experiment of September 03, 1969, in sending packets from his UCLA computer science lab (see Segaller (1998), page 92). It became quickly apparent that protocol consistency and software compatibility were the main problems in making computers communicate. This issue lends itself when one tries to create a connected system combining various (often geographically disperse) networks together in order to make essential resources more universally accessible.

An ad hoc network is a set of nodes and a set of links interconnecting the nodes. The neighbors of a node are the nodes directly attached with links. The number of nodes that can be linked to a given node is dynamic and unlimited. Links are point to point and either symmetric (same level of bandwidth in both directions) or asymmetric (different transmit–receive characteristics for each direction).

A separation can be made between forwarding and routing. These two functions share routing tables. Forwarding is the retransmission of received packets according to the information extracted from the routing tables. The scope of routing is the construction of routing tables. Forwarding and routing are layer three functions, although, forwarding and routing may also exist at the link layer or application layer. Hereafter, we focus on layer three forwarding and routing in ad hoc networks in reference to IP.

Nodes with forwarding capability use the links to deliver packets from one end to another of the network. Forwarding can be conducted in two different ways: source routing or hop-by-hop routing.[1] With source routing, each packet contains, in addition to the source

[1] Although, we are strictly referring to forwarding, we use the term routing here for maintaining consistency with a widely used terminology.

Principles of Ad hoc Networking Michel Barbeau and Evangelos Kranakis

address and destination address, a list of intermediate node addresses defining the path that must be followed to reach the destination. A pointer in the list, updated from hop to hop, is also included as an indicator of the next node to be visited. In the routing tables, a source routing protocol stores, for each destination, a list of intermediate nodes that have to be visited to reach the target. When a packet needs to be sent, the route is either extracted from the routing table or discovered dynamically, with certain protocols. When the route is available, the list of intermediate hops is included in a special section of the packet. The packet is transported to destination according to the specified route.

With hop-by-hop routing, each packet contains, as addressing information, solely the source address and destination address. In the routing tables, a hop-by-hop routing protocol maintains, for each destination, the address of the next node to be used to reach the destination. When a packet is received, if the destination address corresponds to the current node's address, then the packet is accepted and its content is processed locally. If the destination address corresponds to one of the entries in the routing table, then the packet is forwarded to the next hop which address is extracted from the entry. Otherwise, the next hop is discovered dynamically, with certain protocols or rejected with others (an error message may eventually be sent to the source).

There are two well-accepted approaches to routing protocols for ad hoc networks: the reactive approach and proactive approach. With the reactive approach, discovery and maintenance of routes is delayed until necessary. In other words, an attempt to find or repair a route for a destination will not be undertaken until a packet effectively needs to be delivered to that destination. With the proactive approach, nodes are actively engaged in route discovery and maintenance even though they may not have data traffic to deliver. Routes are built, maintained and made ready for eventual data traffic.

The reactive approach generates less control traffic in lightly loaded networks, but packets may experience higher latency at the beginning of the establishment of a session with a new destination because the route needs to be discovered or repaired. The proactive approach generates more control traffic in lightly loaded networks, but packets experience less latency because if a route is theoretically possible it is probably already discovered. The proactive approach works best in relatively stable networks that is, with low mobility and low failure rate. In highly dynamic networks, plenty of routing information needs to be exchanged and performance degrades.

First, we review how normal IP routing works in Section 5.1. The reactive approach is examined in more detail in Section 5.2, while the proactive approach is presented in more detail in Section 5.3. The proactive approach and the reactive approach can be combined and yield a hybrid protocol, which is explored in more detail in Section 5.4. Section 5.5 discusses clustering, a way to structure an ad hoc network to enable hierarchical forms of routing.

Routing protocols may employ geographical routing strategies. All the routing protocols discussed in this section employ nongeographical strategies. Geographical routing strategies are discussed in Chapter 6.

5.1 Normal IP routing

For the sake of comparison, we first review how routing with normal IP works. Each node has a number of network interfaces. Each network interface provides a physical connection to a communication medium, for example, an Ethernet cable. Typically, inside a node each network interface is implemented as an electronic card together with some driver and operating system kernel software. The operating system provides services (e.g. subroutines, system calls) invoked by other components of the nodes that use the network interface to communicate (e.g. the TCP/IP stack). Each interface has a hardware address and an IP address. The hardware address is typically hard coded in the interface by the manufacturer when the card is assembled. For example, each Ethernet card has a 48-bit permanent hardware address. The IP address is configured by the node manager and normally reflects the location of the node. Indeed, a part of the IP address is a network prefix identifying the IP network to which the node is attached. Each interface has an output queue in which frames waiting to be transmitted are buffered. The frames are pushed in the queue by the IP protocol. IP also has a queue in which the incoming packets waiting to be processed are pushed. When network interfaces receive frames, they de-capsulate the embedded IP packets and push them in the IP queue.

Forwarding is decided at the IP layer, but by using the network interfaces that provide access to the communications media. If hop-by-hop routing is used, then each IP packet contains a source IP address and a destination IP address. Each node has a routing table. The entries in the routing table can be inserted by a node manager through the use of a software utility. The entries can also be inserted automatically by a routing protocol that implements an algorithm to obtain the information automatically. Each entry applies to a destination network. There is no entry for every single destination node. It wouldn't be scalable. Forwarding in IP is hierarchical and based on destination network. Each entry specifies the IP address of the next node to be used to reach a destination network. The routing table lists a number of destination networks and there is a default interface for any destination network that cannot be resolved in the routing table. There is also another table that maps IP addresses to hardware addresses. These entries can be inserted either by a node manager or by an IP address to hardware address resolution protocol.

When a packet is taken from the IP incoming queue, the destination IP address is examined. If the destination IP address corresponds to one of the addresses associated with the network interfaces of that node, then the packet has reached its destination and its payload is pushed to higher level protocols in that node (e.g. TCP). Otherwise, the packet needs to be forwarded. With hop-by-hop routing, the path is determined dynamically. When a packet needs to be forwarded by a node, an attempt is made to match the network identifier contained in the destination IP address, contained in the packet, with a network identifier stored in the routing table. The entry with longest (in bits) matching network identifier prefix is used. A next-hop IP address is returned. The IP address of the next hop is resolved to a hardware address. The network identifier of this IP address is used to determine the appropriate network interface that must be used for forwarding. The packet is pushed on the output queue of the network interface. The packet is eventually handled, encapsulated in a frame destined to the next hop and transmitted by the selected network interface.

If source routing is used, then each packet contains, in addition to the source address, a list of IP addresses to be used to reach a destination. The destination IP address is updated from hop to hop and is always that of the next intermediate node. The IP address of the final destination is listed last. In contrast to hop-by-hop routing, source routing determines the entire path to be taken before a packet is injected in the network. When a source routed packet is received by a node, the destination IP address is examined. If the destination IP address corresponds to one of the addresses associated with a network interface of that node, then the packet has reached its destination and its payload is pushed to higher level protocols in that node (e.g. TCP). Otherwise, the packet needs to be forwarded. When a source routed packet needs to be forwarded, the destination IP address is updated using the IP addresses listed. A lookup in the routing table is performed as an attempt to match the network identifier of one of the entries with the network identifier contained in the IP address. The longest matching network identifier is used. The network identifier is used to determine the network interface that needs to be used to reach the next hop. The destination IP address is resolved to a corresponding destination hardware address. The packet is pushed on the queue of the network interface. The IP packet is eventually taken from the queue, encapsulated in a frame destined to this hardware address and transmitted using the network interface.

Because of the dynamic character of ad hoc networks, normal IP routing and forwarding do not work as is. In ad hoc networks, IP addresses are no more related to physical locations. If hop-by-hop routing is used, then the network topology needs to be discovered to fill the routing tables. Routing tables need frequent updates. If source routing is used, then source routes need to be discovered and updated frequently. Ad hoc networks are not by nature hierarchical as the Internet is. Hence, routing is not necessarily hierarchical, which means that the ad hoc networking approach may be applicable on a small to medium scale only.

5.2 The reactive approach

In this section, we discuss the reactive approach using the Dynamic Source Routing (DSR) protocol as a representative example. DSR results from the work of Johnson (1994) and Johnson and Maltz (1996). It has been standardized by Johnson et al. (2004), but thereafter there has been no attempt either for exhaustive coverage or for conformance to standards. DSR uses source routing that is discussed first. Route discovery and maintenance are discussed thereafter.

The source routing concept is introduced using the capabilities of IPv4. The format of an IPv4 packet with a source route option is pictured in Figure 5.1. As usual, the IP header contains the source address and destination address. The source address is the origin of the packet. The destination address is the target, if there is a direct link with it, otherwise it is the address of the first hop on a route leading to the target. If there is a direct link with the target,

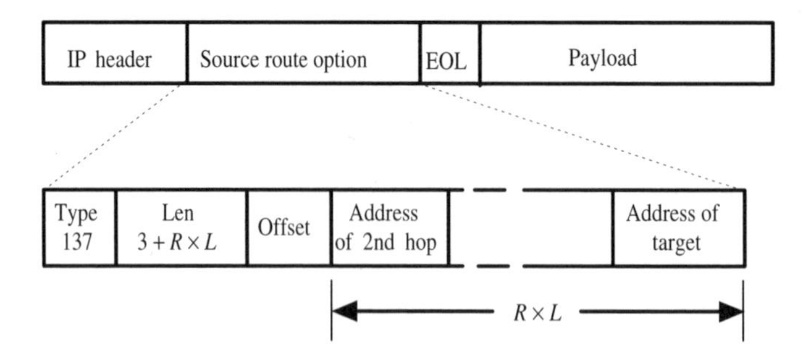

Figure 5.1 Data packet with a source route option.

then there are no addresses listed in the source route option and its length is 3 bytes. Alternatively, the source route option may be omitted. If there is no direct link with the target, then the source route option lists all intermediate hops and the IP address of the target. The source route option is followed by an end of list (EOL)option. The payload is listed next.

The source route option is made of fields: Type, Len, Offset, a list of intermediate hops and Address of target. The *Type* field is 1 byte long and is set to value 137. The *Len* field is 1 byte long and corresponds to the length in bytes of the entire option. If R is the number of listed addresses and L is the length of an IP address, then the *Len* field is set to $3 + R \times L$. The *Offset* field is 1 byte long and gives the index of the byte of the next address to be used. Bytes are numbered from left to right starting with number one. For instance, if the next address to be used is *address of 2nd hop*, then the value of Offset is four. After each hop, the value L is added to the value of *Offset*. When the packet reaches its final destination, the *Offset* field has value $3 + R \times L$ plus one.

Figure 5.2 pictures a source routing scenario. The source (src) is node 1 and the target is node 4. There are five nodes, numbered 1, ..., 5, with five links. Note that the links between nodes 1 and 5, nodes 2 and 3 and nodes 4 and 5 are unidirectional. The contents of the source address (*src*), destination address (*dst*), second hop (*2nd hop*) and target fields are given for each packet. The location determined by the value of the offset is also indicated with a left arrow. From hop-to-hop, the source address field is invariant. The destination address field is updated from hop-to-hop. It successively takes the values 2, 3 and 4. It is always the address of the next node the packet must reach. The offset points to the address of the next hop. When a packet is received by a node, either it has reached its destination

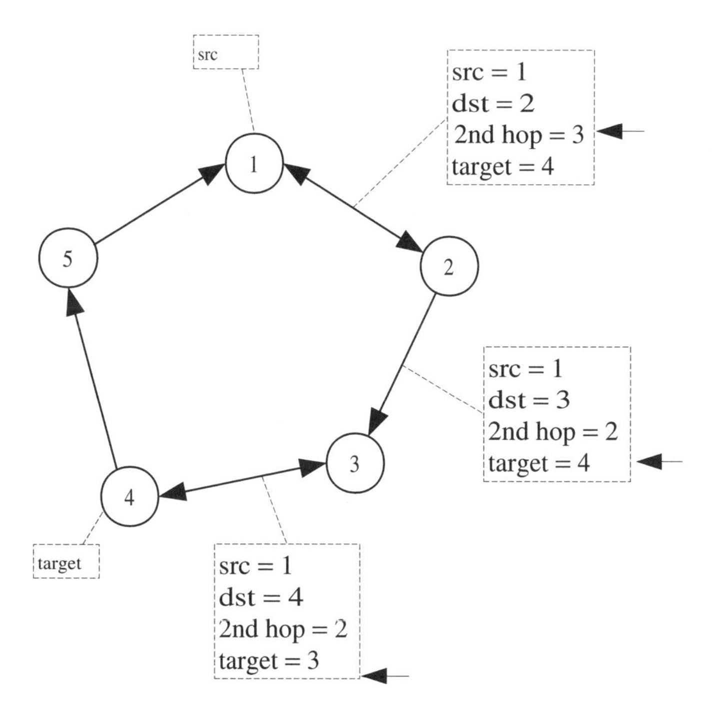

Figure 5.2 Source routing from node 1 to node 4.

(i.e. the destination address is that node address) or it needs to be forwarded. In the latter case, the value in the destination address field is switched with the value determined by the offset. This way the final destination can use the list of addresses contained in the source route option to build a reverse part. This is, although, applicable only if we make use of bidirectional links only. This does not hold in the example, but a path does exist from node 4 to node 1 through node 5.

Source routing increases the overhead associated to data traffic. Applicability makes sense in small diameter networks where routes may consist of a maximum of 5–10 hops.

The route discovery function of DSR can be implemented using IPv4. The format of an IPv4 packet with a route request (RREQ) option is represented in Figure 5.3. The remote address is used as the target address. An EOL is added at the end to make the length of the option a multiple of 4 bytes; 3 no operation (NOP) bytes are appended.

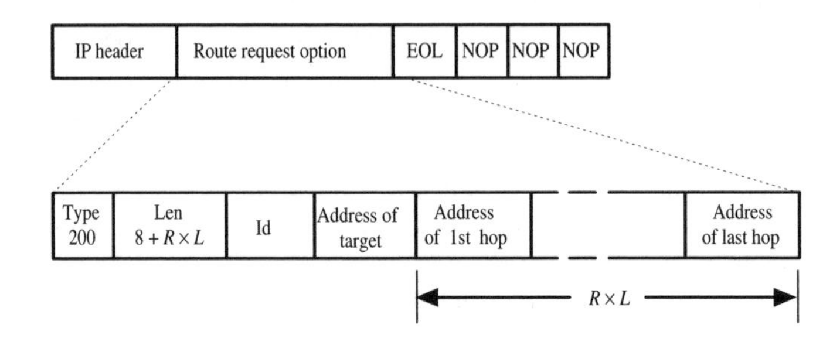

Figure 5.3 Data packet with a route request option.

The source address in the IP header is the home address of the origin. The destination address is the `255.255.255.255` broadcast address. The RREQ option is made of the following fields: Type, Len, Id, Address of target and a list of intermediate hops. The *Type* field is one byte long and is set to value 200. The *Len* field is one byte long and corresponds to the length in bytes of the entire option. If R is the number of intermediate hops and L is the length of an IP address, then the *Len* field is set to $8 + R \times L$. The *Id* field is 2 bytes long (not used in this example).

Upon reception of a RREQ, a node does the following. If this node is the target of the RREQ (i.e. target IP address is equal to the home address), then the life cycle of the RREQ ends and a route reply (RREP) is returned to the origin. If this node is the source of this RREQ, then the latter is ignored. If this node's home address is in the list of addresses in the RREQ, then the latter is ignored. Else, if a new address cannot be appended to the RREQ (because of a lack of space), then the RREQ is ignored. Otherwise, this node's home address is appended to the RREQ and the message is forwarded.

Figure 5.4 illustrates the discovery of a route from node 1 to target node 4. The contents of the source address and destination address fields together with the RREQ option are detailed. The value for the source address is invariant from packet to packet and is node 1. The value of the destination address is also invariant and set to the broadcast address 255.255.255.255. The value of the target field is also fixed and set to node 4. The list of intermediate nodes is collected along the way to the target. The first hop address is added by node 2 and the second hop address is node 3. The RREQ reaches its final destination at node 4.

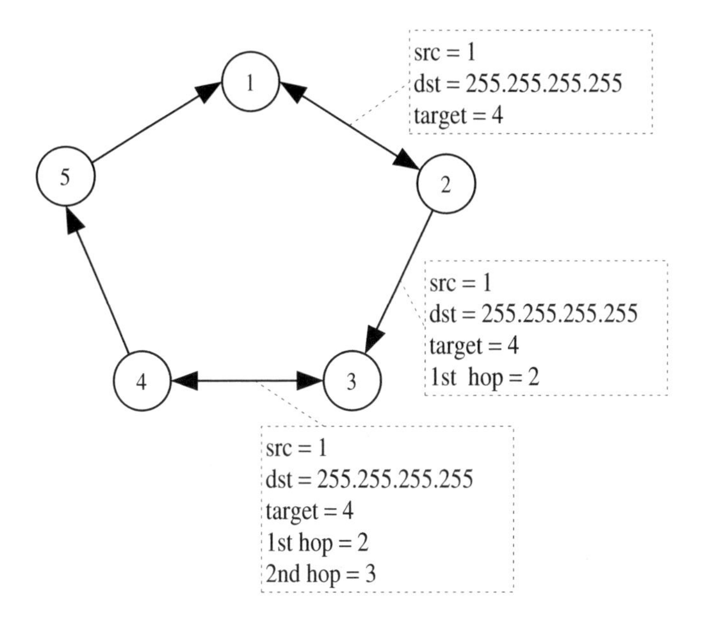

Figure 5.4 Route request from node 1 to node 4.

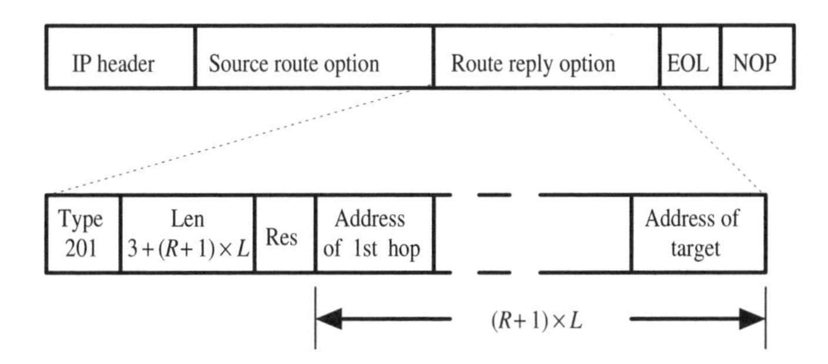

Figure 5.5 Data packet with a route reply option.

A RREP packet is built and sent when a RREQ packet reaches the target node. The format of an IPv4 packet with a source reply option is represented in Figure 5.5. The route is returned using the source route option of IP (if symmetric links are assumed). Therefore, the packet containing the RREP is made of an IP header, a source route option, a RREP option and an end of list option. An NOP is appended to make the header length a multiple of 4 bytes. The payload of the packet is empty.

The source address in the IP Header is the home address of the node sending the RREP, which is also the end of the route and target of the corresponding RREQ. If the length of the route is 0, then the destination address is the IP address of the origin of the route. Else, it is the last IP address listed in the returned route (the first hop).

If the length of the route is 0, no addresses are listed in the source route option. If the length of the route is greater than 0, then the source route option contains all the addresses listed in the returned route, but the last, and listed in reverse order. The target is added.

The RREP option contains the list of IP addresses of the intermediate nodes and home address of the target. The intermediate nodes are all the IP address listed in the route formal parameter. R is the length of the route, zero if is there if direct link and L is the length of an IP address. Therefore, $(R + 1) \times L$ is the length, in bytes, of the data part of the option. The *Len* field includes as well the first 3 bytes: Type, Len and Res (not used). The route always contains at least one IP address: the target. Hence, the value of *Len* is always at least 7.

A RREP packet is not handled until it reaches its final destination. When a RREP reaches its final destination, a source route is constructed from the information contained in the RREP option of the packet. The first hop of that source route is the first IP address listed in the RREP option. The RREP option contains at least the IP address of the target. If the number of IP addresses listed in RREP option is greater than 1, then the listed addresses, but the first, constitute the intermediate hops of the source route. Else, the source route has no intermediate hops.

Figure 5.6 shows the forwarding of a RREP from node 4. Note that the path from node 1 to node 4 acquired with an RREQ (as pictured in Figure 5.4) cannot be reversed to reach node 1 from node 4. Piggy backing a RREP with another RREQ targeted to node 1 is used instead. Figure 5.4 illustrates the contents of both the RREQ option (which target is node 1) and RREP option with detailed route from node 1 to node 4. When this second RREQ reaches node 1, another RREP needs to be returned to node 4 such that it learns the discovered path from node 4 to node 1.

The forwarding of this second RREP, forwarded with source routing, is represented in Figure 5.7. The RREP is routed through the nodes 2 and 3 and reaches its final destination at node 4.

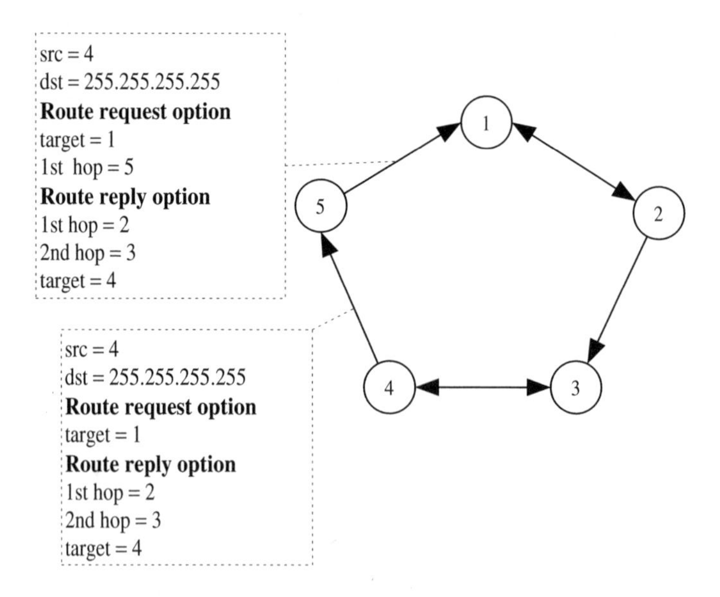

Figure 5.6 Route reply from node 4 to node 1.

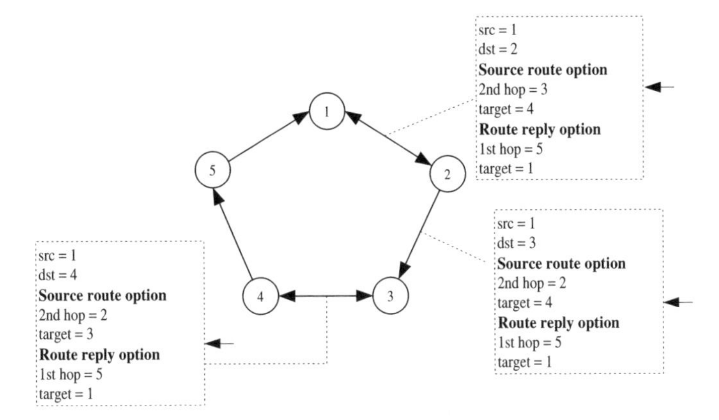

Figure 5.7 Source route reply from node 1 to node 4.

The Ad Hoc On-Demand Distance Vector (AODV) is another example of a reactive protocol, and is now an Internet Engineering Task Force (IETF) standard authored by Perkins et al. (2003).

5.3 The proactive approach

The Optimized Link-State Routing (OLSR) protocol is used to illustrate the proactive approach. We discuss OLSR, according to the definition of Clausen and Jacquet (2003). OLSR is a routing-only protocol. Forwarding is performed as defined by Baker (1995) using routing entries built by OLSR. In combination with OLSR, the standard IP packet format is used for data traffic.

Routing in OLSR is hop-by-hop. OLSR uses symmetric links only, that is, links operating in both directions. A node may have several network interfaces, each identified with a unique IP address. Each node maintains a routing table, represented in Table 5.1 with sample data. Each entry in the table consists of a destination address, a next-hop address, a distance and an interface address. The next-hop address identifies a neighbor on a shortest path to the destination. Paths are made of symmetric links only. The distance to the destination is expressed as a number of hops. The interface address indicates the link to be used to reach the next-hop neighbor. When a received IP packet is considered for forwarding, a lookup is performed on the routing table using the destination address. If the lookup

Table 5.1 OLSR routing table.

Destination address	Next hop address	Distance	Interface address
3	2	2	a1
4	2	2	a1
6	5	2	a2
8	5	2	a2

operation succeeds, the corresponding next-hop address is extracted from the routing table entry.

Note that routing entries are indexed by destination node. In contrast, in normal IP routing, entries are indexed by destination network identifier, which is more abstract. Indeed, a network is an aggregation of several nodes. In other words, OLSR maintains in every node a routing table entry for every other node. The routing table needs to be updated whenever new information is acquired about the topology of the network. This is an impediment to the scalability of OLSR. Lack of scalability here means that the performance is not maintained when the load increases (e.g. number of nodes, number of links or traffic). OLSR can be applied to small-scale networks only.

OLSR uses UDP datagrams to exchange control messages. The use of UDP for control messages facilitates a user-level implementation of OLSR. In other words, OLSR can be implemented as a user process with communication with the operating system kernel, where IP is most of the time placed, for updating routing table entries.

The format of OLSR control messages in illustrated in Figure 5.8. A single UDP datagram may actually embed several OLSR messages. There is a common header, in which the total length of the sequence of messages is stored. Each individual message has a header, in which the type of the message, its size and address of origin can be found. Messages are assembled together to efficiently use the lower-lever protocols. IP and MAC headers are amortized over several OLSR messages. Each control message is, however, processed individually by OLSR. Two types of OLSR messages are discussed in the sequel, namely, the hello message and topology control message.

Figure 5.8 OLSR control message format.

Each node maintains a selector table. Each entry in the selector table is the address of a neighbor node for which the current node is playing the role of *multipoint relay* (MPR). Each node independently selects, among its one-hop neighbors, its multipoint relays. They are, however, selected such that every two-hop neighbor is reachable through at least one of the multipoint relays. Topology control (TC) messages, which propagate routing information, are forwarded by multipoint relays only. If the number of multipoint relays is minimized in some way, then a reduction of duplication in flooding traffic may be expected. An example is pictured in Figure 5.9. The multipoint relays of node 1 are nodes 2 and 5. There is no need to use node 7, since all two-hop neighbors are reachable using nodes 2 and 5. Both nodes 2 and 5 have the address of node 1 entered in their selector table.

OLSR detects and eliminates repeatedly received control messages to avoid duplication of work. OLSR keeps track of the recently received messages using a duplicate table, pictured in Table 5.2. Each entry in the duplicate table consists of the address of the originator of a message, the sequence number of the message, a retransmission indicator, an interface address list and a scheduled removal time. The retransmission indicator is a

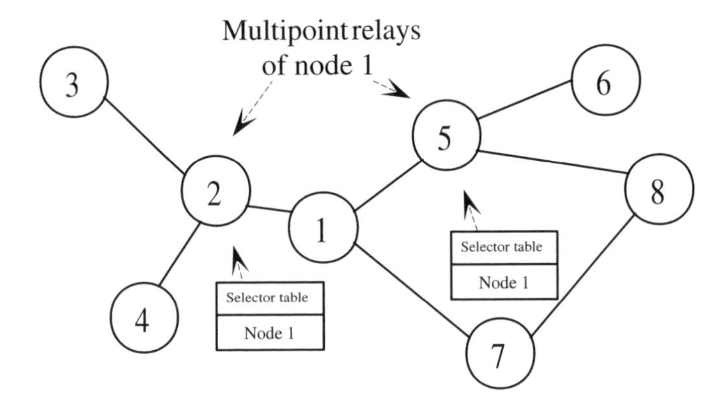

Figure 5.9 Multipoint relays.

Table 5.2 OLSR duplicate table.

Originator address	Message sequence number	Retransmission indicator	Interface address list	Scheduled removal time
1	1	True	$a1$	$T1$

Boolean indicating whether or not the message has been forwarded by the current node. The interface address list contains the IP addresses associated to the interfaces on which the message has been received. The scheduled removal time specifies the moment when the entry must be deleted to recover the space for more recent messages. Whenever an update is needed, a constant value is added to the current time and the result is stored in the scheduled removal time field.

A received message is a candidate for processing, that is, interpretation of the embedded control information, and forwarding, that is, retransmission to neighbors. Any message whose originator address and sequence number match those of an entry in the duplicate table is considered to be a duplicate with respect to processing. Its payload is not examined. Otherwise, the payload of the message is decoded and the extracted information is used to update the routing table.

Any message for which the address of the interface on which it was received, originator address and sequence number match those of an entry in the duplicate table is considered. A duplicate message is forwarded only if the retransmission indicator of the corresponding entry is false. It is set to true after forwarding. When a nonduplicate message is received, if its originator address and sequence number are not contained in the table, then a new entry is created for this message. If the message is a nonduplicate, but an entry already exists for the originator address and sequence number, then the IP address associated to the interface on which the message was received is added to the interface address list in the corresponding entry. The scheduled removal time is updated.

Every node periodically sends hello messages to advertise its presence to other nodes and to inform about its one-hop neighborhood and selected multipoint relays. Hence, through the interception of hello messages, every node learns about its two-hop neighborhood. This provides the information required by a node to select its multipoint relays. Hello messages are not forwarded.

Each node that has been selected by other nodes to play the role of multipoint relay informs the other members of the network by sending TC messages. Each TC message contains the addresses of the neighbors for which the originator is a multipoint relay. TC messages are forwarded over the whole network using the multipoint relays. This provides the information required by a node to build its routing table. Nodes learn the multipoint relays required to reach every other node in the network. Nodes build the entries of their routing table using the shortest-path algorithm of Dijkstra (1959). Topology control messages are forwarded only by multipoint relays. When forwarded, they are sent using broadcast on all interfaces. In Figure 5.9, TC messages generated by nodes 2 and 5 contain the address of node 1.

A forwarding scenario of an OLSR TC (TC) message illustrating the use of the information stored in the duplicate table is depicted in Figure 5.10. In this example, there are three flows of message represented by the graphs (a), (b) and (c). A TC message with a Time To Live (TTL) with value 4 is produced by node 1. This message is new and forwarded by the nodes 1, 2, and 3, but with TTL value decreased by 1 at each step. When it reaches node 4, forwarding ceases because the TTL reaches the value 0. Note that node 1 also sends the message to node 5 (since it must also be a multipoint relay). It is forwarded by node 5 to node 3, but dies there because node 3 detects a duplicate (Table 5.2 for sample data).

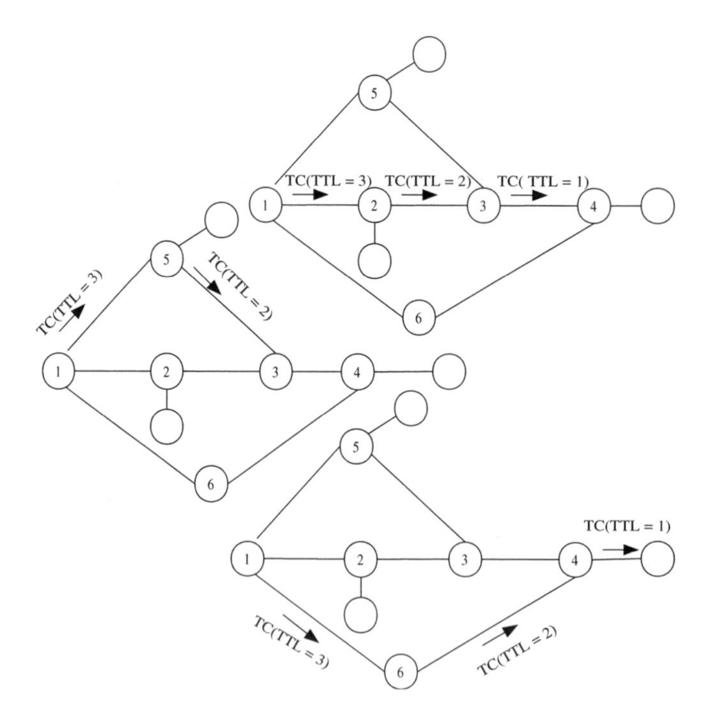

Figure 5.10 Forwarding example of a TC message in OLSR.

Node 1 sends the message to node 6 as well, where node 6 is also needed as an MPR. The message is forwarded by node 6. At node 4, if we assume that it is the latest to arrive and the interfaces to nodes 3 and 6 are different, then it will be forwarded by node 4.

The Destination-Sequenced Distance Vector (DSDV) of Perkins and Bhagwat (2001) and Topology Dissemination Based on Reverse-Path Forwarding (TBRPF) of Ogier et al. (2004) are two other examples of proactive protocols.

5.4 The hybrid approach

A hybrid routing approach integrates ideas from reactive protocols and proactive protocols. The Zone Routing Protocol (ZRP) from Haas and Pearlman (2001) is used in this section to illustrate the hybrid approach.

ZRP is designed under the hypothesis that in ad hoc networks a large portion of the traffic is between nodes that are geographically closed. The proactive approach, which is costly in term of control traffic, together with source routing is applied to deliver packets to short-distance destinations. The reactive approach together with source routing is applied to reach long-distance destinations. Since, communications with long-distance destinations are, by assumption, less likely to occur, savings in the amount of control traffic compensate for the beginning-of-a-session latency associated with the reactive approach. ZRP consists of three subprotocols, namely, the Neighbor Discovery Protocol (NDP), Intrazone Routing Protocol (IARP) and Interzone Routing Protocol (IERP) . IARP deals with short-distance routing, while IERP deals with long-distance routing. These protocols, coordinated together, are discussed in more detail hereafter.

5.4.1 Neighbor discovery protocol

The NDP is used to discover one-hop neighbors. To advertise its presence, each node repeatedly sends, using broadcast, a hello beacon. The message contains the address of the sender. Each NDP entity maintains a table to store the information about the neighborhood. The structure of the neighbor information is represented in Table 5.3, with sample data for node with address A pictured in Figure 5.13. Each entry stores a neighbor's address and corresponding information. The arrival field models a recent reception of a hello beacon from the corresponding neighbor. Upon arrival of a hello beacon from the neighbor, the arrival field is set to the value True. There is a neighbor table update algorithm, which runs regularly. The time between updates is chosen at random. The field arrival is set to value False at the end of the execution of this algorithm. The field last recorded models the logical age of the most recent hello beacon received from the neighbor. It is set to value -1 when a hello beacon is received from a neighbor. It takes the values $0, 1, 2, \ldots$ consecutively

Table 5.3 Neighbor table.

Neighbor address	Arrival	Last recorded
B	True	–
C	True	3
D	False	5

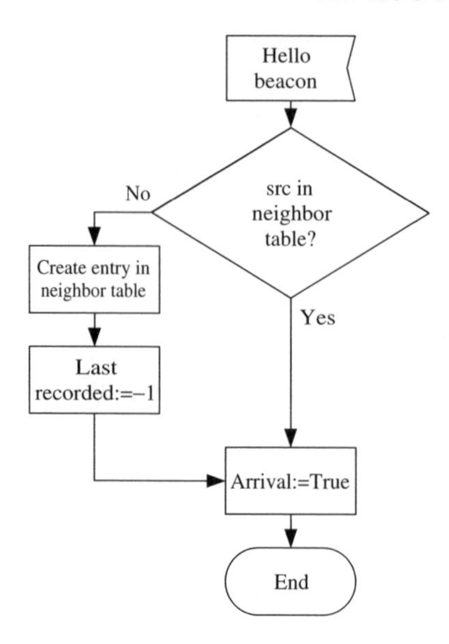

Figure 5.11 Reception of a hello message in NDP.

at the end of each execution of the neighbor table update algorithm. Randomness in this algorithm reduces the risk that all nodes do the same thing at the same time, for example, send a hello beacon. The transmission attempts are hence distributed in time.

The reception of a hello beacon in NDP is pictured in Figure 5.11. The top rectangle with right side enclave represents the reception of a hello beacon from the network. If the address of the source of the beacon is in the neighbor table, then the corresponding arrival field is set to the value True (as for neighbor C). Otherwise, a new entry for that neighbor is created in the table, and the fields arrival and last recorded are set to values True and -1, respectively(as for neighbor B).

The interpretation of the various combinations of the values of the fields in the neighbor table is given in Table 5.4. In accordance with this interpretation, the neighbor table update algorithm is pictured in Figure 5.12. This flow chart is repeatedly applied on each entry of the neighbor table. Given an entry, the value of the arrival field is examined. If the value is False (as for neighbor C), then the value of field Last recorded is inspected. A maximum number of updates (of Max, a parameter of an implementation) on the neighbor

Table 5.4 Interpretation of neighbor table entries.

Arrival	Last recorded	Interpretation	
True	-1	Beacon has arrived since last update	A new neighbor found!
True	$\neq -1$		
False	$\geq$ Max	Beacon has not arrived since last update	Neighbor lost!
False	$<$ Max		

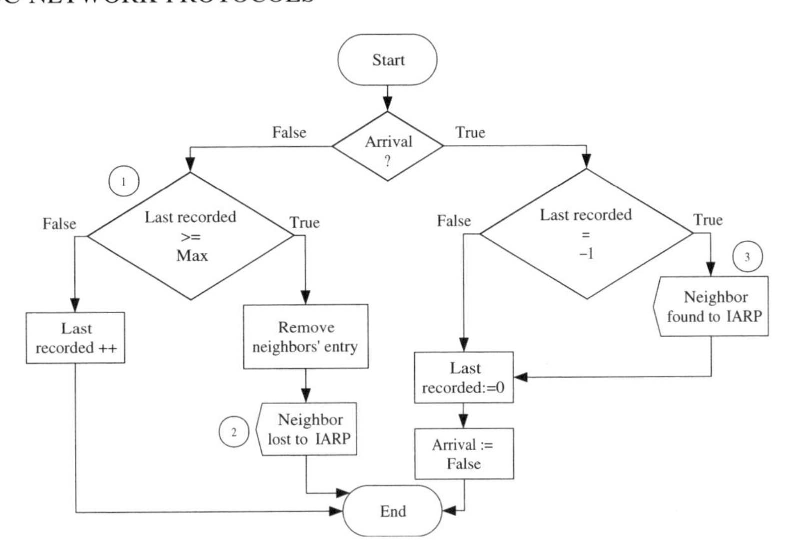

Figure 5.12 Update of the neighbor table in NDP.

table entry can performed while no hello beacon is received from the neighbor (1). When this maximum is reached, the entry is deleted and contact with the neighbor is considered to be lost. This fact is signaled with the neighbor lost message to the IARP protocol (2).

If on the other hand, the value of the arrival field is True and value of the last recorded field is -1 as well, then the neighbor found signal is sent to IARP (3). If the value of the last recorded field is different than -1, then arrival and last recorded are set to values False and 0, respectively.

5.4.2 Intrazone Routing Protocol

The IARP addresses route discovery and route maintenance for short-distance destinations. It is an adaptation of the Open Shortest Path First (OSPF) algorithm defined by Moy (1998).

The concept of *zone* is central in ZRP. A zone has a meaning relative to a given node. It has a radius ρ defined as a number of hops. Let $N_i(v)$ denote the set of distance i neighbors of node v ($i = 0, 1, 2, \dots$). By convention, $N_0(v) = \{v\}$. The zone $Z(v)$ of node v consists of all its $0, \dots, \rho$-hop neighbors, that is $Z(v) = \cup_{i=0}^{\rho} N_i(v)$. Each node has its zone, and zones from different nodes may overlap, particularly if they are close neighbors. In the example pictured in Figure 5.13, ρ is equal to 2. The zone of node A is delimited by a circle. Hence $Z(A) = \{A, B, C, D, E, F\}$. The *peripheral nodes* $P(v)$ of a node v are the nodes reachable through ρ hops, that is, $P(v) = N_\rho(v)$ (nodes E and F in this example). The *interior nodes* of a zone are the nodes within the routing zone (nodes A, B, C and D in the example), but which are nonperipherals.

The IARP does proactive routing and uses a link-state algorithm that maintains the information about the topology of the zone of a node. If the node has a large number of neighbors, then it implies a large number of nodes in the routing zone and heavy update traffic, but limited to the zone in scope. The zone radius can vary from node to node. For the sake of simplicity, it is assumed to be invariant in the sequel.

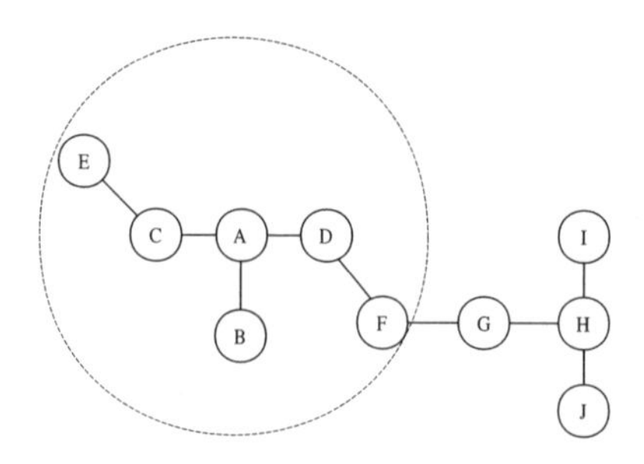

Figure 5.13 Zone of node A.

The IARP nodes exchange link-state packets describing links between nodes. A link-state packet may contain several link descriptions. Each description consists of a source address and a destination address. The source address is the origin of the link. The destination address is the target of the link. Link-state packets are sent using IP broadcast to neighbors. For limiting the scope of the propagation, the TTL field in link-state packets is set to two times the zone radius minus one.

Every IARP node maintains a routing table, a link-state table, a pending-link state table and a peripheral table. They are pictured in Tables 5.5, 5.6, 5.7, and 5.8 with sample data for node A. The routing table contains source routes to destinations inside the zone, except for zero or one-hop neighbors. The link-state table stores information about known links. The pending-link state table contains information extracted from newly received link-state packets, but not yet integrated in the routing table.

The handling by IARP of a link-state packet, a neighbor found signal or a neighbor lost signal is pictured in Figure 5.14. When the discovery of a new neighbor is reported by

Table 5.5 IARP routing table.

Destination address	Source route
E	C
F	D

Table 5.6 IARP link-state table.

Link source address	Link destination address	Insert time
A	C	$T0$
A	D	$T1$
A	B	$T2$
C	E	$T3$
A	D	$T4$

Table 5.7 IARP pending-link state table.

Link source address	Link destination address
D	F
F	G

Table 5.8 IARP peripheral table.

Peripheral	Peripherals of peripheral
E	A
F	A, H

NDP, a new entry is created in the link-state table to represent the link. When a link-state packet is received, an entry is created in the pending-link-state table for every received link description. A control field in a link-state packet is used by its originator to mark the end of a flow of link-state packets constituting together one update. When such a signal is received, the information stored in the pending-link-state table is loaded in the link-state table. Then, the routing table update is performed as follows.

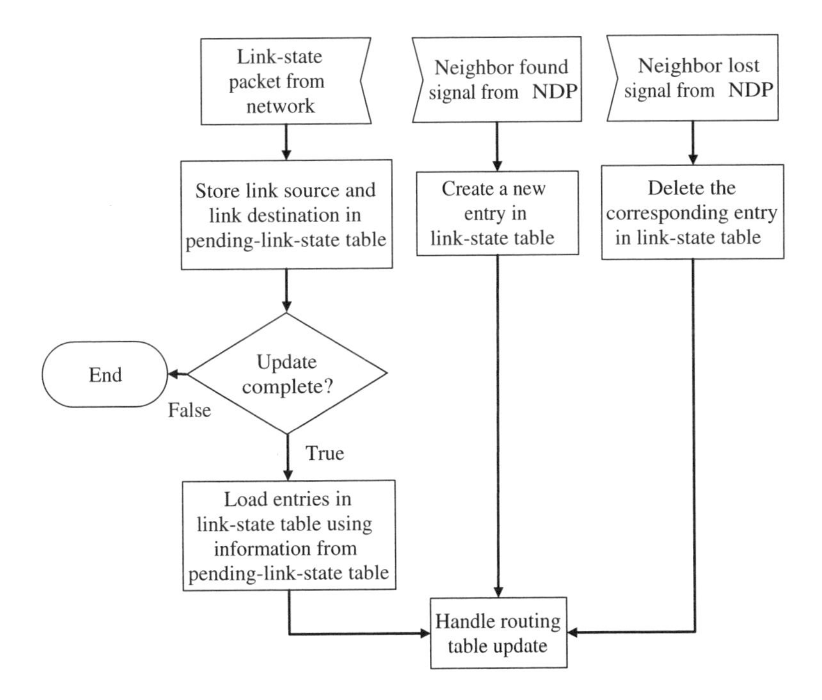

Figure 5.14 Handling of a link-state packet, a neighbor found signal or a neighbor lost signal in IARP.

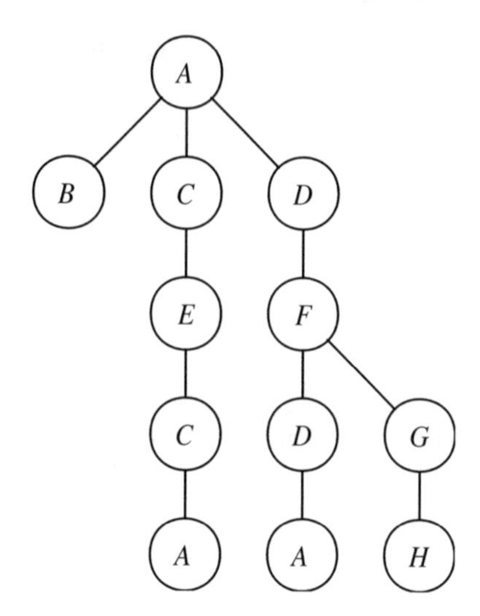

Figure 5.15 Extended zone of node A represented as a spanning tree.

Using the information contained in the link-state table, a spanning tree rooted at the current node is constructed. Branches are pruned such that their length is not greater that two times the zone radius. Such a spanning tree for node A is pictured in Figure 5.15. The coverage of the spanning tree defines the *extended zone*. The extended zone contains the current node's zone and zones of all its peripherals. From the root, each subpath of length $2, \ldots, \rho$ defines a source route to a destination in the zone. All such source routes are stored in the routing table. Nodes in the spanning tree at depth ρ correspond to the peripheral. The peripherals of each peripheral are obtained by following the branches of their corresponding subtrees until depth 2ρ is reached. The addresses of the peripherals are loaded in the peripheral table. Links in the link-state table attached to nodes outside the extended zone are removed. The information remaining in the link-state table is sent to new neighbors, and updated information is broadcast to existing neighbors. The TTL field in the packets is set to two times the zone radius minus one. The propagation of a link-state packet does not extend beyond the extended zone of its originator. When an entry is loaded in the link-state table, the current time is stored with the entry. An implementation-dependent life time is granted to every entry. When this time expires, the entry is deleted. An entry in the link-state table may also be deleted because the loss of the link is reported by NDP.

When a packet needs to be sent, an attempt to resolve the destination is done in the neighbor table. If it works, the packet is delivered using the link with that neighbor. Otherwise, a lookup in the routing table is attempted. If this attempt succeeds, the source route paired with the destination is extracted from the routing table. Source routing is applied to deliver the packet. Otherwise, the IERP is applied.

5.4.3 Interzone routing protocol

Whenever a destination cannot be resolved using the neighbor table or routing table, then a route discovery procedure using the IERP is launched. Route discovery in the IERP is based on a route request-reply model. The procedure is similar to the one used by DSR with slight differences taking into account the concept of zone. Whenever a node receives a RREQ, the zone is first inspected. If the target is in the neighbor table or routing table, then the accumulated path is returned to the origin of the RREQ. Otherwise, a RREQ is sent to every peripheral, while avoiding overlapping with previously queried nodes. This action is termed *bordercasting*.

The IERP maintains a detected request table (Table 5.9). Together, the fields source address and request ID constitute the key of the detected request table. IERP assumes symmetric links. The previous hop address is used as next hop of reverse path when the RREP is returned to the source (in contrast, DSR assumes asymmetric links and uses source routing for the same function). In the table, the sample data corresponds to the reception by node F of a request from node A with ID one. Duplicate RREQs can be detected and ignored. An aging mechanism is used to delete entries and make room for new requests.

Table 5.9 IERP detected request table.

Source address	Request ID	Previous hop address
A	1	D

The transmission of a RREQ is pictured in Figure 5.16. If a lookup in the neighbor table or routing table fails, then the route for the target is unknown (1). A RREQ is assembled for transmission. A unique request ID is assigned to every RREQ by incrementing the value of an ID counter (2). The field bordercast_from contains the address of the bordercaster of the

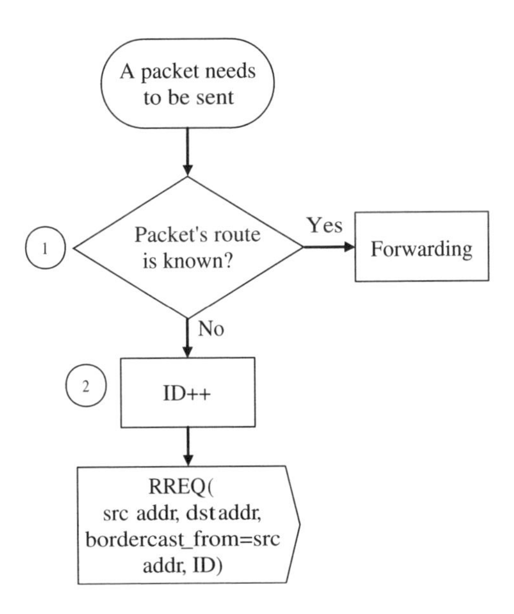

Figure 5.16 Sending a route request in IERP.

request. It is initially the original sender of the request, but it is updated when the request is handed over from a node's zone to another.

Figure 5.17 pictures the relaying of a RREQ. When a RREQ is received for the first time, the embedded information is stored in the detected request table. If the source address and request ID are present, then the RREQ is ignored (1). The test regarding the membership of the destination in the zone is determined by a lookup in the neighbor table and routing table (2). The path is accumulated as the RREQ makes its way in the network. When the destination is finally resolved, the path extracted from the routing tables is returned in a RREP to the origin. If the destination is not yet resolved, then the RREQ follows its path to the peripheral node (3). When a RREQ reaches the peripheral node

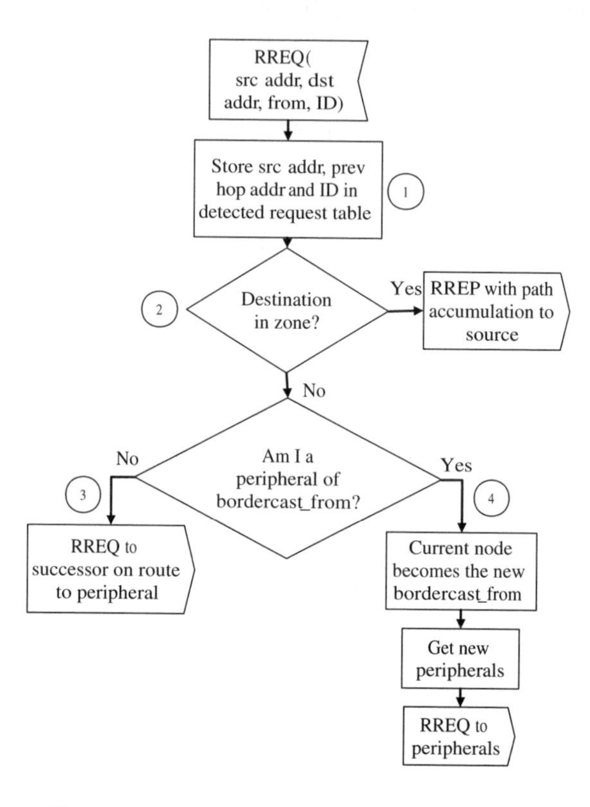

Figure 5.17 Relaying a route request in IERP.

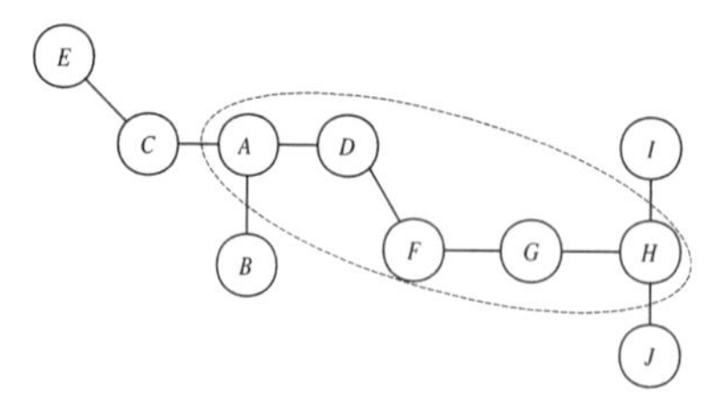

Figure 5.18 Zone of node F.

of the bordercaster (whose address is stored in field bordercast_from), the peripheral node becomes the new bordercaster (4). The new peripherals for the node are determined, but excluding previously consulted nodes. For example, if node A is is the old bordercaster and node F the new bordercaster, then the new peripheral set is $P(F) - Z(A) = \{H\}$ (the zone of node F is pictured in Figure 5.18). A RREQ is sent to every new peripheral node.

5.5 Clustering

A *cluster* is a subset of nodes of a network. *Clustering* is the process of partitioning a network into clusters. Clustering is a way of making ad hoc networks more scalable. A possible scheme is the following. Each cluster elects a cluster head. Each clusterhead knows the members of its cluster and how to reach each of them (may be through a direct link). The cluster heads together run one of the ad hoc network protocols. Cluster heads also forward packets on behalf of the members of their cluster. Cluster members forward all their traffic to their cluster head. The cluster heads act as routers for their cluster members. This section discusses the formation of clusters.

A network is modeled as a graph $G = (N, E)$ where $N = \{1, 2, \ldots, n\}$ is a set of n nodes and $\{i, j\} \in E$ if and only if there is a link between node i and node j. The set of nodes to which a node i is linked is denoted as $N(i)$. The number of neighbors of node i is denoted as $|N(i)|$ and called the *degree of* i. Given a node i, the function $W(i)$ is the weight of node i. The weight may simply correspond to the address of the node.

Given a node i, predicate $Ch(i)$ is *true* if node i is a clusterhead; otherwise it is false. Given nodes i and j, function *Join*(i) is equal to j if node j is a clusterhead (i.e. $Ch(j)$ is true), node i is a neighbor of node j (i.e. $i \in N(j)$) and node i has joined j's cluster; otherwise it is 0. Note that by definition a node cannot join itself, that is, *Join*$(i) = i$ is an invalid statement.

Basagni (1999) defines a cluster formation algorithm, called *Distributed and Mobility-Adaptive Clustering* (*DMAC*), addressing the initial setup and maintenance of the clusters in a mobile and ad hoc network. Mobility changes the topology of the network: new links can be created and existing links can break. Each node has a unique time-invariant weight. By definition, $W(i)$ equals $W(j)$ if and only if i equals j.

The following predicate compares the weight of every neighbor j of a node i with the weight of a third node h.

$$p_1(i, h) \equiv (\forall j \in N(i))[W(j) > W(h) \Rightarrow (\exists k \in N - \{i\}) Join(j) = k] \qquad (5.1)$$

Predicate $p_1(i, h)$ is true if every i's neighbor j heavier than node h has joined k's cluster, for some k. It is a necessary, but not sufficient condition, for h to be the clusterhead of i (the other condition being that h is a clusterhead). If predicate $p_1(i, i)$ is true, then every i's heavier neighbor has joined a cluster and i can become a clusterhead, that is, $Ch(i)$ is set to true. If $\{i, h\} \in E$ (i.e. nodes i and h are neighbors), $Ch(h)$ is true (node h is a clusterhead) and predicate $p_1(i, h)$ is true, then *Join*(i) is defined as h. In other words, node i joins h's cluster.

Whenever a node i becomes a clusterhead, it sends the message $Ch(i)$ to all its neighbors. Whenever a node i joins node j's cluster, it sends the message *Join*(i, j) to all its

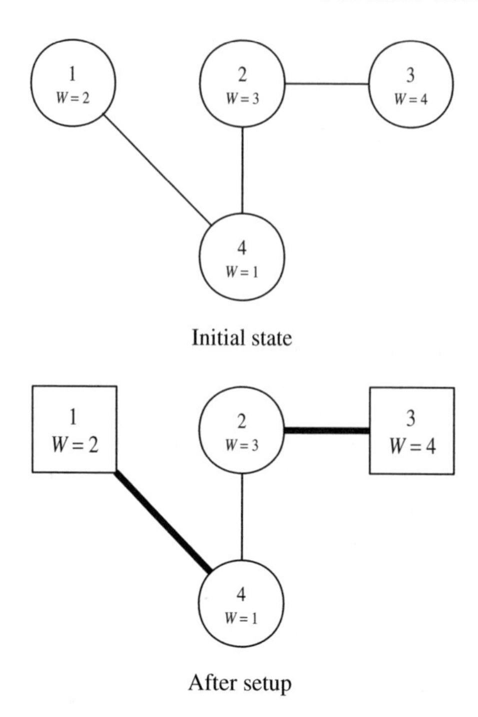

Figure 5.19 Initial setup.

neighbors. Hence, a node can track the status of all its neighbors and acts accordingly. We define the message complexity as the total number of *Ch*(_) and *Join*(_, _) messages transmitted to set up and maintain the organization of a network.

Figure 5.19 pictures a network of four nodes, $1, 2, 3, 4$. Each node i is pictured as a circle embedding i and its weight is shown as $W = W(i)$. Initially, for all i, *Ch*(i) is false and *Join*(i) is 0. Node 3 is the heaviest node and initially predicate p_1 is true for $i = h = 3$. Node 3 becomes a clusterhead, *Ch*(3) becomes true (pictured as a square), and message *Ch*(3) is sent to node 2. Predicate p_1 then becomes true for $i = 2$ and $h = 3$. *Join*(2) becomes defined as 3, pictured as a thick line, and message *Join*(2, 3) is sent, using broadcast, to nodes 4 and 3. Initially, predicate p_1 is true as well for $i = h = 1$. Node 1 becomes a clusterhead, *Ch*(1) becomes true, and message *Ch*(1) is sent to node 4. Predicate p_1 then becomes true for $i = 4$ and $h = 1$. *Join*(4) becomes defined as 1 and message *Join*(4, 1) is sent to node 1. The message complexity is 4 in this example and always corresponds to the number of nodes, when the topology is fixed. Each node takes one decision and uses one message to communicate the decision to its neighbors.

Figure 5.20 pictures the network whose topology has changed because of a movement of node 1. Predicate p_1, for $i = h = 1$, is retracted and *Ch*(1) becomes false, that is, node i stops playing the clusterhead role. However, predicate p_1, for $i = 1$ and $h = 3$, is asserted. *Join*(1) becomes defined as 3, pictured as a thick line, and message *Join*(1, 3) is sent, using broadcast, to node 3. Predicate p_1 then becomes false for $i = 1$ and $h = 4$. However, predicate p_i then becomes true for $i = h = 4$ node 4 becomes a clusterhead, *Ch*(4) becomes

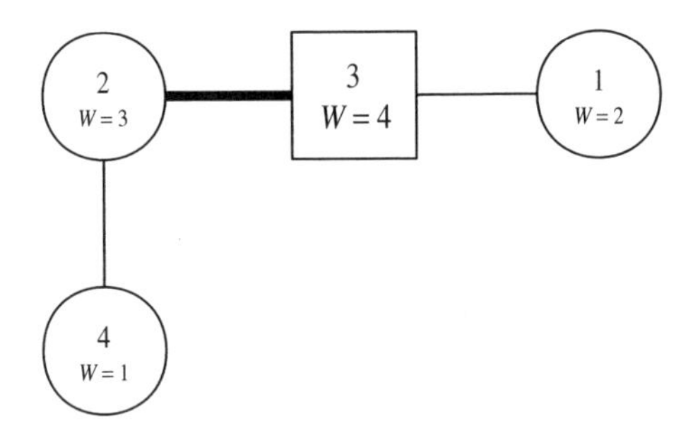

Location of node 1 is changed

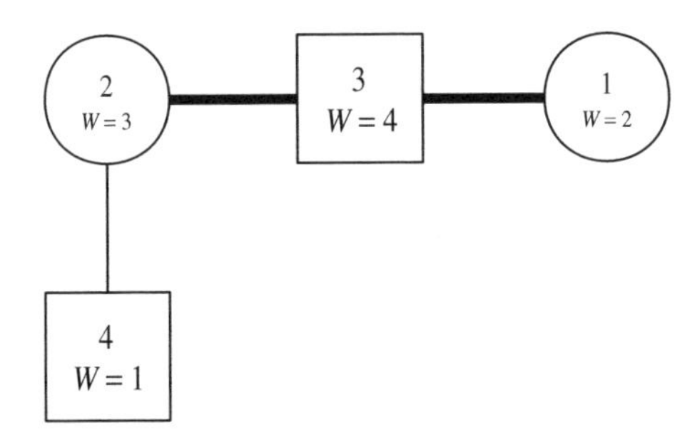

After maintenance

Figure 5.20 Cluster maintenance.

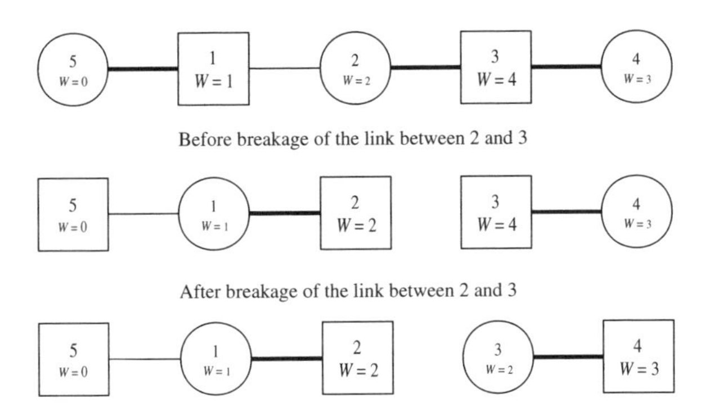

After breakage of the link between 2 and 3 and changing the weight of node 3 to 2

Figure 5.21 Cascaded cluster re-formation.

true and message $Ch(4)$. The message complexity of cluster maintenance in this example is 2.

Creation/breakage of a link between two nodes can trigger a costly cascaded cluster re-formation. This is pictured with an example in Figure 5.21. Before a breakage, node 2 is member of node 3's cluster. The link between nodes 2 and 3 breaks. Predicate $p_1(1, 1)$

becomes false, predicate $p_1(2, 2)$ becomes asserted and nodes 1, 2 and 5 switch their roles. It is easy to imagine a chain reaction from one end to the other in a wide network when an update of that type occurs. Note that on the side of node 3 nothing changes because *breakage of a link* does not change the status of being the heaviest node relative to neighbors. It is worth mentioning that if neither node i nor node j is a clusterhead or none of them is serving as the other's clusterhead, then cluster maintenance is not required when weighting is invariant. If, however, the weighting is not fixed and determined as a function of the degree of a node, then losing weight for a clusterhead can mean losing its status, because predicate $p_1(i, i)$ is retracted. Then a possible cascaded cluster re-formation is also possible on the side of the clusterhead. For example, in Figure 5.21 if after the breakage the weight of node 3 is reduced to 2, then it switches role with node 4. Variable weighting requires higher maintenance cost. Similar observations can be made when a new link is created.

5.6 Quality of service

Quality of service refers to requirements that must be satisfied by routing algorithms. For example, if multimedia traffic needs to be transported, then routes must satisfy bandwidth, jitter, delay and loss rate requirements. Besides, the requirements must be satisfied while achieving an efficient utilization of the resources in order to serve a maximum number of users. The challenge is particularly difficult for ad hoc networks because of their dynamic nature.

Zhang and Mouftah (2005) capture the delay and bandwidth requirements together with efficiency through a notion of cost with the following model. It is assumed that every node has the capacity to measure the amount of bandwidth available on its links (i.e. the amount of available resources) and to reserve bandwidth on a link. Given a network modeled as a graph $G = (V, E)$, each edge $e = (u, v) \in E$ is characterized by a cost $C(e)$, a delay $D(e)$ and a bandwidth $B(e)$. The cost is uniformly and statically defined as one unit for each link. The delay is the time it takes for a frame to move from node u to node v. The bandwidth is the residual capacity available on edge e.

Given a source s, a destination d and a path

$$p \equiv s = v_0, v_2, \ldots, v_{n-1}, v_n = d \text{ with } (v_i, v_{i+1}) \in E \text{ for } i = 0, 1, \ldots, n-1$$

the cost of using p is

$$\sum_{(v_i, v_{i+1}) \in p} C(v_i, v_{i+1}) \text{ units}$$

from s to d, and the packet trip delay using p is

$$\sum_{(v_i, v_{i+1}) \in p} D(v_i, v_{i+1}) \text{ time units}$$

Zhang and Mouftah (2005) define the On-Demand Delay Constrained Unicast Routing Protocol (ORDP)), which can be used, in conjunction with other routing protocols, to address delay constraints. ODRP reserves bandwidth on links according to end-to-end delay requirements. ODRP works on the following assumptions. First, it is assumed that the residual bandwidth on links can be measured. Secondly, available bandwidth on links can be reserved. Thirdly, the delay required to transfer a packet on a link can be measured. Fourthly, broken links can be detected by nodes at both ends.

Each ODRP node maintains a routing table and a bandwidth reservation table. The routing table contains one entry for every other node in the network. Each entry consists of the address of a destination node, the length (in hops) of the shortest path to that destination node and the first on hop on that shortest path (a neighbor node). ODRP does not explicitly address the construction of the routing tables, which needs to be done by a companion routing protocol. Each entry in the bandwidth reservation table refers to a link, connected to the local node, an amount of bandwidth and a source address, destination address pair, to which the bandwidth is granted.

There are three ODRP bandwidth reservation specific messages: Probing, RREQ and Acknowledgment. The Probing message contains the addresses of a source and a destination, and information on the required amount of bandwidth (in bps), the required delay (in time units) and the accumulated time delay (in time units). The acknowledgment contains the addresses of a source and a destination and information on the required amount of bandwidth. The RREQ contains the addresses of a source and a destination, an accumulated delay, a hop count and a TTL (in hops).

The execution of ODRP is triggered by a request to establish a channel of capacity C from a source S to a destination D that satisfies a delay constraint δ. ODRP works in two rounds. The routing tables contain, by construction, the shortest path from every possible node to every possible node. The network in Figure 5.22 is used to illustrate the operation of ODRP. In this example, the shortest route from S to D, in hops, goes through node 1. During the first round, the Probing message is used to test satisfaction of the bandwidth and δ constraints by the route stored in the routing tables. A Probing message is constructed with the appropriate addresses of source and destination, required amount of bandwidth, required delay and null accumulated time delay. The Probing message is forwarded hop by hop to the destination using the shortest path represented in the routing tables. From hop to hop, the accumulated delay is augmented with the time required to transfer the Probing message on each link. Availability of bandwidth on the links is assessed. When the Probing message reaches the destination, the accumulated delay is compared with the required delay. If the former is less than or equal to the latter, then an Acknowledgment message is returned to the source. The Acknowledgment message is loaded with the source and destination addresses and required amount of bandwidth, all available from the Probing

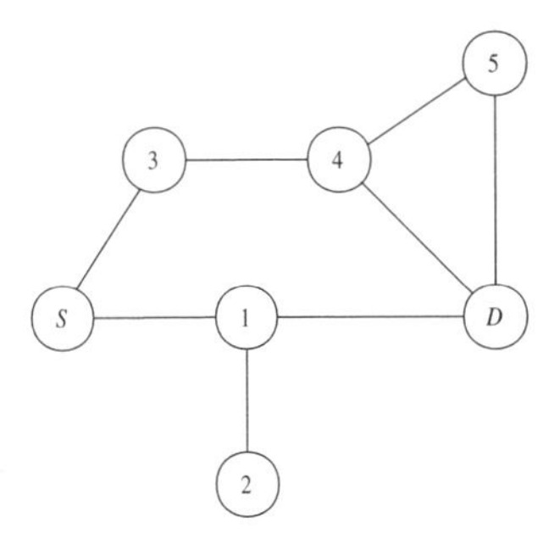

Figure 5.22 Probing and route discovery in ODRP.

message. The Acknowledgement message is forwarded through the path followed by the Probing message, but reversed. The Acknowledgement is forwarded hop by hop. While visiting each node on the path, bandwidth is reserved on the incoming and outgoing links and an appropriate entry is created in the bandwidth reservation table.

The test performed by a probing round may fail because the shortest path may not satisfy the bandwidth and delay constraints. It is, however, possible that another path exists with the available required bandwidth and ability to satisfy the delay constraint. For example, in Figure 5.22, the route S, 1, D may not be suitable, even though it is the shortest, maybe because node 2 loads node 1 with heavy traffic. The second round of ODRP explores other options. In case of failure of the probing round, a route discovery round is attempted. After the detected failure of the probing round, the source constructs a RREQ message. The message is loaded with the addresses of the source and destination. The accumulated delay is set to null. The hop count is also set to null. The TTL limits the scope of the search for an alternate route. In this example, a value of three hops is used. The network is flooded with the RREQ message. Each node retransmits the RREQ. There is a mechanism that avoids retransmission of duplicate RREQs. From hop to hop, the accumulated delay is augmented with the time required to transfer the Probing message on each link. Availability of bandwidth on the links is assessed. When a RREQ message reaches the destination, the accumulated delay is compared with the required delay. If the former is less than or equal to the latter, then an Acknowledgement message is returned to the source. The sequel is as in the Probing round. In the example of Figure 5.22, the route $S, 3, 4, D$ may be discovered. If such a route is successfully discovered, then the routing tables are updated (by the Acknowledgement message). The shortest route is replaced by the route that satisfies the quality of service requirements. In the example of Figure 5.22, the route $S, 3, 4, 5, D$ cannot be discovered because the life of the RREQ terminates in node 5 after three hops.

Because of the dynamic nature of ad hoc networks, route maintenance needs to be addressed. Whenever a link is broken, the nodes at both ends need to inform the sources and destinations nodes using the bandwidth on that link. A Route Error message is sent to every source. While being forwarded to the source, the bandwidth reserved on each link is released by each visited node. When the Route Error message reaches the source, another RREQ round can be launched to explore other available options. A Release message is sent to every destination. On it way to a destination, the reserved bandwidth is released on each visited node.

5.7 Sensor Network Protocols

Sensor networks are a special kind of ad hoc networks geared toward data collection. Networked sensors collect, for example, environmental or biological information, which is concentrated toward a base station. The information is then stored or processed in real time to achieve useful goals. Examples of applications are environment monitoring and intrusion detection in wireless networks. Sensor measurements include water temperatures at different depths, concentrations of various chemical elements, water movements at various depth levels, temperature and wind above water.

Sensor networks are characterized by low availability of energy, limited communication bandwidth, low processing power and storage, a model of deployment where either the nodes are fixed or, if there is mobility, the speed is low, high rate of failures and large number of nodes. Sensor networks typically consist of hundreds or thousands of devices.

To cope with the sensor network specific constraints, a number of routing techniques have been devised. The main approaches follow a flat, hierarchical or location-based model Al-Karaki and Kamal (2004).

5.7.1 Flat routing

The routing protocols following the flat model assume an unstructured sensor network topology. The Sensor Protocols for Information via Negotiation (SPIN), proposed by Heinzelman, Kulik and Balakrishnan, is an example of flat routing protocol (Heinzelman et al. 1999). SPIN is based on a three-way handshake exchange protocol (Figure 5.23). There are three kinds of messages: new data advertisement (ADV), data request (REQ) and data message (DATA). A sensor node, that is ready to supply new data, sends an ADV message using broadcast an ADV message. The ADV message contains a specification of the available data, termed a *metamodel*. One-hop neighbors interested in consuming the offered data reply to the supplier with an REQ message. The actual value of the data is communicated from the supplier sensor to the consumer sensor with a DATA message.

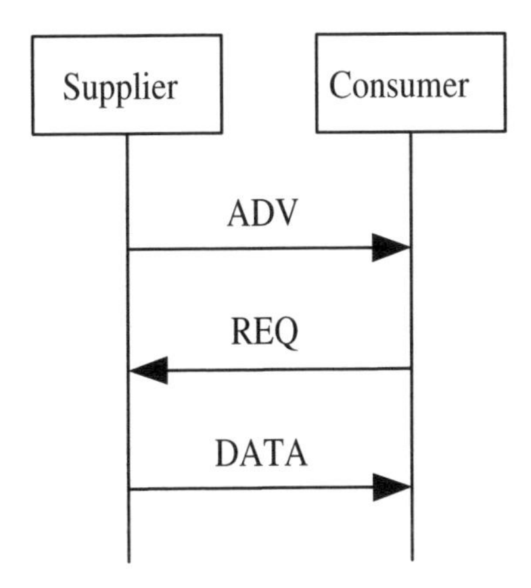

Figure 5.23 Three-way handshake of SPIN.

Figure 5.24 pictures a data dissemination scenario in SPIN. Nodes consuming the data progressively are shown in gray. The origin supplier is node 1. An ADV message is sent using broadcast to nodes 3 and 5. Both are interested, reply with an REQ message and consume with a DATA message. In the next cycle, both nodes 2 and 5 send an ADV message to their neighbors using broadcast. Only node 6 shows interest and consumes the data from node 5. The data is progressively disseminated from the original supplier node to the interested consumer nodes. In the pictured scenario, node 3 creates a barrier. Because of its lack of interest, node 3 blocks the eventual propagation of the data up to node 4. Full network coverage is not guaranteed. SPIN is a best-effort protocol.

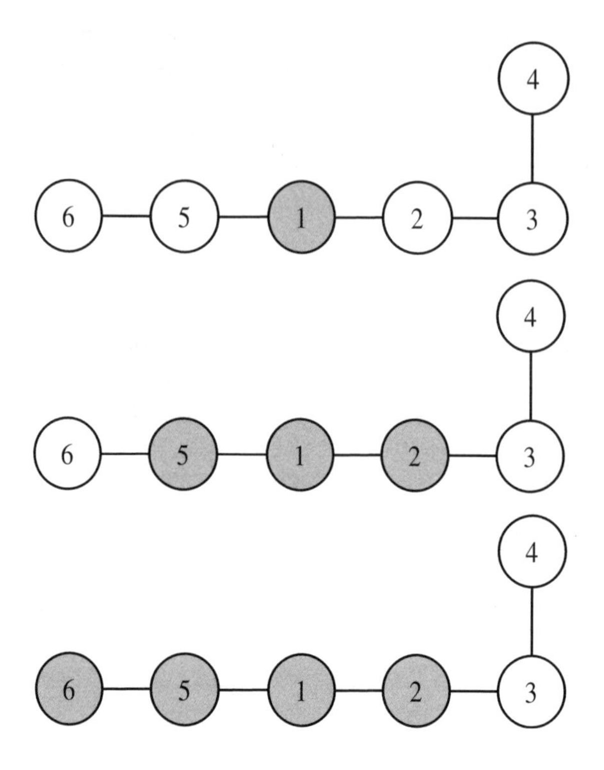

Figure 5.24 Progressive dissemination of SPIN.

5.7.2 Hierarchical routing

The Low-Energy Adaptive Clustering Hierarchy (LEACH) of Heinzelman et al. (2000) is an example of hierarchical sensor network protocol. There is the strong requirement that all sensors must be able to communicate directly with the BS. LEAC is cluster-based. It means that the network is structured dynamically into partitions called *clusters*. Each cluster as a head whose role consists of aggregating the data generated by the cluster members and forwarding the data to the BS. The cluster head also determines a timed division multiple access (TDMA) schedule used by the members and head to transmit in a round-robin manner. The head communicates the TDMA schedule to the members using broadcast. Each cluster uses a *unique* code division multiple access (CDMA) for intracluster communications, hence avoiding interference with other clusters. The code is selected from a predetermined list.

The LEACH is a two-phase protocol, repeated to insure a rotation in cluster head role playing. In the first phase, each node self-determines its role, member or head, using a randomized algorithm. In the second phase, in each cluster, data generated by the members is sent to the head and redirected by the head to the BS.

5.7.3 Zigbee

ZigBee is a wireless personal area networking technology (ZigBee Alliance (2005)). The envisioned applications include sensor networks and wireless alarm systems. ZigBee clearly identifies the two worlds of applications: commercial applications and residential applications. ZigBee has a five-layer model of reference consisting of a physical layer, a link layer, a network layer, an application support layer and an application layer. The physical layer supports relatively low data rate communication channels (i.e. 20 , 40 and 250 kbps). The link layer may follow either a polling medium access scheme or carrier sense multiple access with collision avoidance scheme. Hereafter, the focus is on the network layer, namely, the network architecture and routing mechanism.

ZigBee defines four different types of nodes; coordinator, trust center, router and end device. The *coordinator* determines the level of security of the network and address of the device that plays the role of trust center. The responsibilities of the *trust center* are to authorize devices that join a ZigBee network and distribute keys to other network devices (ZigBee security is discussed in more detail in Section 7.3.2). The role of coordinator and trust center can be combined and assumed by the same single device. A *router* sponsors network access to end devices. *End devices* have no network management role except for themselves. Between the different types of nodes, there are two kinds of links: star link and mesh link. A *star link* connects an end device to a router. A mesh link is a router to router (or coordinator) link. An example ZigBee network is pictured in Figure 5.25.

A ZigBee node joins a network using a direct link with the coordinator or through a multihop route leading to the coordinator. In the former case, the coordinator is the node's parent. In the latter case, the first router on the route is the node's parent. More details about the procedure followed to join a network can be found in Section 7.3.2. During the network entry procedure, the joiner receives an address from its parent. A block of addresses is granted to the coordinator and every router. The address space is hierarchical. A hierarchical network structure is induced. Routing is done following the parent–child links.

The hierarchical network structure is a tree that is characterized by a number of parameters (determined by the network administrator). The tree has the maximum depth of L. The

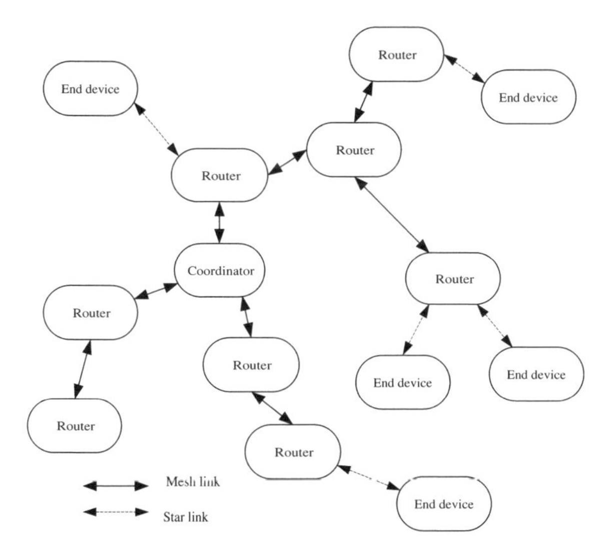

Figure 5.25 An example ZigBee network.

coordinator is at depth 0 and leafs nodes can be at a maximal depth of L. Each router in the tree is allowed a maximum number of children C. Among these C children, a maximum of R nodes can play the role of routers. For example, L, C and R may be equal to 3, 4 and 4. These parameters are selected as a function of the application.

A ZigBee router grants addresses to it children, which may include other routers. A router is granted an address bank by its parent. An important question is how many addresses does a router need in its bank? At depth L, the leaf end devices do not allocate addresses and do not need to be granted an address bank, but require one address for themselves. At depth $L-1$, each router needs a bank of C addresses for its children plus one for itself. At depth, $L-2$, the router needs

$$R(C+1)+C-R+1 \text{ addresses.}$$

In general, a router a depth d needs a bank of addresses of size:

$$S(d) = R(S(d-1)) + C - R + 1 \text{ addresses}$$

with $S(L) = 1$. An example is detailed in Table 5.10. Note that this network may have up to 85 nodes in total.

Table 5.10 Sizes of ZigBee address spaces.

Depth (in hops)	Size (in addresses)
0	85
1	21
2	5
3	1

The addresses are assigned (initially the coordinator and eventually other routers) in a sequential manner. An example is pictured in Figure 5.26. The coordinator has address 0. The first router that joins the coordinator receives the next available address, that is, address 1, and a consecutive chunk of $S(1)-1$ addresses for its descendent that is, addresses 2 to 21. In this example, address 2 is granted to an end device. The next router to join the coordinator receives the next unallocated address, that is, address 22, and the following chunk of $S(1)-1$ addresses for its children, that is, addresses 23 to 42. In general, if the address of the parent is a_p, and a router, at depth d, is rth router to join the parent, then the router receives the address

$$a_r = a_p + S(d)(r-1) + 1$$

and the range of addresses of its children is

$$a_p + S(d)(r-1) + 2, \ldots, a_p + S(d)(r-1)$$

Hence the second router to join router 22 at depth 2 receives the address 28 and the range of addresses $29, \ldots, 32$ for its children. The end devices also receive their address sequentially. The first end device to join the router with address 28 receives the address 29.

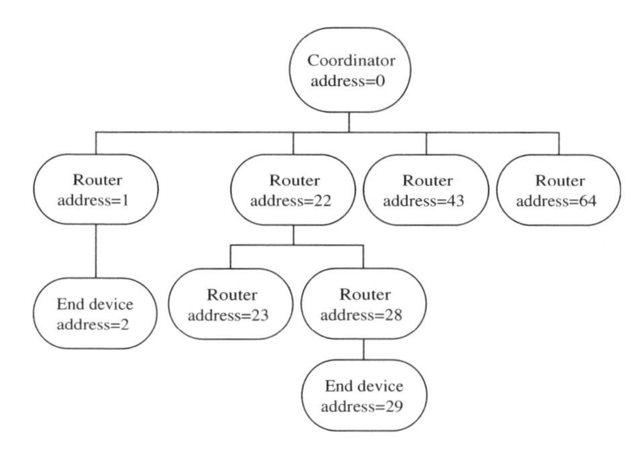

Figure 5.26 ZigBee tree.

Routing in a ZigBee tree is of great simplicity. When a message needs to be received, the destination address is examined. If it is equal to the local address, then the message has reached its final destination and its payload is pushed to the upper layer protocol. If the destination address is greater than the local address, but lower than the local address plus $S(d)$, where d is the depth of the current router, then the message needs to be routed to the children. The exact children may be an end device or a subrouter responsible for the subrange of addresses in which the destination address is contained. Otherwise, the message is forwarded to the parent of the local router.

ZigBee also supports the AODV protocol, defined by Perkins et al. (2003). AODV is a reactive (RREQ/reply-based), table-driven, hop-by-hop routing protocol. The tree algorithm is examined in more detail hereafter.

5.8 Exercises

1. What are the problems of the forwarding mechanisms of DSR regarding security?

2. In OLSR, explain why it is necessary for a node to forward again a TC message when it is received on an interface not included in the *interface address list* field of the duplicate table.

3. What are the security vulnerabilities of NDP?

4. Formally demonstrate that, by construction, the routing table model used by ODRP capture, for every pair of node, the shortest path.

5. In ODRP, explore different ways failure of a probing test or a route discovery round can be detected.

6. Explain how received duplicate RREQs in ODRP can be detected and ignored.

7. Introduce extensions to ODRP to address jitter and loss rate quality of service requirements.

8. In ODRP, the several RREQ may succeed to find an adequate path. Devise a strategy to identify and select the best possible option.

9. Introduce in ORDP an extended route discovery that expands the scope of the search of a route discovery from one round to another in case of repeated failures.

10. In Figure 5.26, complete the subtree under the router with address 64 with two subrouters and one end device under each subrouter.

6

LOCATION AWARENESS

> This is as far as the computer scientist is going to get.
> Now let's be mathematicians again for a while.
> Knuth (1985)

The interaction of mathematical models and computer science simulations is nowhere more evident than in studies of ad hoc network localization and performance. Ad hoc networks provide the paradigms that influence the development of mathematical models, which in turn become a guide to further paradigms. In a sense, Donald Knuth's quotation attests to the need of *becoming mathematicians again*, even if only *for a while*, in order to establish mathematical models that are representative of the needs of ad hoc networking (see Knuth (1985, page 190)).

Ad hoc networks are networked systems that facilitate communication among peer hosts without any preestablished infrastructure. They may be established in remote areas, for example, to handle emergency communications and should not require any previous network support. In fact, in such cases, cost may also be an argument against infrastructure thereby limiting battery resources, available bandwidth and number of network hosts. In addition, not every sensor can hear every other sensor and data needs to be forwarded in a multihop manner. Therefore, it becomes a significant challenge to provide for efficient information dissemination in an ad hoc network consisting of energy limited hosts operating in a bandwidth limited communication environment.

Protocols and algorithms in ad hoc networks must enable global communication using only local resources. In addition, information dissemination should be local in the sense that a message emanating from a host should propagate for only a constant number of steps. The important property guiding algorithmic design in ad hoc networking is called *locality* and was introduced in a different context by Linial (1992). It refers to the property of distributed computations that the algorithm must terminate in constant time that is independent of the size of the input network.

Locality is combined with two important paradigms that are being used in the design of efficient geographically based algorithms in ad hoc networks. We refer to the first paradigm

Principles of Ad hoc Networking Michel Barbeau and Evangelos Kranakis

as *local to global* and to the second as *global to local*. Note that, although only local information and resources can be used to maintain the infrastructure, communication should be attained even in a possibly large diameter ad hoc system. Data should be processed in a distributed manner and an algorithm is good for local computation provided it attains good global connectivity and spanning characteristics. Therefore, the first paradigm means that you should use only local resources in order to attain global communication. However, communication can only be attained if connectivity is maintained in the ad hoc system as a whole and therefore we may have to use principles with established global, spanning characteristics. As a matter of fact, any global computation algorithm can be used if restricted to local computation provided it attains connectivity, whereby the degree of locality of a node is determined by the set of nodes at distance k hops away from that node, for some $k \geq 1$. Therefore, the second paradigm means that you are allowed to use global algorithms, but restricted to a geographically local vicinity. In this chapter, we will see several instances of application of these two principles in the design of communication algorithms in ad hoc networks.

The main theme of this chapter is the study of geometric principles guiding communication in ad hoc networks. Section 6.1 is devoted to defining geographic proximity in a set of sensor nodes. Section 6.2 looks at the issue of preprocessing the ad hoc network in order to construct a spanner, that is, an underlying backbone over which routing can be guaranteed delivery. Section 6.3 is on information dissemination: in particular, it looks at several routing and traversal algorithms in unit disk graphs (UDGs). Section 6.4 delves into the problem of geographic location determination. Random UDGs, which are studied in Section 6.5, are basic in understanding the average case behavior of UDGs and are very important in ad hoc network simulations. Finally, Section 6.6 looks at the particular problem of coverage and connectivity in networks with directional sensors.

6.1 Geographic proximity

An essential requirement of information dissemination in communication networks is that the hosts be endowed with a consistent addressing scheme. IP-addresses, very typical of traditional internet networking, can be either set up at initialization or else acquired by means of an address configuration protocol. Unfortunately, none of these techniques are suitable for wireless networking and addresses have to be based on the geographic coordinates of the hosts. The latter can be acquired either directly from a participating satellite of an available Global Positioning System (GPS) (Global Position System) or via radiolocation by exchanging signals with neighboring hosts in the wireless system. Thus, the coordinates of the hosts form the foundation of any addressing scheme in such infrastructureless networks. Ad hoc networks are infrastructureless, have no central components and the hosts may have unpredictable mobility patterns. In order to produce dynamic solutions that take into account the changing topology of the backbone, the hosts should base their decision only on factors whose optimality depends on geographically local conditions.

A fundamental concept that will play a significant role in the sequel is the unit disk graph (also abbreviated as UDG). It is defined over a given pointset (Figure 6.1). The set of

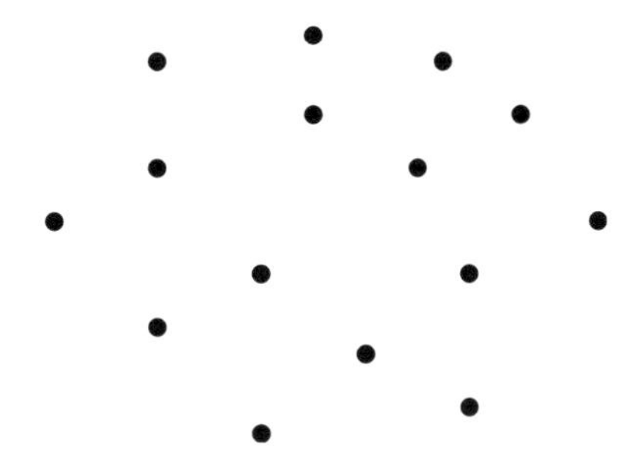

Figure 6.1 A set P of points in the plane.

vertices is exactly the point set while edges take into account the distance between nodes, that is, two points p, q are adjacent if and only if their Euclidean distance is at most one, (where *one* is the transmission range of the nodes when considered as sensors).

6.1.1 Neighborhood graphs

Suppose that P is a set of n points in the plane. Every pair $\{p, q\}$ of points is associated with a planar region $S_{p,q}$ that specifies the neighborhood association of the two points. Let $\mathcal{S}$ be the set of neighborhood associations in the pointset. Formally, the neighborhood graph $G_{\mathcal{S},\mathcal{P}}$ of the planar pointset P is determined by a property $\mathcal{P}$ on the neighborhood associations:

1. $\{p, q\} \to S_{p,q} \subseteq R^2$, for $p, q \in P$.
2. $\mathcal{P}$ is a property on $\mathcal{S} := \{S_{p,q} : p, q \in P\}$.

The graph $G_{\mathcal{S},\mathcal{P}} = (P, E)$ has the pointset P as its set of vertices and the set E of edges is defined by $\{p, q\} \in E \Leftrightarrow S_{p,q}$ has property $\mathcal{P}$. There are several variants on this model including nearest neighbor graph (NNG), relative neighbor graph (RNG), and Gabriel graph (GG), which we define in the sequel.

6.1.1.1 Nearest Neighbor Graph (NNG)

The edges of an NNG are determined by minimum distance (Figure 6.2). More precisely, for two points p and q, $(p, q) \in E \Leftrightarrow p$ is the nearest neighbor of q. If $S_{p,q}$ is the union of the two disks centered at points p and q, respectively, and radius is the distance between p and q then it is easy to see that the NNG is the same as the graph $G_{\mathcal{S},\mathcal{P}}$, where the property $\mathcal{P}$ is defined by "$S_{p,q}$ contains no points other than p or q". Although the NNG incorporates a useful notion, it has the disadvantage that the resulting graph may be disconnected. A generalization of the NNG is the k-Nearest Neighbor Graph (k-NNG), for some $k \geq 1$. In this case, $(p, q) \in E \Leftrightarrow p$ is kth nearest neighbor of q or q is kth nearest neighbor of p.

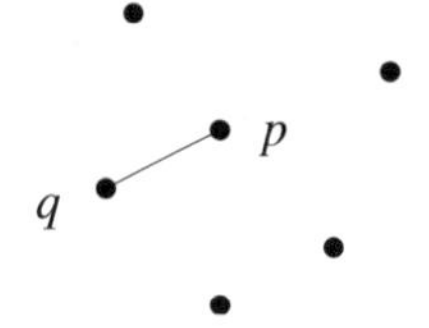

Figure 6.2 q is the nearest neighbor of p.

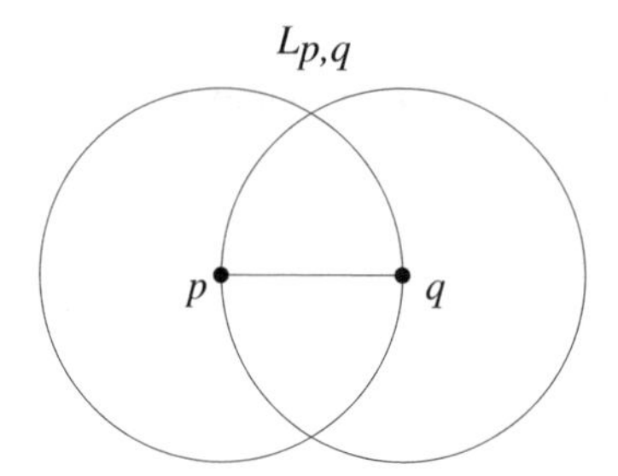

Figure 6.3 The lune defined by points p and q.

6.1.1.2 Relative Neighbor Graph (RNG)

The neighborhood association in the RNG is determined by the lune. Formally, the *lune* $L_{p,q}$ of p and q is the intersection of the open discs with radius equal by as the distance $d(p, q)$ between p and q and centered at p and q, respectively (Figure 6.3). The set of edges of the graph is defined by $(p, q) \in E \Leftrightarrow$ the lune $L_{p,q}$ does not contain any point from the pointset P other than p or q.

6.1.1.3 Gabriel Graph (GG)

GGs (also known as *least squares adjacency graphs*) were introduced in Gabriel and Sokal (1969) and have been used for geographic analysis and pattern recognition. The region of association between two points p, q is specified by the disc with diameter determined by p and q (Figure 6.4). Formally, the set of edges of the GG is defined by $(p, q) \in E \Leftrightarrow$ the disc centered at $\frac{p+q}{2}$ and radius $\frac{d(p,q)}{2}$ does not contain any point from the pointset P other than p or q.

Lemma 6.1.1 summarizes the important property of the GG that will be used in the sequel.

Lemma 6.1.1 *GG is planar.*

Proof. Assume that AB is a Gabriel edge. Observe that a point X is inside the circle with diameter AB if and only if the angle AXB is bigger than $\pi/2$ (Figures 6.5, 6.6). Moreover, a point X is inside the circle with diameter AB if and only if its distance from the center of the circle is bigger than $|AB|/2$. Other than p or q Assume on the contrary that the

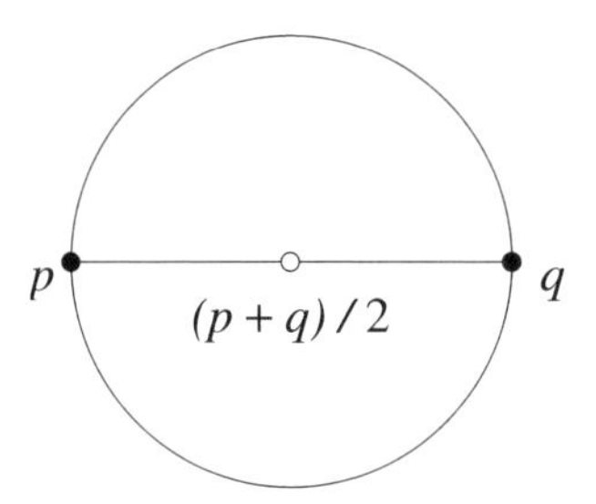

Figure 6.4 The circle with diameter equal to the line segment determined by p, q and centered at the point $\frac{p+q}{2}$.

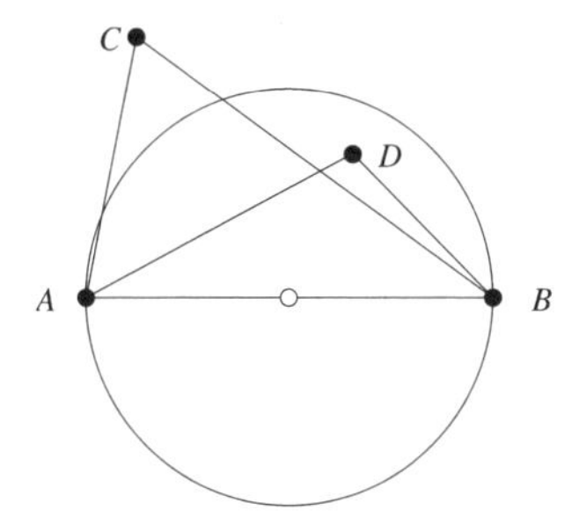

Figure 6.5 When is a point $X \in \{C, D\}$ inside the circle.

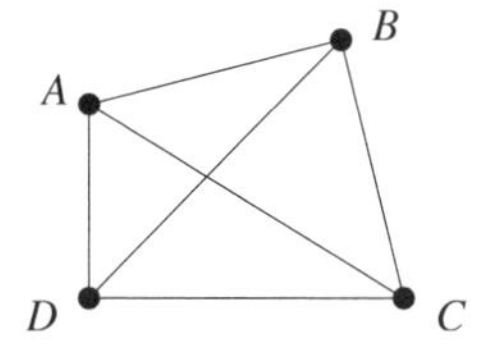

Figure 6.6 Proving the planarity of GG.

Gabriel edges AC and BD intersect, A, B, C, D must form a convex four vertex polygon. Since AC (respectively, BD) is Gabriel edge none of the points B, D (respectively, A, C) can lie inside the circle with diameter AC (respectively, BD). Hence $\angle ABC$, $\angle BCD$, $\angle CDA$, $\angle DAB < \pi/2$, contradicting the fact that $\angle ABC + \angle BCD + \angle CDA + \angle DAB = 2\pi$. This completes the proof of the lemma. ■

6.1.1.4 Minimum Spanning Tree (MST)

A spanning tree of a graph is a subgraph that is a tree (i.e. contains no cycles) and includes all the nodes of the network. We are familiar with constructions like Depth First Search (DFS) and Breadth First Search (BFS). If the edges of the network are weighted (e.g.

representing power consumption, expected delay), a minimum weight spanning tree is one with minimum sum of edge weights. The minimum spanning tree (MST) considered in this chapter is assigned with Euclidean distances as weights. Two standard algorithms for computing MST are Prim's algorithm and Kruskal's algorithm (see Cormen et al. (1990) for additional details).

6.1.1.5 Delaunay Triangulation (DT)

The Delaunay triangulation(DT) is another form of proximity graph. It differs from the previous one in that the neighborhood association is defined over triangulated triples of points from the given pointset. Consider a triangulation $\mathcal{T}$ of the point set P. For any triple (p, q, r) of points let $S_{p,q,r}$ be the circle formed by the points p, q, r (Figure 6.7). $\mathcal{T}$ is a DT (see Preparata and Shamos (1985)) of the pointset P if for any triangle (p, q, r) of the triangulation $S_{p,q,r}$ contains no points from P other than p, q, r.

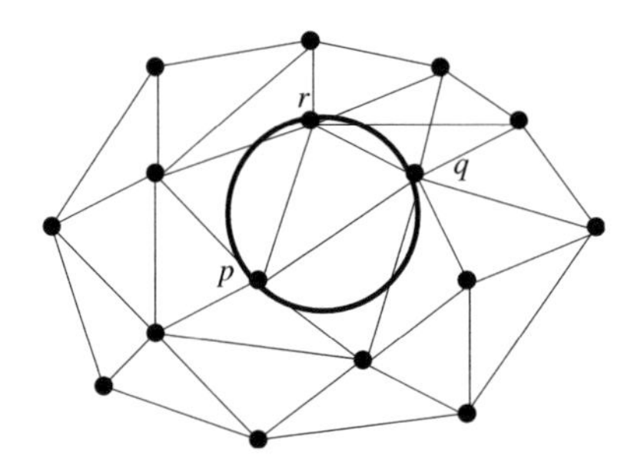

Figure 6.7 Triangulation of a pointset.

6.1.2 Relation between the neighborhood graphs

We can prove the following theorem:

Theorem 6.1.2 (O'Rourke and Toussaint (1997); Toussaint (1980)) *The following inclusions are satisfied for the previously mentioned neighborhood graphs*

$$NNG \subset MST \subset RNG \subset GG \subset DT,$$

where MST denotes the Minimum spanning tree and DT the Delaunay triangulation of the pointset P.

Proof. Figure 6.8 outlines the proof of MST being a subset of RNG. Figure 6.9 outlines the proof of RNG is a subset of GG. The proof of GG is a subset of DT is depicted in Figure 6.10 and uses the fact that the DT is the dual of the Voronoi diagram (see Preparata and Shamos (1985) for additional details). Details of the proofs of $NNG \subset MST$ and $GG \subset DT$ are left to the reader as Exercise 2. This completes the proof of the theorem. ■

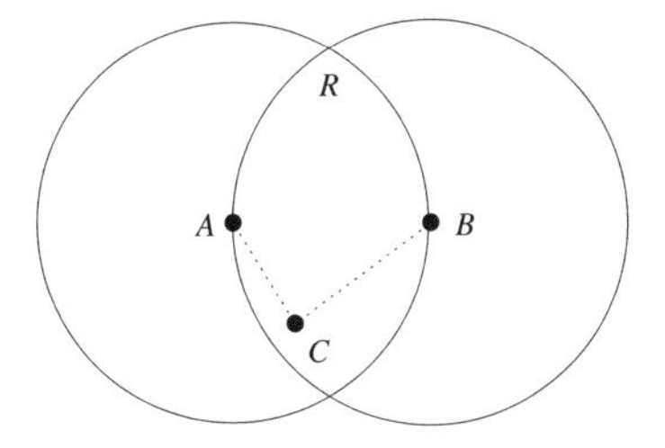

Figure 6.8 Proving that MST is a subset of RNG.

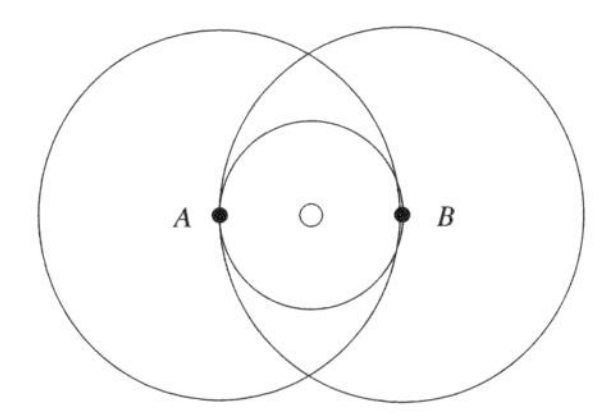

Figure 6.9 RNG is a subset of GG.

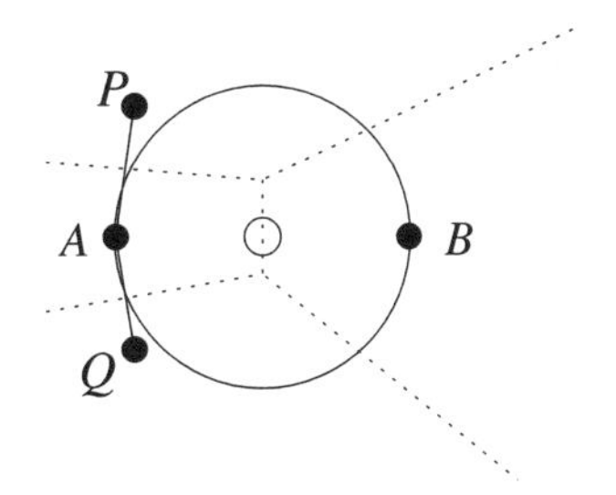

Figure 6.10 Proving that GG is a subset of the DT.

6.2 Constructing spanners of ad hoc networks

Since, in practice, the original ad hoc wireless network is never planar with many links crossing each other, we need to abstract from the original network a planar connected network spanning the entire underlying network. The reason for doing this will be clear in subsequent sections. In this section, we describe several tests for preprocessing the ad hoc network: the Gabriel test, the Morelia test, the Half-Space proximal, and the local MST.

6.2.1 Gabriel test

One of the most important tests for eliminating crossings in a wireless network is called *Gabriel test*, which is applied to every link of the network. Assume that all nodes have the same transmission range R. Let A, B be two nodes whose distance is less than the transmission range R of the network. In the Gabriel test, if there is no node in the circle with diameter AB then the link between A and B is kept. If, however, there is a node C

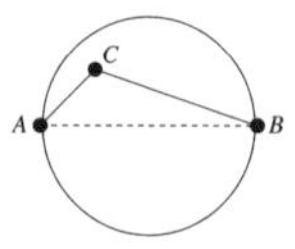

Figure 6.11 Eliminating an unnecessary link (dashed line AB) with the Gabriel test.

in the circle with diameter AB, as depicted in Figure 6.11, then nodes A and B remove their direct link. In particular, when A (respectively, B) is queried on routing data to B (respectively, A) the routing table at A (respectively, B) forwards the data through C or some other similar node if more than one node is in the circle with diameter AB. The advantage of doing this rerouting of data is that the resulting graph is a planar spanner on which we can apply the face routing algorithm for discovering a route from the source to destination.

Theorem 6.2.1 (Bose et al. (2001)) *If the original network is connected, then the Gabriel test produces a connected planar network.*

6.2.2 Morelia test

The Gabriel test does not prevent the multiplehop effect. For example, consider a set of nodes as depicted in Part (a) of Figure 6.12. All nodes are mutually reachable. However, when we apply the Gabriel test, it results in the configuration in the Part (b) of Figure 6.12. We can see that although nodes A and B could reach each other directly in a single hop instead they must direct their data through a sequence of many hops.

The Morelia test preserves links whenever possible and the resulting planar spanner will most likely keep the contour very similar to the original contour. Thus, the resulting planar network will have smaller diameter and routes from the source to destination will require less hops. It is similar to the Gabriel test in that, given two nodes A and B it eliminates links based on the inspection of the circle with diameter AB. Unlike the Gabriel test, it does not necessarily eliminate the direct link AB when it finds another node inside the circle with diameter AB. Instead, it verifies whether the nodes inside the circle create any crossing of the line AB. If no crossing is created, the line AB is kept and it is removed otherwise. The verification of the existence of crossing is done in most cases by inspecting only the neighborhood of nodes A and B at the transmission distance R. In a few cases, the neighborhood of some of the nodes in the circle around AB is inspected. In all cases,

Figure 6.12 Part (a) depicts a set of nodes within reach of each other. Applying the Gabriel test, results in the configuration as depicted in Part (b), which produces a multiplehop effect when eliminating the link AB.

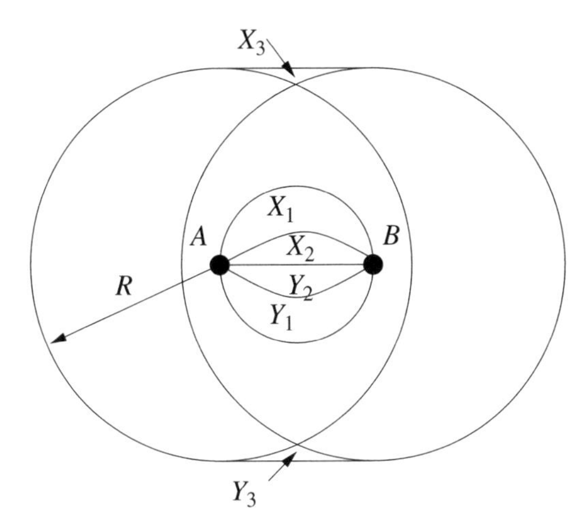

Figure 6.13 Defining the geometric regions of the Morelia Test

it is a local test that computes the neighborhood of nodes at distance at most two hops of each node A and B.

In the Morelia test of a link AB, we subdivide the area of the circle with diameter AB into X_1, X_2, Y_1 and Y_2 as in Figure 6.13. The areas X_2 and Y_2 are determined by an arc of radius R. Furthermore, in the testing we use areas X_3 and Y_3 as indicated in the Figure 6.13 that are outside the transmission radius R of the nodes A and B and within distance R from the link AB. For each node A, let $N(A)$ be the set of nodes Z such that $d(A, Z) \leq R$. The precise specification of the Morelia test applied to a link AB is given below (Figure 6.13).

1. *Elimination Rule:* If there is at least one node in $X_1 \cup X_2$ and at least one node in $Y_1 \cup Y_2$, then the link AB is removed.

2. *Crossing Rule (a):* If there is at least one node in X_1 and no node in $X_2 \cup Y_1 \cup Y_2$, then the node A checks whether any node in $N(A)$ creates a link with nodes in X_1 that crosses AB. If such a crossing occurs, link AB is removed and A sends a message to B to remove the link as well. (Similarly for node B.)

3. *Crossing Rule (b):* If there is at least one node in Y_1 and no node in $Y_2 \cup X_1 \cup X_2$ (symmetric to Rule 2), then the node A checks whether any node in $N(A)$ creates a link with nodes in Y_1 that crosses AB. If such a crossing occurs, link AB is removed and A sends a message to B to remove the link as well. (Similarly for node B.)

4. *Neighborhood Rule (a):* If there is at least one node in X_2 and no node in $Y_1 \cup Y_2$, then the node A inspects the nodes in $N(A)$ to check whether any node there creates a link with nodes in $X_1 \cup X_2$ that crosses AB. If such a crossing occurs, link AB is removed and A sends a message to B to remove the link as well. Furthermore, A sends a message to nodes in X_2 with a request to send back information whether there is a node in the region Y_3. If A receives a message that a node exists in Y_3 then AB is removed and node B is informed to remove the link as well.

5. *Neighborhood Rule (b):* If there is at least one node in Y_2 and no node in $X_1 \cup X_2$ (symmetric to Rule 4), the node A inspects the nodes in $N(A)$ for a possible crossing of AB. If such a crossing occurs, link AB is removed and A sends a message to nodes in Y_2 with a request to send back information whether there is a node in the region X_3. If A receives a message that a node exists in X_3, then AB is removed and node B is informed to remove the link as well.

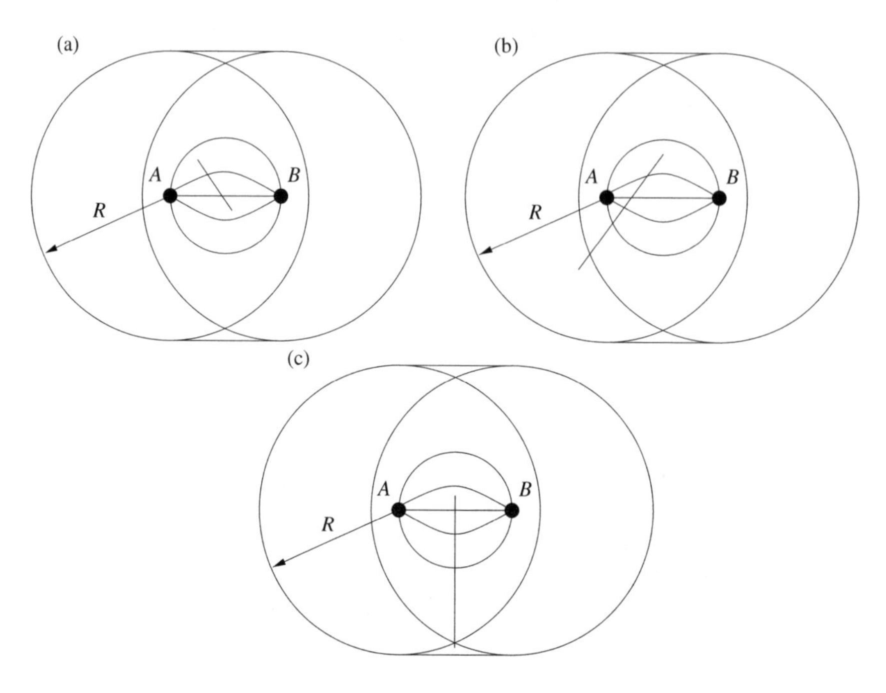

Figure 6.14 Examples of the Morelia test situations: (a) AB deleted by rule 1, (b) AB deleted by rule 2, (c) AB deleted by rule 4.

Figure 6.14 illustrates how the Morelia test will check for crossings prior to eliminating a link. It will eliminate link AB because it detects crossings, but it will not eliminate it when no crossing exists. The following theorem can be proved:

Theorem 6.2.2 (Boone et al. (2004)) *Application of the Morelia test produces a connected planar spanner if the original network is connected. Furthermore, this spanner contains the GG as its subgraph.*

Proof. Let the original connected network be N and suppose that the application of the Morelia test to all links of N produces a network N'. Every edge in N that is kept by the Gabriel test is also kept by the Morelia test. Thus, N' contains the GG of N as its subgraph. Since the Gabriel test produces a connected spanning subgraph of N, the subnetwork N' is connected.

Assume that there is a link e in N' that crosses a link AB of N'. Since the length of any link in N is at most R, both ends of e are in $N(A) \cup N(B) \cup X_3 \cup Y_3$. If one of the ends of e is in the circle with diameter AB, then AB would be deleted by the Morelia test.

If both ends of e are outside the circle with diameter AB, then one of the ends of the edge AB must be inside the circle with diameter e, since the edges cross each other. However, the Morelia test applied to e would eliminate e because of it being crossed by AB. Thus, there can be no crossing in N'. ■

6.2.3 Half-space proximal test

The idea in this test is the following: For each point u select the point closest to it, say v, and draw the line L perpendicular to uv at its midpoint. Next, remove from consideration all vertices in the half-plane determined by L that contain v and iterate.

An example of vertex selection is depicted in Figure 6.15. First, u selects edge uv_2 but rejects uv_1, uv_3, uv_4. Next, u selects uv_5 but rejects uv_6. Finally, it selects the remaining edge uv_7. More precisely, on input an UDG G the test produces $\overrightarrow{HSP}\ (G)$ as given in the following algorithm:

Input: a vertex u of a geometric graph and a list L_1 of edges incident with v.

Output: A list of directed edges L_2, which are retained for the $\overrightarrow{HSP}\ (G)$ graph.

1. Set the forbidden area $F(u)$ to be $\emptyset$.
2. Repeat the following while L_1 is not empty.
   (a) Remove from L_1 the shortest edge, say $[u, v]$, (any tie is broken by smaller end-vertex label) and insert in L_2 directed edge (u, v) with u being the initial vertex.
   (b) Add to $F(u)$ the open half-plane determined by the line perpendicular to the edge $[u, v]$ in the middle of the edge and containing the vertex v. (Notice that the points of the line do not belong to the forbidden area)
   (c) Scan the list L_1 and remove from it any edge whose end-vertex is in $F(u)$.

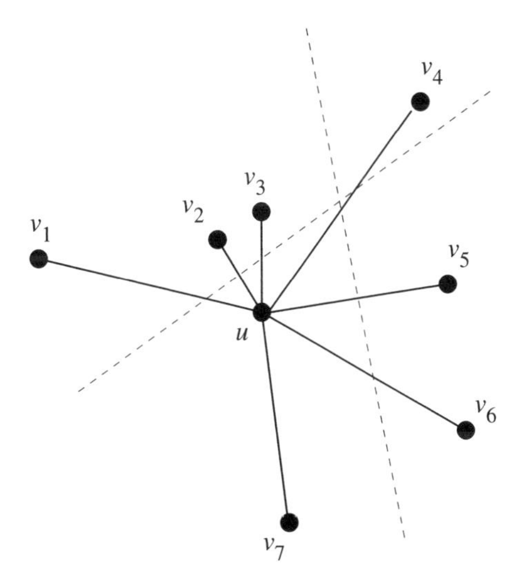

Figure 6.15 Example of selection of the neighbors of u in the Half-space proximal test.

Theorem 6.2.3 (Chavez et al. (2005)) *If G is a connected UDG, then the digraph $\overrightarrow{HSP}$ (G) has out-degree at most 6, is strongly connected and its stretch factor is at most $2\pi + 1$*

.

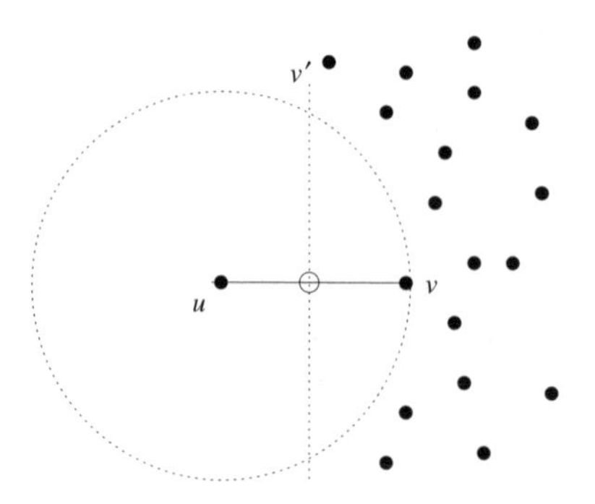

Figure 6.16 Proving that the half-space proximal test produces a graph of degree at most 6.

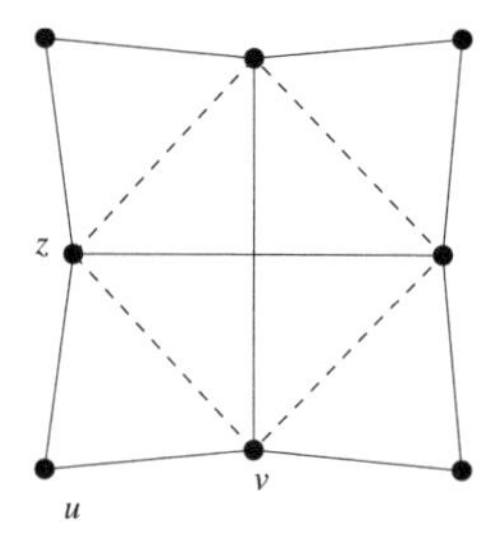

Figure 6.17 Proving that the half-space proximal is not planar.

To prove that the Half-space proximal test produces a bounded degree graph, take uv to be the shortest edge from u and let v' be the next shortest edge (Figure 6.16). It follows from the above-mentioned definitions that the angle $\angle vuv'$ must be $\geq \pi/3$. Iterating this, it follows that the degree of u can be at most 6. In fact, the algorithm can be modified slightly (see Chavez et al. (2005)) so that the resulting graph has degree at most 5.

Unfortunately, the graph thus produced is not planar (namely there are two edges from vertices v, z, respectively that cross each other) but it has stretch factor at most $2\pi + 1$ (Figure 6.17). For additional details and complete proof of the above-mentioned theorem, the reader is referred to Chavez et al. (2005).

6.2.4 Spanner for nodes with irregular transmission ranges

A UDG is a representation of a wireless network in which all nodes have the same circular transmission range. Considering that the nodes in a network are not exactly identical (e.g. some obstacles in the terrain containing the nodes may result in the transmission ranges of nodes to be irregular) in this section we are assuming that the nodes have irregular transmission ranges (Figure 6.18).

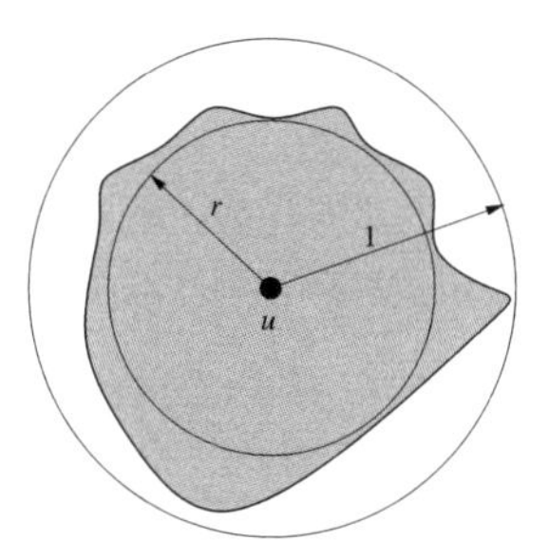

Figure 6.18 The irregular transmission area of node u.

We assume that there is an additional parameter r, a positive real number less than or equal to 1. The transmission range of a node in the network is assumed to be a region contained within the unit disk around the node, but this region contains all points at distance less than r. Thus, any two nodes at distance at most r can communicate directly, but no nodes at distance more than 1 can communicate directly. Two nodes at distance more than r and at most 1 may or may not communicate directly. An example of a transmission range of node u is shown in Figure 6.18 as the darker area. We call the geometric representation of such a network a *unit disk graph with irregularity r*. The problem of constructing a spanner for UDGs with irregular transmission ranges is more complex, for example, the usual planarization algorithms like the Gabriel test or the relative neighborhood graph algorithm do not work for them.

Edge selection in the main algorithm will depend on a linear order on edges of the input geometric graph G. Let $|u, v|$ denote the Euclidean distance between nodes u and v. Intuitively, we can define a linear order $\prec$ on the edges of G by first considering the Euclidean length, if two edges have the same length, the one with rightmost, topmost endnode is larger, and finally if two edges of same length share their rightmost, topmost node, then their second endnode is considered; the edge with the right most, top most second endnode is defined as larger.

Notice that in the order $\prec$, we first consider the Euclidean length of edges and the coordinates are used for ordering edges of the same length. The input graph G may have many MSTs when the Euclidean length of edges is the cost function. However, if we break the ties by the linear order $\prec$, then G has a unique MST $T^{\prec}$ which can be computed for example by Kruskal's algorithm (see Cormen et al. (1990)). For a given geometric graph H, define $cost(H)$ as the sum of Euclidean lengths of the edges of H. Let $\Delta(H)$ denote the maximum degree of H. Given a graph G and a vertex v of G, we denote by $N_k[v]$ the distance k closed neighborhood of v, that is, the nodes of G reachable from v by a path with at most k edges.

Consider algorithm **LocalMST**$_k$, for $k \geq 2$, which was presented by Li et al. (2004). On input a connected geometric graph G with the linear order $\prec$, it outputs the graph $G_k^{\prec}$ by running the following algorithm at each node v of G:

1. Learn your distance k neighborhood $N_k[v]$.

2. Construct locally the unique MST $T_k(v)$ of $N_k[v]$.

3. Broadcast in $N_1[v]$ the edges of $N_1[v]$, which have been retained in $T_k(v)$ (i.e. $N_1[v] \cap T_k(v)$).

4. The output graph $G_k^{\prec}$ is defined as follows: an edge is selected into $G_k^{\prec}$ if and only if it was retained by both of its incident nodes.

Clearly, this is a distributed algorithm. To learn its distance k neighborhood, v first broadcasts its coordinates to all its neighbors. After having learnt its distance k neighborhood, it broadcasts it to all its neighbors. It can then construct the unique MST $T_k(v)$ (which is selected using Kruskal's algorithm (see Cormen et al. (1990)) and the linear order $\prec$) of $N_k[v]$ and broadcasts edges in $N_1[v] \cap T(v)$ to all nodes in $N_1[v]$. The parameter k determines the desired locality of our algorithm, and thus the resulting graph $G_k^{\prec}$ is constructed "locally," each node v uses only knowledge of $N_k[v]$ and the results of its neighbors.

Let G be a unit geometric graph with irregularity r and $k \geq 2$. We show that the graph $G_k^{\prec}$ constructed by the above algorithm has interesting properties summarized in the following theorem:

Theorem 6.2.4 (Chavez et al. (2006)) *If G is a connected geometric UDG with irregularity r and $k \geq 2$, then the graph $G_k^{\prec}$ has the following properties:*

1. *$G_k^{\prec}$ is connected;*

2. *if the distance between any two nodes of the graph is at least $\sqrt{1 - r^2}$, then the graph $G_k^{\prec}$ is planar;*

3. $\Delta(G_k^{\prec}) \leq \begin{cases} 5 & \text{if } r = 1, \\ 3 + \frac{6}{\pi r} + \frac{r+1}{r^2} & \text{if } 0 < r < 1; \end{cases}$

4. *If $G_k^{\prec}$ is planar and $kr > 1$, then $cost(G_k^{\prec}) \leq \frac{kr+1}{kr-1} \times cost(T^{\prec})$.*

We will not give the proof of this result. Instead we refer the reader to Chavez et al. (2006) for details.

Here, we will prove only the connectivity of $G_k^{\prec}$. This clearly follows from the claim $T^{\prec} \subseteq G_k^{\prec}$. To prove this, we argue by contradiction. Let the edge $\{u, v\}$ be retained in $T^{\prec}$, but rejected in $G_k^{\prec}$. Without loss of generality we may assume it was rejected in $T_k(v)$. Since $\{u, v\}$ was retained in $T^{\prec}$, there is no other path in $T^{\prec}$ joining u and v. Since $\{u, v\}$ was rejected by $T_k(v)$, there exists a path, say p, in $T_k(v)$ joining u and v and using only edges smaller than $\{u, v\}$.

Let $\{w, w'\}$ be an edge in p such that $\{w, w'\} \notin T^{\prec}$ (Figure 6.19). It follows that there is a path in $T^{\prec}$ joining w and w' and using only edges smaller than the edge $\{w, w'\}$. As

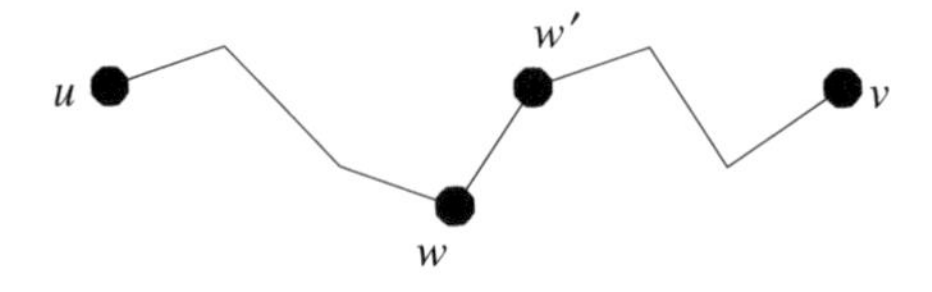

Figure 6.19 Path p from u to v.

this argument applies to each such edge of p, there must be a path in $T^{\prec}$ joining u and v using only edges smaller then $\{u, v\}$. This contradicts the fact that the edge $\{u, v\}$ was retained in $T^{\prec}$.

6.3 Information dissemination

The underlying wireless network is a rather complex graph consisting of vertices (hosts) and edges (representing host connectivity). An important question is how to discover "routes" and guarantee real-time message delivery in a dynamically changing system. For example, an issue in routing is how to guarantee delivery by selecting a path that satisfies three rather contradictory goals: (1) avoids flooding the network, (2) path is optimal, (3) use only geographically local information. Although, it is generally accepted that flooding must be avoided in order to increase the network lifetime, limitations on the knowledge of the hosts make it necessary that one may need to drop the second goal in favor of the third. Traditional methods employed discover routes using a greedy approach in which the "next hop" is determined by iteratively (hop-to-hop) reducing the value a given function of the distance and/or angle determined by the source and destination hosts. Unfortunately, this does not always work either because of loops (return to the source host without finding the destination) or voids (hosts with no neighbors closer to the destination). To guarantee delivery, it seems necessary that one has to reduce the wireless system to a simpler geometric spanner over which communication is simpler. In this section, we look at the problem of whether geographically local routing with guaranteed delivery is possible in arbitrary graphs.

A *planar subdivision* is a partitioning of the plane into a set V of vertices (points), a set E of edges (line segments) and a set F of faces (polygons). In this paper, we always consider only finite partitions. Furthermore, we assume that no edge passes through any vertex except its end-vertices. A combinatorial abstraction of a planar subdivision is the planar graph $G = (V, E)$ together with its straight line embedding into the plane. We will often identify the planar subdivision with its planar graph G in this paper. A subdivision is *connected* if its graph is connected.

6.3.1 Compass routing in undirected planar graphs

The first routing algorithm we consider is greedy in the sense that it always selects its next hop based on the smallest angle between the "current edge" and the "straight line" formed by the current node and the target node. More precisely, we have the following algorithm:
Input: Connected Planar Network $G = (V, E)$
Starting vertex: s
Destination vertex: t

1: Start at source node $c := s$.
2: **repeat**
3: Choose edge incident to current position c position and forming the smallest angle with straight line $c - t$ connecting c to t.
4: Traverse the chosen edge.
5: Go back to 3
6: **until** target t is found

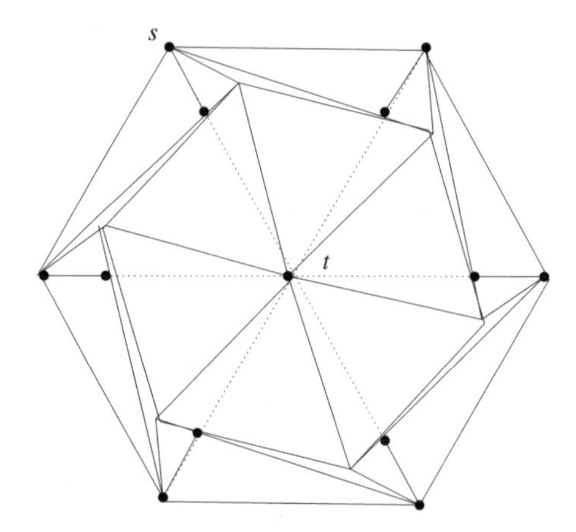

Figure 6.20 An example of a planar graph over which compass routing fails to find the destination.

It can be shown that in a random setting, whereby the points of the set P are uniformly distributed over a given convex region, compass routing can reach the destination with high probability (Exercise 6). In general, however, compass routing cannot guarantee delivery. This is not only due to possible voids (i.e. hosts with no neighbors closer to the destination) along the way but also inherent loops in the network. For example, in the planar graph depicted in Figure 6.20, a host starting at s with destination t and using compass routing will be looping on the vertices of the outer hexagon without ever reaching its target (Exercise 7).

6.3.2 Face routing in undirected planar graphs

Failure of compass routing to guarantee delivery is due to its "greedy" nature in order to minimize memory storage. Face routing overcomes this problem by remembering the straight line connecting the source s to the target node t.

Input: Connected Planar Network $G = (V, E)$
Starting vertex: s
Destination vertex: t

1: Start at source node $c := s$.
2: **repeat**
3:    **repeat**
4:       Determine face f incident to current node c and intersected by the straight line $s - t$ connecting s to t.
5:       Select any of the two edges of f incident to c and traverse the edges of f.
6:    **until** we find the second edge, say xy, of f intersected by $s - t$.
7:    Update face f to the new face of the graph containing edge uv.
8:    Update current vertex c to either of the vertices x or y.
9: **until** target t is found.

An example of the application of face routing is depicted in Figure 6.21. Face routing always advances to a new face. We never traverse the same face twice and, as a matter of fact,

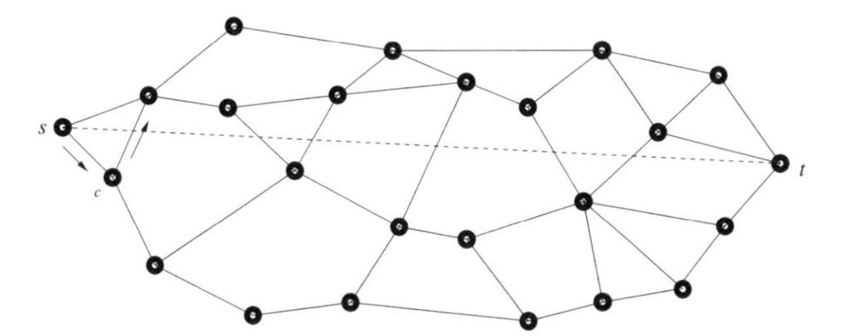

Figure 6.21 Face routing on a planar graph: s is the starting node, t is the target node, and c is the current node.

the routing algorithm is optimal in the number of faces between the source and destination nodes. The distance from the current position c to t gets smaller with each iteration. Since each link is traversed a constant number of times, we can prove the following theorem:

Theorem 6.3.1 (Kranakis et al. (1999)) *Face routing in a planar graph always guarantees delivery and traverses at most $O(n)$ edges.*

6.3.3 Traversal of quasi-planar graphs

Traversal is a fundamental task performed on planar subdivisions. Typically, it involves reporting each vertex, edge and face of G exactly once, in order to apply some operation to each of them. DFS (Depth First Search) of the primal (vertices and edges) or dual (faces and edges) of the graph is the usual approach followed, but unfortunately, this technique cannot be implemented without using mark bits on the vertices, edges or faces, and a stack or queue. If the data structure used to represent the subdivision G does not have extra memory allocated (as might be the case for the hosts in ad hoc wireless network), then an auxiliary array must be allocated and some form of hashing is required to map vertex/edge/face records to array indices. The DFS approach also has the drawback that the traversal cannot be performed simultaneously by more than one thread of execution without some locking mechanism and, of course, the memory requirements are increasing. In this section, we report on a traversal technique that is applicable to a class of non-planar networks.

A *quasi-planar subdivision* is a graph $G = (V, E)$ with vertices embedded in the plane and partitioned into $V_p \cup V_c = V$ so that

- vertices in V_p induce a connected planar graph P,
- the outer face of P does not contain any vertex from V_c or edge of $G - P$, and
- no edge of P is crossed by any other edge of G.

An example of a quasi-planar subdivision is depicted in Figure 6.22.

We will refer to the graph P as an *underlying planar subgraph* and to its faces as *underlying faces*. The notion of vertices and edges is explicit in the definition of quasi-planar subdivision, however, the notion of faces is not. To define the notion of a face, we need to introduce some basic functions on quasi-planar subdivisions.

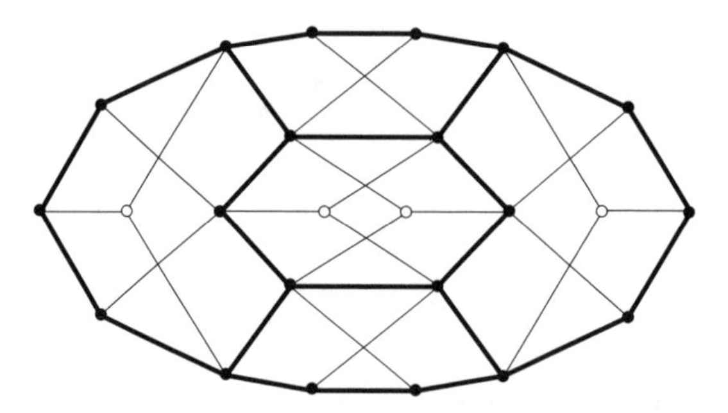

Figure 6.22 An example of a quasi-planar subdivision that satisfies the Left-Neighbor Rule. The filled vertices are in V_p and bold edges are edges of the underlying planar subgraph P.

6.3.3.1 Basic Functions on Quasi-Planar Subdivisions

A vertex u is uniquely determined by a pair $[x, y]$, where x is its horizontal coordinate and y is its vertical coordinate. The representation of the graph G is such that every edge $e = uv$ is stored as two oppositely directed edges (u, v) and (v, u). If we need to specify a direction of an edge e determined by nodes u and v, we write either $e = (u, v)$ or $e = (v, u)$, and if the direction is irrelevant, we write $e = uv$. For a vertex v, the function **xcor(v)** will return the horizontal coordinate of the vertex v, while the function **ycor(v)** will return the vertical coordinate of v. For an edge $e = (u, v)$, the function **rev(e)** will return a pointer to the edge (v, u). Similarly the function **succ(e)** will return a pointer to the edge (v, x) so that (v, x) is the first edge counterclockwise around v starting from the edge (v, u), and the function **pred(e)** will return a pointer to the edge (y, u) so that (u, y) is the first edge clockwise around u starting from the edge (u, v). For an illustration of these functions see Figure 6.23.

Since the functions **succ**() and **pred**() are injective, for every (directed) edge $e = (u, v)$ of G, we can define a closed walk by starting from $e = (u, v)$ and then repeatedly applying the function **succ**() until we arrive at the same edge $e = (u, v)$. Such a walk is called a *quasi-face* of G. The set of all quasi-faces of G is denoted by F. The function **qface(e)** will return a pointer to the quasi-face determined by the (directed) edge $e = (u, v)$. Note that for a planar subdivision G the quasi-faces become (regular) faces, and hence the notion of quasi-planar subdivision generalizes the notion of the connected planar subdivision.

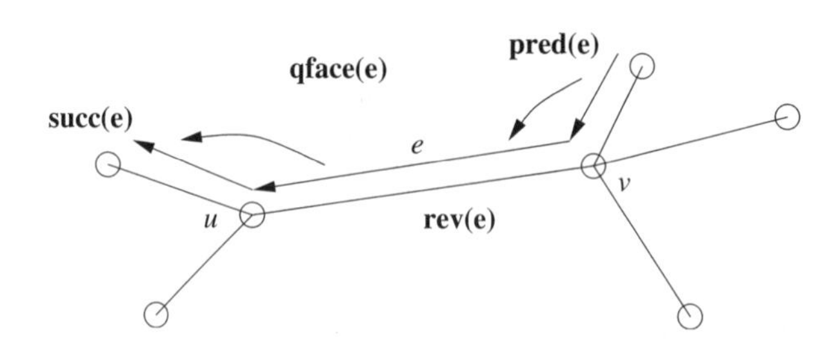

Figure 6.23 Illustration of basic functions on quasi-planar subdivisions.

The task of *traversing a quasi-planar subdivision* is to report every vertex, (undirected) edge, and quasi-face exactly once in some order. For general quasi-planar subdivisions this seems to be a hard task if we want to perform it without using mark bits and a stack. A quasi-planar subdivision G satisfies a *Left-Neighbor Rule* if every vertex $v \in V_c$ has a neighbor u so that $\mathbf{xcor(u)} < \mathbf{xcor(v)}$. For an example of G that satisfies the Left-Neighbor Rule see Figure 6.22.

6.3.3.2 Quasi-planar subdivision traversal algorithm

The general idea of the algorithm is to define a total order $\preceq$ on all edges in E. Using this order, we define a unique predecessor for every quasi-face in F such that the predecessor relationship imposes a virtual directed tree $G(F)$. The algorithm (due to Chavez et al. (2004a)) will search for the root of $G(F)$ and then will report quasi-faces of G in DFS order on the tree $G(F)$. For this, a well-known tree-traversal technique is used in order to traverse $G(F)$ using $O(1)$ additional memory. Note that the tree $G(F)$ is never stored in memory and at any given time the algorithm will remember only a constant number of edges (at most two) of this tree.

Input: $e = (u, v)$ of $G(V, E)$;

Output: List of vertices, edges and quasi-faces of G.

1: **repeat** {find the minimum edge e_0}
2: $e \leftarrow \mathbf{rev(e)}$
3: **while** $e \neq \mathbf{entry(qface(e))}$ **do**
4: $e \leftarrow \mathbf{succ(e)}$
5: **end while**
6: **until** $e = e_0$
7: $p \leftarrow \mathbf{left\check{t}(e)}$
8: **repeat** {start the traversal}
9: $e \leftarrow \mathbf{succ(e)}$
10: let $e = (u, v)$ and let $\mathbf{succ(e)} = (\mathbf{v}, \mathbf{w})$
11: **if** p is contained in $cone(u, v, w)$ **then** {report u if necessary}
12: **report** u
13: **end if**
14: **if** $|up| < |vp|$ or $(|up| = |vp|$ and $\overrightarrow{up} < \overrightarrow{vp})$ **then** {report e if necessary}
15: **report** e
16: **end if**
17: **if** $e = \mathbf{entry(qface(e))}$ **then** {report e and return to parent of $\mathbf{qface(e)}$}
18: **report** $\mathbf{qface(e)}$
19: $e \leftarrow \mathbf{rev(e)}$
20: **else** {descend to children of $\mathbf{qface(e)}$ if necessary}
21: **if** $\mathbf{rev(e)} = \mathbf{entry(qface(rev(e)))}$ **then**
22: $e \leftarrow \mathbf{rev(e)}$
23: **end if**
24: **end if**
25: **until** $e = e_0$
26: **report** $\mathbf{qface(e_0)}$

Note that our algorithm reports each (undirected) edge $e = uv$ exactly once. We have the following theorem:

Theorem 6.3.2 (Chavez et al. (2004a)) *The Quasi-Planar Traversal Algorithm reports each vertex, (undirected) edge, and quasi-face of a quasi-planar subdivision G that satisfies the Left-Neighbor Rule exactly once in* $O(|E|\log|E|)$ *time.*

6.3.4 Routing in eulerian directed planar graphs

A *planar geometric network* is a planar graph G with vertex set V, edge set E and the face set F together with its straight line embedding into the plane. An *orientation* $\vec{\ }$ of a planar geometric network G is an assignment of a direction to every edge e of G. For an edge e with endpoints u and v, we write $e = (u, v)$ if its direction is from u to v. The geometric network together with its orientation is denoted by $\vec{G}$.

Consider a connected planar geometric network $\vec{G} = (V, E)$. We say $\vec{G}$ is *Eulerian* if for every vertex $u \in V$, the size of $N^+(u) = \{x,\ (u, x) \in E\}$ equals the size of $N^-(u) = \{y,\ (y, u) \in E\}$; that is, the number of edges outgoing from u equals the number of edges ingoing into u.

First consider a planar geometric network G without any orientation. Given a vertex v on a face f in G, the boundary of f can be traversed in the counterclockwise (clockwise if f is the outer face) direction using the well-known right hand rule which states that it is possible to visit every wall in a maze by keeping your right hand on the wall while walking forward. Treating this face traversal technique as a subroutine, face routing as given by Kranakis et al. (1999) provides an elegant algorithm for routing in an undirected planar geometric network.

If we impose an orientation $\vec{\ }$ on G, then this algorithm will not work since some edges may be directed in an opposite direction while traversing a face. In this section, we describe a simple technique on how to overcome this difficulty. In particular, we propose a method for routing a message to the other end of an oppositely directed edge in Eulerian geometric networks.

Now suppose $\vec{\ }$ is such that $\vec{G}$ is an Eulerian planar geometric network. For a given vertex u of $\vec{G}$, we order edges (u, x) where $x \in N^+(u)$ clockwise around u starting with the edge closest to the vertical line passing through u. Similarly we order edges (y, u) where $y \in N^-(u)$ clockwise around u (Figure 6.24). Clearly, these orderings are unique and can be determined locally at each vertex.

Let $e = (y, u)$ be the ith ingoing edge to u in $\vec{G}$. The function **succ(e)** will return a pointer to the edge (u, x) so that (u, x) is the ith outgoing edge from u. For an illustration of the function see Figure 6.25. Again, this function is easy to implement using only local information. Obviously, the function **succ()** is injective, and thus, for every edge $e = (u, v)$ of $\vec{G}$, we can define a closed walk by starting from $e = (u, v)$ and then repeatedly applying the function **succ()** until we arrive at the same edge $e = (u, v)$. Since $\vec{G}$ is Eulerian, the walk is well defined and finite. We call such a walk a *quasi-face* of $\vec{G}$.

The following is the route discovery algorithm from Kranakis et al. (1999) for planar geometric networks. We modify it so that it will work on Eulerian planar geometric networks. For this, we only need to extend the face traversal routine so that whenever the face traversal routine wants to traverse an edge $e = (u, v)$ that is oppositely directed, we traverse the following edges in this order: $\mathbf{succ(e)}, \mathbf{succ(e)}^2, \ldots, \mathbf{succ(e)}^k$, so that

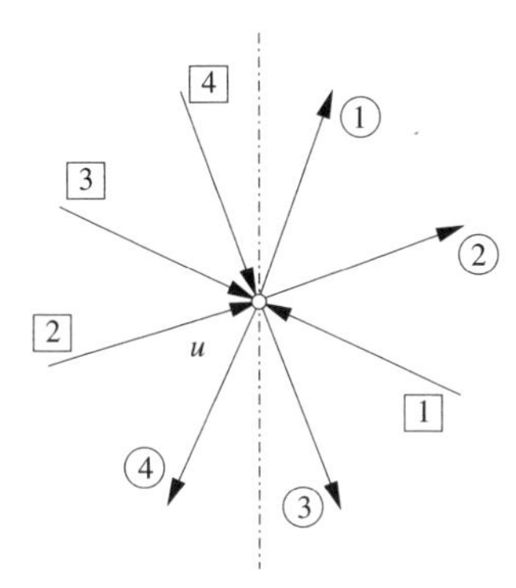

Figure 6.24 Circled numbers represent the ordering on outgoing edges, squared numbers on ingoing ones.

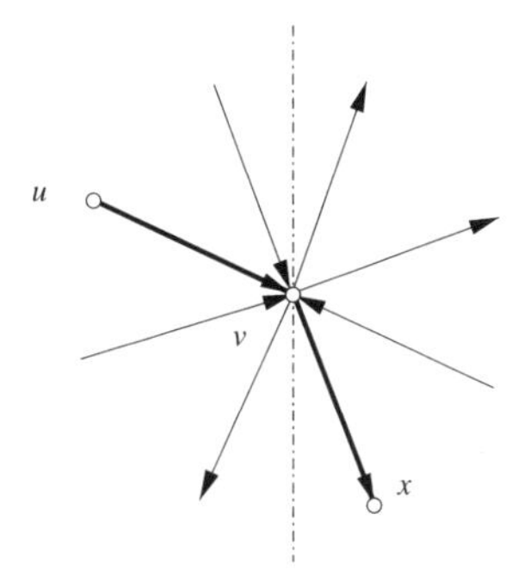

Figure 6.25 In this example the ingoing edge (u, v) is third, so the chosen outgoing edge (v, x) is also third. Both these edges are depicted bold.

$\mathbf{succ(e)^{k+1}} = (\mathbf{u}, \mathbf{v})$. After traversing $\mathbf{succ(e)^{k}}$, the routine resumes to the original traversal of the face. Formally, the algorithm is as follows:
Input: Connected Eulerian geometric network $\vec{G} = (V, E)$
Starting vertex: s
Destination vertex: t

1: $v \leftarrow s$ {Current vertex = starting vertex.}
2: **repeat**
3: Let f be a face of G with v on its boundary that intersects the line v-t at a point (not necessarily a vertex) closest to t.
4: **for all** edges xy of f **do**
5: **if** $xy \cap v\text{-}t = p$ and $\mathbf{dist(p, t)} < \mathbf{dist(v, t)}$ **then**
6: $v \leftarrow p$
7: **end if**
8: **end for**
9: Traverse f until reaching the edge xy containing the point p.
10: **until** $v = t$

We can prove the following theorem:

Theorem 6.3.3 (Chávez et al. (2004b)) *The Eulerian Directed Graph Algorithm will reach t from s in at most $O(n^2)$ steps.*

6.3.5 Routing in outerplanar graphs

A planar geometric network G is *outerplanar* if one of the elements in F contains all the vertices (also called *outer face*). Without loss of generality, we may assume that this face is a convex polygon in $\mathbb{E}^2$. For a given triple of vertices x, y, and z, let $V_{\curvearrowleft}(x, y, z)$ (respectively, $V_{\curvearrowright}(x, y, z)$) denote the ordered set of vertices distinct from x and z that are encountered while moving from y counterclockwise (respectively, clockwise) around the outer face of G until either x or z is reached; see Figure 6.26.

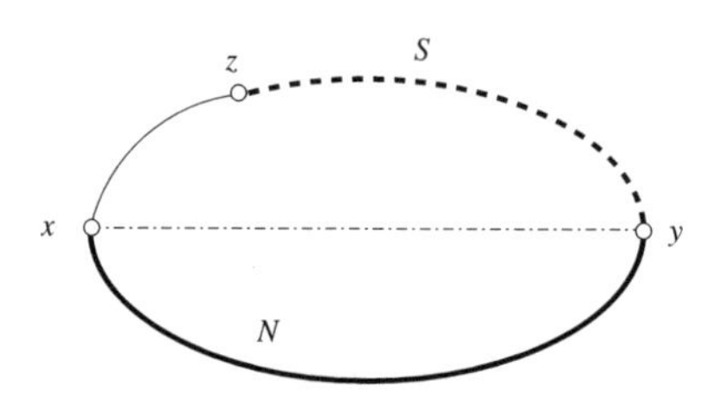

Figure 6.26 The dashed part of the outer face represents the vertices in $V_{\curvearrowleft}(x, y, z)$ and the bold solid part represents vertices in $V_{\curvearrowright}(x, y, z)$, respectively. Note that y belongs to both these sets and is in fact the first element of those sets.

Now consider an orientation $\vec{\ }$ of the geometric network G. Define

$$N_{\curvearrowleft}(x, y, z) = V_{\curvearrowleft}(x, y, z) \cap N^+(x)$$

$$N_{\curvearrowright}(x, y, z) = V_{\curvearrowright}(x, y, z) \cap N^+(x).$$

If $N_{\curvearrowleft}(x, y, z) \neq \emptyset$, let $v_{\curvearrowleft}(x, y, z)$ denote the first vertex in $N_{\curvearrowleft}(x, y, z)$. Similarly we define $v_{\curvearrowright}(x, y, z)$ as the first vertex in $N_{\curvearrowright}(x, y, z)$, if it exists. A geometric network with fixed orientation is *strongly connected* if for every ordered pair of its vertices, there is a (directed) path joining them.

The algorithm starts at the source node s. It specifies the two nodes v_1, v_2 adjacent to the current node c such that the straight line determined by nodes c and t lies within the angle $v_1 c v_2$. It traverses one of the nodes and backtracks if necessary in order to update its current position. Formally, the algorithm is as follows:

Input: Strongly connected outerplanar geometric network $\vec{G} = (V, E)$
Starting vertex: s
Destination vertex: t

1: $v \leftarrow s$ {Current vertex = starting vertex.}
2: $v_{\curvearrowleft}, v_{\curvearrowright} \leftarrow s$ {counterclockwise and clockwise bound = starting vertex.}
3: **while** $v \neq t$ **do**
4:     **if** $(v, t) \in E$ **then**

5: $\quad v, v_{\curvearrowleft}, v_{\curvearrowright} \leftarrow t$ {Move to t.}
6: **else if** $N_{\curvearrowleft}(v, t, v_{\curvearrowleft}) \neq \emptyset$ and $N_{\curvearrowright}(v, t, v_{\curvearrowright}) = \emptyset$ **then** {No-choice vertex; greedily move to the only possible counterclockwise direction toward t.}
7: $\quad v, v_{\curvearrowleft} \leftarrow v_{\curvearrowleft}(v, t, v_{\curvearrowleft})$
8: **else if** $N_{\curvearrowleft}(v, t, v_{\curvearrowleft}) = \emptyset$ and $N_{\curvearrowright}(v, t, v_{\curvearrowright}) \neq \emptyset$ **then** {No-choice vertex; greedily move to the only possible clockwise direction toward t.}
9: $\quad v, v_{\curvearrowright} \leftarrow v_{\curvearrowright}(v, t, v_{\curvearrowright})$
10: **else if** $N_{\curvearrowleft}(v, t, v_{\curvearrowleft}) \neq \emptyset$ and $N_{\curvearrowright}(v, t, v_{\curvearrowright}) \neq \emptyset$ **then** {Decision vertex; first take the "counterclockwise" branch but remember the vertex for the backtrack purpose.}
11: $\quad b \leftarrow v; \quad v, v_{\curvearrowleft} \leftarrow v_{\curvearrowleft}(v, t, v_{\curvearrowleft})$
12: **else if** $N_{\curvearrowleft}(v, t, v_{\curvearrowleft}) = \emptyset$ and $N_{\curvearrowright}(v, t, v_{\curvearrowright}) = \emptyset$ **then** {Dead-end vertex; backtrack to the last vertex where a decision has been made. No updates to $v_{\curvearrowleft}$ and $v_{\curvearrowright}$ are necessary.}
13: $\quad$ **if** $v \in V_{\curvearrowleft}(t, b, t)$ **then**
14: $\quad\quad$ **while** $v \neq b$ **do**
15: $\quad\quad\quad v \leftarrow v_{\curvearrowleft}(v, b, v)$
16: $\quad\quad$ **end while**
17: $\quad$ **end if**
18: $\quad$ **if** $v \in V_{\curvearrowright}(t, b, t)$ **then**
19: $\quad\quad$ **while** $v \neq b$ **do**
20: $\quad\quad\quad v \leftarrow v_{\curvearrowright}(v, b, v)$
21: $\quad\quad$ **end while**
22: $\quad$ **end if**
23: $\quad v, v_{\curvearrowright} \leftarrow v_{\curvearrowright}(v, t, v_{\curvearrowright})$ {Take the "clockwise" branch toward t.}
24: **end if**
25: **end while**

Theorem 6.3.4 (Chávez et al. (2004b)) *Given a start node s and a target node t, in an outerplanar graph of n nodes, Outer Planar Directed Graph Algorithm will reach t from s in at most $2n - 1$ steps.*

6.4 Geographic location determination

Supporting nodes without GPS capability, in wireless ad hoc and sensor networks, have numerous applications in guidance and surveying systems in use today. At issue is that a procedure be available so that the subset of nodes with GPS capability succeed in supporting the maximum possible number of nodes without GPS capability and as a result enable the highest connectivity of the underlying network infrastructure. In this section, we first discuss radiolocation methods and then describe several techniques for improving on radiolocation.

6.4.1 Radiolocation techniques

The position of wireless nodes can be determined using one of four basic techniques, namely time of arrival (TOA), time difference of arrival (TDOA), angle of arrival (AOA) or signal strength.

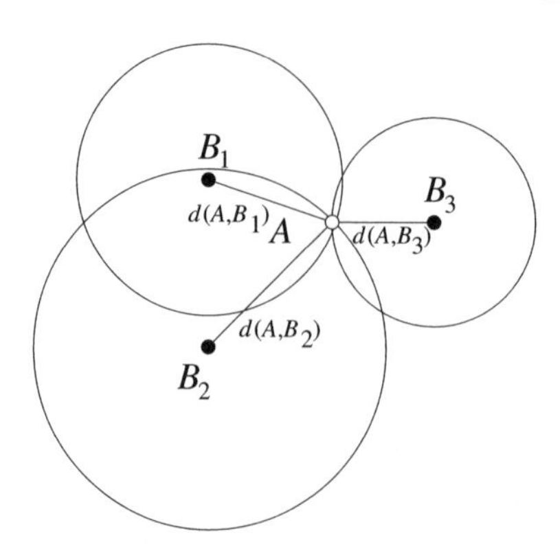

Figure 6.27 A sensor at A not equipped with a GPS device can determine its position from the positions of its three neighbors B_1, B_2, B_3.

6.4.1.1 Time of arrival

The TOA technique is pictured in Figure 6.27. Node A is unaware of its position. Three position aware nodes are involved, let us say B_1, B_2 and B_3. Each position aware node B_i sends a message to A and the trip time of the signal is measured. The trip time multiplied by the propagation speed of signals (i.e. the speed of light) yields a distance d_i. The distance $d(A, B_i)$ defines a circle around node $d(A, B_i)$. The position of A is on the circumference of this circle. In a two-dimensional model, the position of A is unambiguously determined as the intersection of three such circles.

Trip time measurement from each node B_i to node A requires synchronized and accurate clocks at both locations. If round-trip time is measured instead (halved to obtain trip time), then this requirement is relaxed. No clock synchronization is required and accurate clocks are required only at the B_i's.

6.4.1.2 Time difference of arrival

TDOA is pictured in Figure 6.28. Nodes B_1, B_2 and B_3 are aware of their position while node A is not. B_1 and B_2 simultaneously send a signal. Times of arrivals t_1 and t_2 of signals from B_1 and B_2, respectively, are measured by A. Node A has the capability to recover the identity of the sender of a signal and its position. The TDOA is calculated, that is, $\delta_t = t_2 - t_1$. The time difference δ_t multiplied by the speed of light is mapped to the distance difference δ_d. The position (x_1, y_1) of B_1, position (x_2, y_2) of B_2 and δ_d define a hyperbola h with equation:

$$\sqrt{(x - x_1)^2 + (y - y_1)^2} - \sqrt{(x - x_2)^2 + (y - y_2)^2} = \delta_d. \tag{6.1}$$

The positions of B_1 and B_2 are at the foci of the hyperbola. The position of node A is a solution of Equation 6.1. The geometrical properties of the hyperbolas are such that all points located on the line of h are of equal time difference δ_t and equal distance difference δ_d.

Two such hyperbolas can be defined by involving two different pairs of nodes (B_1, B_2) and (B_2, B_3), which produce two time differences of arrival δ_1 and δ_2. In a two-dimensional

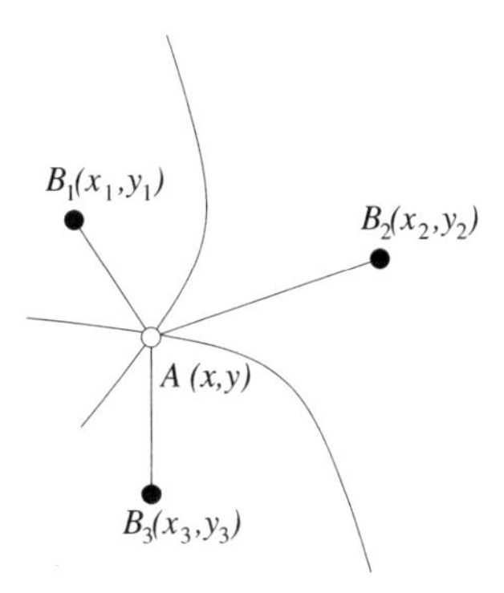

Figure 6.28 The TDOA technique.

model, the observer of the times of arrival δ_1 and δ_2, that is, node A, is at the position corresponding to the intersection of the two hyperbolas. Note that there are cases in which the two hyperbolas intersect at two points. In these cases, a third independent measurement is required to resolve the ambiguity.

6.4.1.3 Angle of arrival

The AOA technique is pictured in Figure 6.29. Two position aware nodes, let us say B_1 and B_2, are required to determine the position of a node A. Nodes B_1 and B_2 have to be able to determine the direction from which a signal is coming. This can be achieved with an array antenna Cooper (2003). An imaginary line is drawn from B_1 to A and another imaginary line is drawn from B_2 to A. The AOA is defined as the angle that each of these lines make with a line directed towards a common reference. The intersection of these two lines unambiguously determines the position of A. Note however, that if A, B_1 and B_2 all lie on the same straight line, another independent measurement is required to resolve the ambiguity. Accuracy of the AOA technique is largely dependent on the beamwidth of the antennas. According to Pahlavan and Krishnamurthy (2002), the TOA technique is superior to the AOA technique. In CDMA cellular networks, Caffery and Stüber (1998b) come to similar conclusions.

6.4.1.4 Signal Strength

The signal strength based technique exploits the fact that a signal loses its strength as a function of distance. Giving the power of a transmitter and a model of free-space loss, a

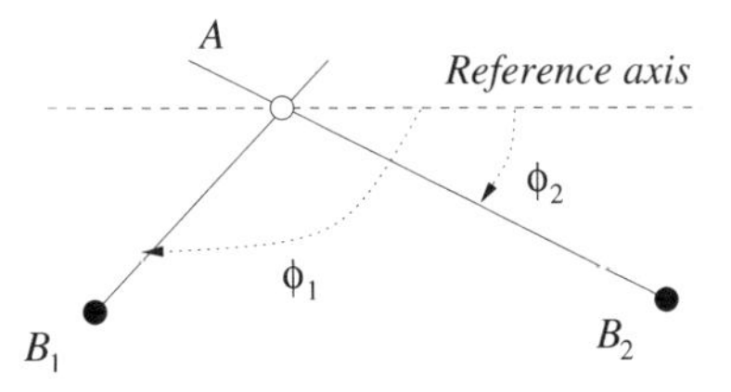

Figure 6.29 The AOA technique.

receiver can determine the distance traveled by a signal. If three different such signals can be received, a receiver can determine its position in a way similar to the TOA technique. Application of this technique for cellular systems has been investigated by Figuel et al. (1969) and Hata and Nagatsu (1980). The main criticism of Caffery and Stüber (1998a) is about the accuracy of the technique. This is due to transmission phenomena such as multi path fading and shadowing that cause important variations in signal strength.

All these techniques require line-of-sight propagation between the nodes involved in a signal measurement. Line-of-sight means that a nonobstructed imaginary straight line can be drawn between the nodes. In other words, the accuracy is sensitive to radio propagation phenomena such as obstruction, reflection and scattering. With all the distance-based techniques (i.e. TOA, TDOA, signal strength), three position aware neighbors are required to determine the location of a position unaware node, in a two-dimensional model (e.g. latitude and longitude are determined) and four neighbors are required in a three-dimensional model (e.g. altitude is determined as well). In the sequel, we augment the distance-based techniques with an algorithmic component that can resolve ambiguity in a two-dimensional model when only two position aware neighbors are available. The ambiguity can also be resolved using knowledge about the trajectory when the wireless nodes are mobile or using the AOA technique. When the nodes are fixed and the technology required to apply the AOA technique is not available, the algorithm described in this paper can be used. We note that a similar algorithm is also possible in a three-dimensional model.

6.4.2 Computing the geographic location

Consider a set S of n sensors in the plane. Further, assume that the sensors have the same reachability radius r. We divide the set S of sensors into two subsets. A subset E of S consists of sensors equipped with GPS devices that enable them to determine their location in the plane. The set U of remaining sensors, that is, $S \setminus E$, consists of sensors not equipped with GPS devices. (In pictures, the former are represented with solid circles and the latter with empty circles.) In this section, we consider the problem of determining the positions of sensors in a sensor network. In the beginning, we review the well-known *three neighbor algorithm* (3-NA) and conclude with an example illustrating why the algorithm does not necessarily compute the positions of the maximum possible number of nodes. Subsequently, we present an improvement, the *three/two neighbor algorithm* (3/2-NA). Essentially, this is an iteration of the 3-NA followed by an algorithm that uses only two neighbors as well as their distance one neighborhood.

6.4.2.1 Three Neighbor Algorithm and its Deficiencies

If a sensor at A (Figure 6.31) is not equipped with a GPS device, then it can determine its (x, y) coordinates using three neighboring nodes. After receiving messages from B_1, B_2 and B_3, that include their position, node A can determine its distance from these nodes using a distance-based radiolocation method. Its position is determined as the point of intersection of three circles centered at B_1, B_2 and B_3 and distances $d(A, B_1)$, $d(A, B_2)$ and $d(A, B_3)$, respectively.

The well-known 3-NA is as follows. Each node that is not equipped with a GPS device sends a position request message. A sensor that knows or can compute its position sends it to all its neighbors. A sensor that receives position messages from three different nodes, say B_1, B_2 and B_3 can calculate its position as in Figure 6.27.

6.4.2.2 Computing the Position of the Maximum Number of Nodes

The 3-NA does not necessarily compute the positions of the maximum possible number of sensors without GPS devices. We illustrate this with a simple example (Figure 6.30). The Part (a) of the picture depicts five nodes A, A', B_1, B_2 and C. Node A' is within the communication range of nodes B_1, B_2 and C. Node A is within the communication range of nodes B_1 and B_2. Node A is out of the range of both nodes C and A'. Assume that nodes B_1, B_2 and C know their (x,y)-coordinates. Application of the 3-NA will equip A' with its (x,y) coordinates (because it will receive messages from all three of its neighboring nodes B_1, B_2, C). This is depicted in the right side of the picture in Figure 6.30. However, node A will never receive three position messages and therefore will never be able to discover its (x,y)-coordinates. Nevertheless, we will see in the next algorithm that node A can indeed discover its position if additional information (i.e. the distance one neighborhood of its neighbors) is provided.

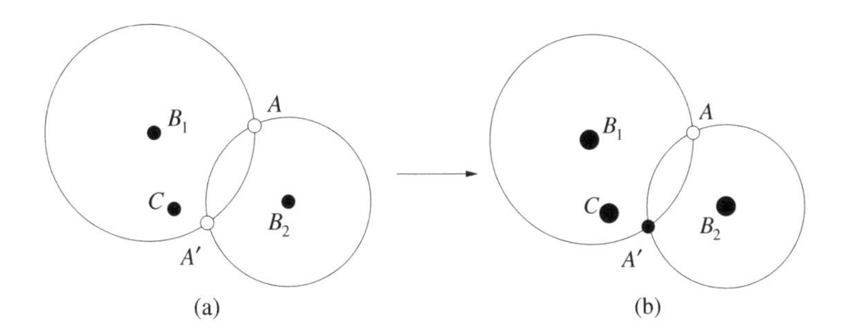

Figure 6.30 Application of 3-NA will equip node A' with its (x,y)-coordinates, but it fails to do it with node A.

6.4.3 Three/two neighbor algorithm

We now consider an extension of the 3-NA for the case where all the sensors have exactly the same reachability radius, say r.

6.4.3.1 On Utilizing the Distance One Neighborhood

Assume that we have concluded the execution of the 3-NA. For each node P, let $N(P)$ be the set of neighbors of P, that is, the set of sensors within communication range of P. Suppose we have two nodes B_1 and B_2 (depicted in Figure 6.31) that know their position. The solid circles are determined by the reachability radius of the nodes. The dashed circles are centered at B_1 and B_2, respectively. After using radiolocation, they specify that the inquiring node must be located at one of their points of intersection (in this case, either A or A'). Further assume that a given node X receives the positions of nodes B_1 and B_2 by radiolocation. On the basis of this information, X may be located in either position A or A'.

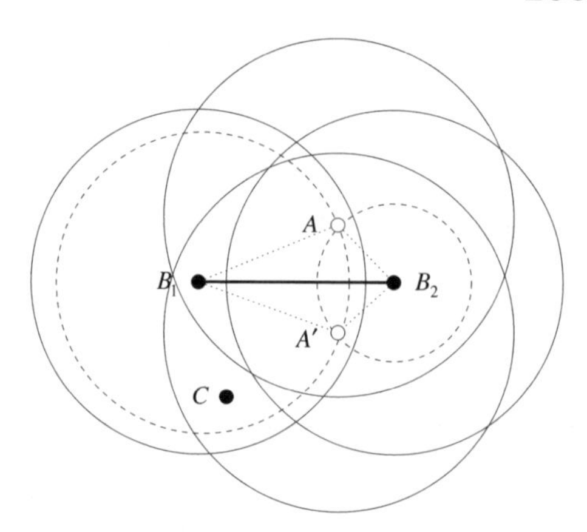

Figure 6.31 A sensor not equipped with a GPS device can determine its coordinates using the positions of two neighbors B_1 and B_2 and their neighborhood information.

Lemma 6.4.1 *Suppose that both nodes A and A' receive from B_1 and B_2 the set $N(B_1) \cup N(B_2)$. If there is a sensor $C \in (N(B_1) \cup N(B_2)) \cap (N(A) \cup N(A'))$ that knows its position, then both nodes A and A' can determine their position.*

Proof. We must consider two cases. In the first case, assume that C is within the range of both nodes A and A'. In this case, the two nodes will receive position messages from all three nodes B_1, B_2 and C and can therefore determine their position. In the second case, assume that C is within the range of only one of the two nodes. Without loss of generality assume that it is within the range of A' but outside the range of A, that is, $d(C, A') \leq r < d(C, A)$ (Figure 6.31). Then A' can determine its position. Also, A can determine its position because it knows it must occupy one of the two positions A or A' and can also determine that it cannot be node A' since its distance from C (a node whose position it has received) is outside its range. ∎

It is now clear that Lemma 6.4.1 gives rise to the following 3/2-NA for computing sensor positions.

6.4.3.2 3/2-NA (Three/Two Neighbor Algorithm):

1. For each node as long as new nodes determine their position **do**
2. Each sensor executes the 3-NA algorithm and also collects the coordinates of all its neighbors.
3. **If** at the end of the execution of this algorithm a sensor has received the coordinates from only two neighbors, say B_1, B_2 **then**
   (a) it computes the distances from its current location to the sensors B_1, B_2 and also computes the coordinates of the two points of intersection A, A' of the circles centered at B_1, B_2, respectively;
   (b) it requests the coordinates of all the neighbors of B_1 and B_2 that are aware of their coordinates;

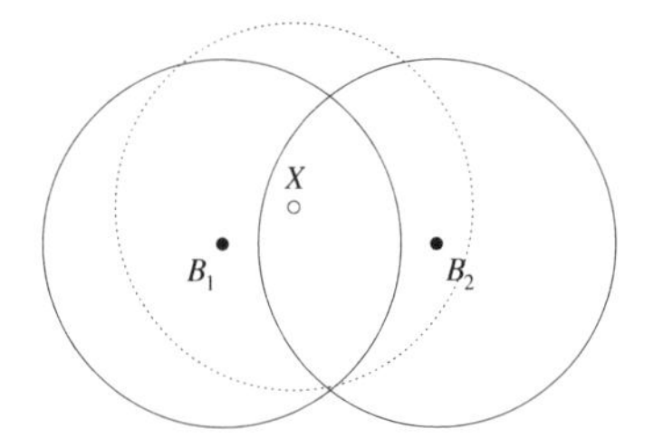

Figure 6.32 The termination condition for the 3/2-NA.

(c) **if** $N(B_1) \cup N(B_2) \neq \emptyset$ then take any node $C \in N(B_1) \cup N(B_2)$ and compute $d(C, A)$, $d(C, A')$; then the sensor occupies the position $X \in \{A, A'\}$ such that $d(X, C) > r$;

For any node P let $D(P; r)$ be the disc centered at P and with radius r, that is, the set of points X such that $d(P, X) \leq r$. We can prove the following theorem:

Theorem 6.4.2 *The Three/Two Neighbor Algorithm terminates in at most $(n - e)^2$ steps, where e is the number of GPS-equipped nodes.*

Proof. First consider the three neighbor algorithm. At each iteration of the algorithm, a sensor either waits until it receives three messages (of distances of its neighbors) or else it receives the coordinates of at least three neighboring sensors, in which case it computes its own coordinates and forwards it to all its neighbors. Let E_t be the number of sensors that know their coordinates by time t. Observe that initially $E_0 = E$ and $E_t \subseteq E_{t+1}$. If, at any given time, no new sensors that are not GPS-equipped determines their coordinates (i.e. $E_t = E_{t+1}$) then no new non-equipped sensor will ever be added. It follows that the algorithm terminates in at most $n - e$ steps. After this step, no new sensors will be equipped with their coordinates.

Now consider the three/two neighbor algorithm. Concerning correctness, we argue as follows. Consider a sensor as above that has received position messages only from two neighbors, say B_1 and B_2. The sensor knows that it is located at one of the points of intersection of the two circles (Figure 6.31). Since $N(B_1) \cup N(B_2) \neq \emptyset$ and the sensor received no position message from any sensor in $N(B_1) \cup N(B_2)$ after the execution of the 3-NA algorithm it concludes that it must occupy position X, where $X \in \{A, A'\}$ such that $d(C, X) > r$. Since within each iteration at one new sensor computes its position the running time is as claimed. ■

6.4.4 Beyond distance one neighborhood

When the 3/2-NA terminates no new node can compute its position. There is an improvement that can be made to the 3/2-NA. In Figure 6.31, this may happen when for some $k \geq 1$ the node C is at distance k hops from either B_1 or B_2. If the entire distance k neighborhood is being transmitted and C is within A''s range but outside A's range, then the node A can determine its position. This gives rise to the 3/2-NA(k) algorithm which is similar to 3/2-NA algorithm except that now the nodes transmit their distance k neighborhood.

We define the distance k hop(s) neighborhood of a node P as follows. First, $N_1(P)$ is defined as $N(P)$. For $k \geq 2$

$$N_k(P) = \{A | A \in N_{k-1}(P) \vee \exists B \in N_{k-1}(P) \wedge A \in N(B).\}$$

The specific algorithm is as follows:

6.4.4.1 3/2-NA(k) (Three/Two Neighbor Algorithm):

1. For each node as long as new nodes determine their position **do**
2. Each sensor executes the 3-NA algorithm and also collects the coordinates of all its neighbors.
3. **If** at the end of the execution of this algorithm a sensor has received the coordinates from only two neighbors, say B_1, B_2 **then**
   (a) it computes the distances from its current location to the sensors B_1, B_2 and also computes the coordinates of the two points of intersection A, A' of the circles centered at B_1, B_2, respectively;
   (b) it requests the coordinates of all the distance k neighbors of B_1 and B_2 that are aware of their coordinates;
   (c) **if** $N_k(B_1) \cup N_k(B_2) \neq \emptyset$ then take any node $C \in N_k(B_1) \cup N_k(B_2)$ and compute $d(C, A)$, $d(C, A')$; then the sensor occupies the position $X \in \{A, A'\}$ such that $d(X, C) > r$;

We have the following theorem:

Theorem 6.4.3 *Algorithm* 3/2-*NA(k) terminates in at most* $(n - e)^2$ *steps, where e is the number of GPS-equipped nodes.* ■

Note however, that nodes have to transmit their distance k neighborhood which will produce higher overhead.

6.5 Random unit disc graphs

Let P be a random set of n points, selected randomly and independently with the uniform distribution in a unit interval. Uniform means that the number of points inside an interval I should depend on the length of I and the density n but not on the position of the interval within the unit interval (Figure 6.33). Since the points are chosen at random, the probability that exactly k points lie in a given interval I is equal to

$$\binom{n}{k} |I|^k (1 - |I|)^{n-k}, \tag{6.2}$$

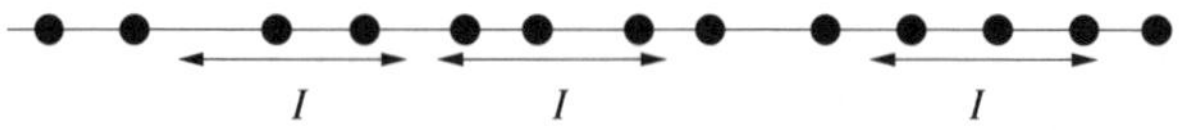

Figure 6.33 n points on a unit interval and three intervals of identical length.

since the length $|I|$ of I is the probability that a random point will be inside I and $1 - |I|$ the probability that it will be outside. This formula arises from the Binomial distribution that was studied in Chapter 2. If now np remains fixed, $n \to \infty$ and k is small, then the probability in Formula 6.2 can be shown to be identical to

$$\frac{(n|I|)^k}{k!} e^{-n|I|}, \tag{6.3}$$

which is the well-known Poisson distribution. The same analysis can, of course, be made for planar domains.

Consider the following setup arising in the study of sensor networks: n sensors, say directional antennas, are dropped randomly and independently with the uniform distribution over the interior of a unit length segment (Figure 6.34). How do we achieve full coverage of the line segment, as well as connectivity of the resulting network? What is the probability that n sensors cover the line segment and/or the network is connected? The sensors drop in the interior of the unit length line segment. We parametrize the characteristics of the sensors as a function of n: $\alpha := \alpha(n)$ is the coverage angle, and $r := r(n)$ the reachability radius of the sensor, respectively. Let $p := p(n)$ be the probability that a given sensor is active, in which case $1 - p(n)$ is the probability that the sensor is inactive. We want to ensure that the network of directional sensors is connected, that is, there is a communication path between any pair of sensors in the network. Since the placement and orientation of the sensors is random, there is a reasonable probability that the two sensors will be unable to communicate with any other sensors. It may be the case that rather than achieve full connectivity we require that a connected subnetwork exists that achieves full coverage. Therefore we can formulate the following question. n directional sensors are dropped randomly and independently with the uniform distribution over the interior of a unit square. What relation between p, α, n must exist in order to ensure the existence of a connected subnetwork of sensors that provides full coverage? Random UDGs are a valuable tool in studying the average case as well as in experimenting with ad hoc networks. In the sequel, we layout the foundations and discuss the basic properties of such graphs.

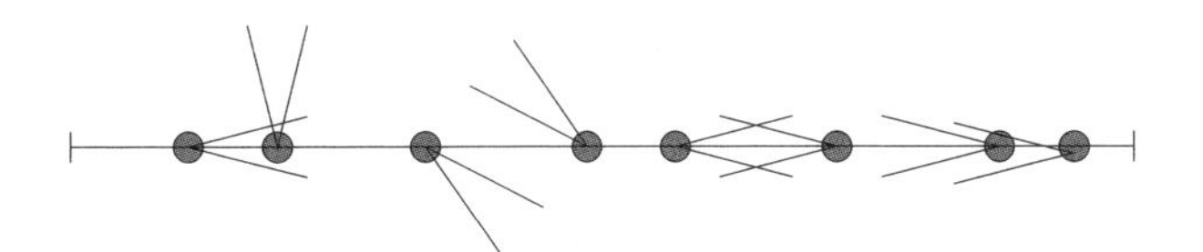

Figure 6.34 n sensors thrown randomly with the uniform distribution on a unit interval.

6.5.1 Poisson distribution in the plane

Suppose that P is a random set of n points, whereby the points are selected randomly and independently with the uniform distribution in a unit square. Thus, n is the *density* of the pointset in the unit square. The points represents the hosts of a wireless network and have identical radius r_n. Let $G(P, r_n)$ denote the unit disc graph on a set P of n nodes, where each node has radius r_n. The probability that a region R of area $|R|$ has exactly l nodes from the random pointset obeys the Poisson distribution and is equal to

$$\frac{(n|R|)^l}{l!}e^{-n|R|}. \tag{6.4}$$

6.5.1.1 Number of Points in a Given Region

The next proposition confirms our intuition that the expected number of points in a planar region R should be proportional to its area, for the given density.

Proposition 6.5.1 *The expected number of points in a given region R is equal to $n|R|$.*

Proof. Given a region R let X_R be the random variable that counts the number of points in R. Then we can prove that

$$\begin{aligned} E[X_R] &= \sum_{k=1}^{\infty} k\,Pr[X_R = k] \\ &= \sum_{k=1}^{\infty} k\frac{(n|R|)^k}{k!}e^{-n|R|} \\ &= e^{-n|R|}n|R|\sum_{k=0}^{\infty} k\frac{(n|R|)^k}{k!} \\ &= n|R|. \end{aligned}$$

■

6.5.1.2 Largest Clique in the Unit Disc Graph

Proposition 6.5.2 *The expected number of points of the largest clique of the unit disc graph is at least $n\pi r_n^2/4$ and at most $n\pi r_n^2$.*

Proof. For any point $x \in P$ consider the disc $D(x, r_n/2)$ is a clique. By Proposition 6.5.1 we have that the expected number of points in $D(x, r_n/2)$ is equal to $n\pi r_n^2/4$. Now assume that K is a clique and let $x \in P$ be an arbitrary point. It follows that $K \subseteq D(x, r_n)$. ■

6.5.2 Connectivity and k-connectivity

The behavior of a random UDG should depend on the relative positions of the set of points P and their density n. In turn, the density depends on the radius r_n which should be chosen in such a way that the UDG is connected, fault-tolerant, etc. Two important questions to ask are:

- Is it possible to provide sufficient conditions on the radius r_n in order to guarantee the k-connectivity of $G(P, r_n)$?
- Is it possible to provide sufficient conditions on the radius r_n in order to guarantee coverage of the unit square?

We note that coverage is not the same as connectivity, since the former means the whole area (e.g. a unit square) is covered by disks while the latter that the nodes form a connected system (Figure 6.35). It is not difficult to determine the conditions for coverage and connectivity. Partition the unit square into subsquares with side $s_n = \Theta(r_n)$ (Figure 6.36).

Choose s_n so that all nodes in a subsquare are at distance at most r_n from the hosts in all adjacent subsquares. The unit square must have $1/s_n^2$ subsquares. Now observe that if every subsquare contains at least one point then the UDG $G(P, r_n)$ is connected. Given that we have n points how do we guarantee that every subsquare contains at least one point with high probability? The coupon collector's problem (see Motwani and Raghavan (1995)) says that if

$$n \geq \left(\frac{1}{s_n^2}\right) \log \left(\frac{1}{s_n^2}\right). \tag{6.5}$$

then with probability at least $1 - \frac{1}{n}$ every subsquare contains at least one point. Hence, Equation 6.5 tells you what s_n should be as a function of n.

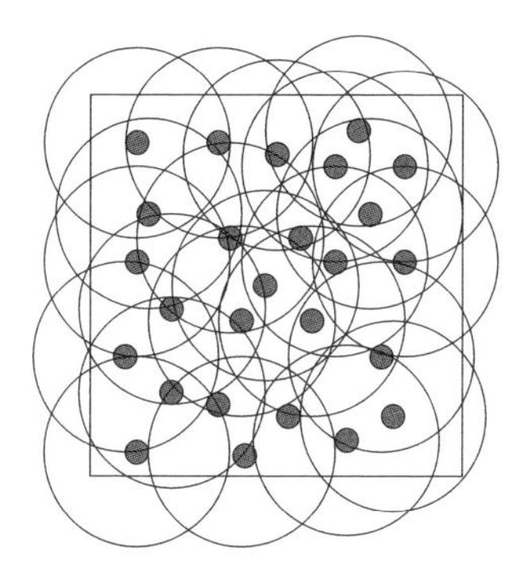

Figure 6.35 Covering a unit square with disks.

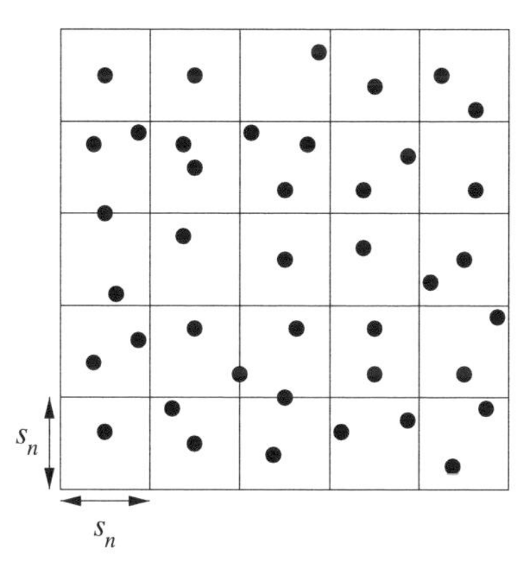

Figure 6.36 Partitioning a unit square into subsquares with side s_n.

6.5.2.1 Precise Formulation of the Problem

More generally, a network is called $(k+1)$-connected if it cannot be disconnected after the removal of any $k+1$ nodes. Thus connected is the same as 1-connected. The main result of Penrose (1999) provides sufficient conditions on the radius r_n in order to guarantee the k-connectivity of $G(P, r_n)$. In fact, we have that for any real number c if

$$r_n \geq \sqrt{\frac{\ln n + c}{n\pi}},$$

then the probability that the network $G(P, r_n)$ is connected is at least $\geq e^{-e^{-c}}$, as $n \to \infty$. If we substitute $e^c = s$ and recall that $e^{-e^{-c}} \approx 1 - e^{-c}$, then we see that if

$$r_n \geq \sqrt{\frac{\ln n + \ln s}{n\pi}}, \tag{6.6}$$

then

$$Pr[G(P, r_n) \text{ is connected}] \geq 1 - \frac{1}{s}. \tag{6.7}$$

More generally, a network is called $(k+1)$-connected if it cannot be disconnected after the removal of any $k+1$ nodes. Thus connected is the same as 1-connected. The main result Penrose (1999) also gives a threshold value for k-connectivity in the "toroidal" unit square. For any integer $k \geq 0$ and real number constant s let the nodes have identical radius $r_n^{(k)}$, satisfying the identity

$$r_n^{(k)} \geq \sqrt{\frac{\ln n + k \ln \ln n + \ln(k!) + \ln s}{n\pi}}. \tag{6.8}$$

Then

$$Pr[G(P, r_n^{(k)}) \text{ is } (k+1)\text{-connected}] \geq 1 - \frac{1}{s} \tag{6.9}$$

and

$$Pr[\text{min degree of } G(P, r_n^{(k)}) \text{ is } k+1] \geq 1 - \frac{1}{s}. \tag{6.10}$$

Thus, if the radius satisfies Inequality 6.8 not only is the network $(k+1)$-connected but within distance $r_n^{(k)}$, each node will have $k+1$ neighbors with high probability as indicated by Inequalities 6.9 and 6.10.

6.5.3 Euclidean MST

Let t_n be the length of longest edge of the MST. Inequality 6.6 and 6.7 are also valid for the MST. In fact, we have that if

$$t_n \geq \sqrt{\frac{\ln n + \ln s}{n\pi}}, \tag{6.11}$$

then

$$Pr[\text{Length of longest edge of MST is} \leq t_n] \geq 1 - \frac{1}{s}. \tag{6.12}$$

This follows from the main result of Penrose (1997).

6.5.4 NNG and k-NNG

In the k-NNG, each point is connected to its kth nearest neighbor out of the points of the pointset P. The main result of Penrose (1997) gives a threshold value for the length $l_n^{(k)}$ of the longest edge of the k-NNG in the "toroidal" unit square. We have that if

$$l_n^{(k)} \geq \sqrt{\frac{\ln n + k \ln \ln n + \ln(k!) + \ln s}{n\pi}}, \tag{6.13}$$

then

$$Pr[\text{Longest edge of } k\text{-NNG is} \leq l_n^{(k)}] \geq 1 - \frac{1}{s}. \tag{6.14}$$

6.5.5 Delaunay triangulations

It has been proved by Li et al. (2003) that if the reachability radius of the nodes is chosen so as to satisfy Inequality 6.7, then with high probability the DT is the same as the localized DT.

Proposition 6.5.3 *Let d_n be the longest edge of the DT. Then*

$$Pr\,[d_n < r_n] \geq 1 - \frac{1}{n},$$

for

$$r_n \geq \sqrt{\frac{9 \ln n}{n\pi}}.$$

Proof. Let d_n be the random variable that denotes the length of the longest edge of the DT of the random pointset. If $d_n \geq d$, then there is a triangle in the triangulation at least one of whose edges is $\geq d$ and whose circumcircle contains no other points of the random pointset: note that the area of this circumcircle is at least $\pi d^2/4$. Since the DT of a set of n points has at most $3n$ triangles, we conclude that $Pr[d_n \geq d] \leq 3ne^{-n\pi d^2/4}$. If we put $1/t = 3ne^{-n\pi d^2/4}$, solve for d and substitute in the last inequality then we see that

$$Pr\left[d_n < \sqrt{\frac{4(\ln n + \ln t + \ln 3)}{n\pi}}\right] \geq 1 - \frac{1}{t}. \tag{6.15}$$

If we now put $s = n^8$ in Inequality 6.7, and $t = n$ in Inequality 6.15 then we see that d_n is smaller than r_n with probability at least $1 - \frac{1}{n}$. ■

6.5.5.1 Arboricity of Delaunay Triangulations

Given a graph G, the arboricity is the minimum number of edge-disjoint trees whose union is G. In the sequel consider the convex subdivision formed by the DT. Consider the following subgraphs of $G(P, r_n)$. In the Max distance left to right (MaxDLTR) tree each node is connected to a max distance neighbor to its right; in the Min distance left to right (MinDLTR) tree each node is connected to a min distance neighbor to its right; finally, in the Mid distance left to right (MidDLTR) tree each node is connected to a neighbor other than a Max or Min neighbor to its right (if it has one) else to the Max. It is easy to see that they are spanning trees of the unit disc graph. We can prove the following result (see Georgiou et al. (2005) for additional details).

Proposition 6.5.4 *Under the assumption that $r_n \geq \sqrt{\frac{9 \ln n}{n\pi}}$, the union of the trees MaxDLTR, MaxDLTR and MidDLTR is the DT of the pointset P with probability at least $1 - 1/n$, asymptotically in n.*

Proof. Assume $r_n \geq \sqrt{\frac{9 \ln n}{n\pi}}$ and consider the trees MinDLTR, MaxDLTR and MidDLTR. It is enough to show that the probability that any two among these trees have an edge in common is at most $1/n$, asymptotically in n.

First consider the trees MinDLTR, MaxDLTR. Consider a given edge e, say $e := \{u, v\}$ that is common to both trees MaxDLTR and MinDLTR. It follows that node u, say, has only one neighbor to its right, namely v. Since the reachability radius of the nodes is r_n, it follows from the definition of the tree MaxDLTR that v is the max distance neighbor of u and therefore the region determined by the semicircle centered at u and radius r_n (call this region S) contains exactly one point from the given pointset P. Hence,

$$\begin{aligned} Pr[e \text{ occurs in both trees}] &= n|S|e^{-n|S|} \\ &= \frac{n\pi r_n^2}{2} e^{-n\pi r_n^2/2} \\ &\leq \frac{n\pi}{n^{9/2}} \\ &= \frac{\pi}{n^{7/2}}. \end{aligned}$$

Since the DT has at most $3n$ edges it follows that the probability the two trees have an edge in common is at most $1/n^2$, asymptotically in n.

A similar proof will work for any other pair of trees. For example, for the trees MaxDLTR and MidDLTR if a given edge $e := \{u, v\}$ is common to both then the semicircle centered at u and radius r_n will have at most two points from the pointset. Therefore an upper bound on the probability that an edge is common can be obtained easily as in the previous analysis using the Poisson distribution. We leave the details as an Exercise to the reader.

Since the DT has $3n - 3$ edges, it turns out that with high probability these three trees contain all the edges of the DT with high probability, asymptotically in n. This completes the proof of Proposition 6.5.4. ■

6.6 Coverage and connectivity with directional sensors

In this section, we study the problem of providing full coverage and connectivity of a planar region with sensors. Consider n sensors in the interior of a unit square. The characteristics of the sensors are parametrized as a function of n. More specifically, let $\alpha := \alpha(n)$ be the coverage angle and $r := r(n)$ the reachability radius of the sensor, respectively. Let $p := p(n)$ be the probability that a given sensor is active, in which case $1 - p(n)$ is the probability that the sensor is inactive. We are interested in using sensors in order to cover every point of the given region, which for simplicity is assumed to be a unit square in the plane over which sensors are dropped, say from an airplane. The placement and orientation of the sensors is random and there is a reasonable probability that some sensors will be unable to communicate with any other sensors. Rather than achieve full connectivity it is required that a connected subnetwork exists that achieves full coverage. In particular, we are interested in the following problem. What relation must exist among the sensor parameters p, α, n in order to ensure the existence of a connected subnetwork of sensors that provides full coverage? The subsequent discussion closely follows the approach of Kranakis et al. (2004).

6.6.1 Covering circles with sensors

Consider a sensor at A. We would like that a circle of radius R fits inside the coverage range of A (Figure 6.37) and is visible from A with an angle of size $\alpha/4$. Let d be the distance of the sensor from the center K of this circle. Since the reachability radius of the sensor is r, the circle at K is entirely visible from the sensor at A if $d + R \leq r$. Since $d = R/\sin(\alpha/4)$ the above inequality is equivalent to $\frac{R}{\sin(\alpha/4)} + R \leq r$. It is now easy to see that the following result is true.

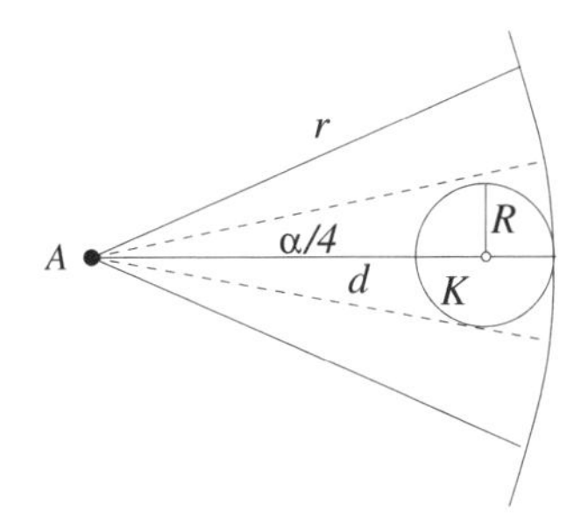

Figure 6.37 A circle of radius R within the coverage range of the sensor at A.

Lemma 6.6.1 *A circle of radius R may lie within the coverage range of sensor if and only if*

$$R \leq \frac{\sin(\alpha/4)}{1 + \sin(\alpha/4)} \cdot r. \tag{6.16}$$

Moreover, the probability that a given sensor at distance d from a given circle of radius R such that $d + R \leq r$ is active and covers the circle, is $\Theta(p\alpha)$. ■

6.6.2 Achieving coverage

We are interested in specifying conditions on the three main parameters $\alpha := \alpha(n)$, $r := r(n)$, $p := p(n)$ so that the active sensors in the unit square provide full coverage. Using Lemma 6.6.1, we are interested in finding sensors that with high probability cover a given

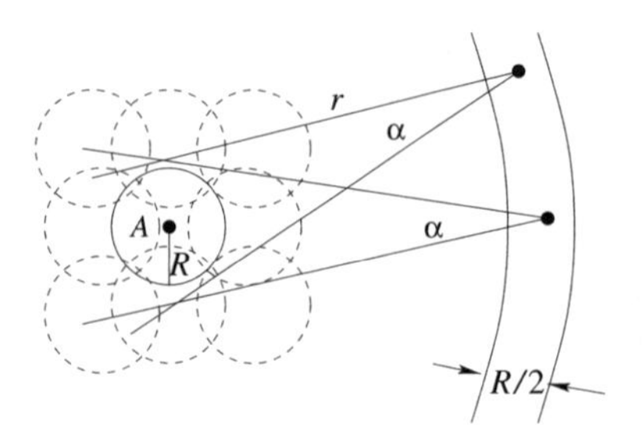

Figure 6.38 Covering a given circle at A from a sensor within the strip, with high probability.

set of circles of radius $R = \frac{\sin(\alpha/4)}{1+\sin(\alpha/4)} \cdot \Theta(r)$. As depicted in Figure 6.38, any sensor within a strip of thickness $R/2$ at distance $\Omega(r)$ can potentially cover a circle of radius R. The probability that a given sensor within this strip will cover fully such a circle is at least $\Omega(p\alpha)$. The expected number of sensors within the strip at distance $\Omega(r)$ from A is at least $\Theta(nrR)$. It follows that up to a constant

$$
\begin{aligned}
&Pr[\text{An active sensor in the strip covers the circle at } A] = \\
&1 - Pr[\text{No active sensor in the strip covers the circle at } A] \geq \\
&1 - \prod_{i=1}^{\Theta(nrR)} (1 - \Omega(p\alpha)) \geq \\
&1 - e^{-\Theta(np\alpha rR)}.
\end{aligned}
$$

It is now easy to see that in order to ensure that $1 - e^{-\Theta(np\alpha rR)} \geq 1 - n^{-O(1)}$ it is enough to assume that $p = \Omega\left(\frac{\log n}{n\alpha rR}\right)$.

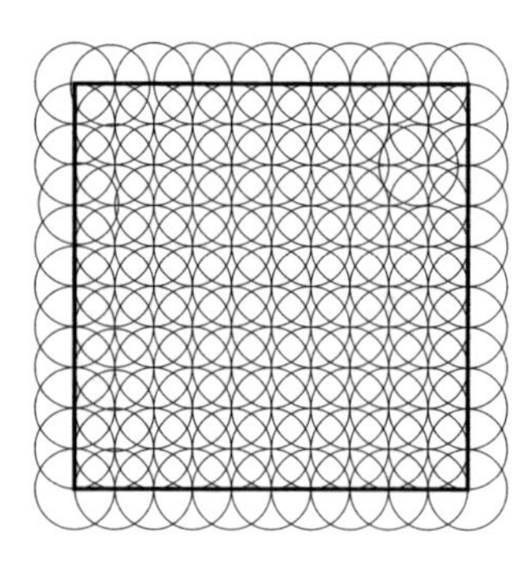

Figure 6.39 Decomposition of the unit square region (square with thick perimeter) with smaller overlapping circles of equal radius.

To obtain coverage of the whole unit square we decompose the given region into circular overlapping subregions $C_1, C_2, \ldots, C_N$ each of radius R, where

$$N = \frac{1}{R^2} = \frac{1}{\Theta(r^2 \sin^2(\alpha/4)/(1+\sin(\alpha/4))^2)} \approx \Theta\left(\frac{1}{r^2 \sin^2(\alpha/4)}\right)$$

(Figure 6.39). This overlap is necessary in order to guarantee that if each such circle is fully covered by a sensor, the whole unit square will then be also covered by the sensor network. Each circular subregion can be covered by a sensor with high probability. Let E_i be the event that C_i can be covered by a sensor. It follows that

$$\begin{aligned}
Pr[\text{Sensors provide full coverage}] &\geq Pr[\text{Each } C_i \text{ is covered by a sensor}] \\
&= Pr\left[\bigcap_{i=1}^{N} E_i\right] \\
&= 1 - Pr\left[\bigcup_{i=1}^{N} \overline{E_i}\right] \\
&\geq 1 - \sum_{i=1}^{N} Pr[\overline{E_i}] \\
&\geq 1 - \sum_{i=1}^{N} e^{-\Theta(np\alpha r R)} \\
&\geq 1 - N e^{-\Theta(np\alpha r R)} \\
&\geq 1 - \frac{1}{\Theta(r^2 \sin^2(\alpha/4))} e^{-\Theta(np\alpha r R)}.
\end{aligned}$$

It is now easy to see that in order to ensure that $1 - \frac{1}{\Theta(r^2 \sin^2(\alpha/4))} e^{-\Theta(np\alpha r R)} \geq 1 - n^{-O(1)}$ it is enough to assume that $p(n) = \Omega\left(\frac{\log(n/r^2 \sin^2(\alpha/4))}{nr^2 \alpha \sin(\alpha/4)}\right)$. We have proven the following theorem:

Theorem 6.6.2 *Suppose that n sensors with coverage angle $\alpha(n)$ and reachability radius $r(n)$ are thrown randomly and independently in the interior of a unit square. If the probability $p(n)$ that a given sensor is active satisfies $p(n) = \Omega\left(\frac{\log(n/r(n)^2 \sin^2(\alpha(n)/4))}{nr(n)^2 \alpha(n) \sin(\alpha(n)/4)}\right)$ then the probability the sensors provide full coverage of the unit square is at least $1 - n^{-O(1)}$.* ■

As a special case we also obtain the result of Shakkottai et al. (2003) for $\alpha = 2\pi$. If $p(n) \geq \Omega\left(\frac{\log n}{nr(n)^2}\right)$ then $Pr[\text{Sensors provide full coverage of the unit square}] \geq 1 - n^{-O(1)}$.

It is also interesting to look at the reverse problem: if coverage is assured with high probability asymptotically in n, what is a lower bound on the probability $p(n)$ that a sensor

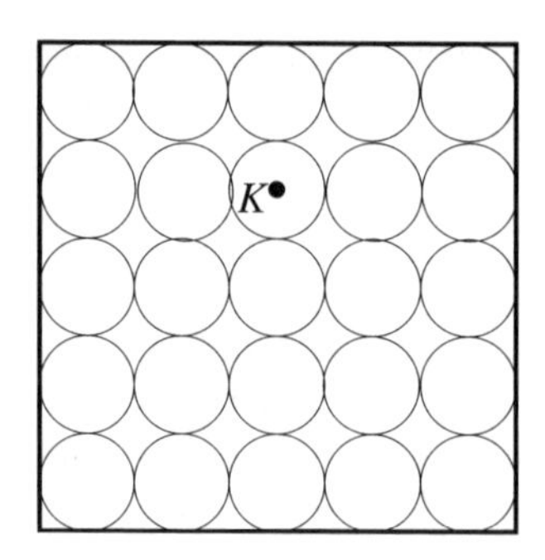

Figure 6.40 Partitioning of the unit square region (square with thick perimeter) with smaller circles of equal radius.

is active? Partition the square into N pairwise disjoint circles each of radius r. If any given of these circles has no active sensor (say the circle centered at K), then K cannot be covered by any sensor lying in a neighboring circle (Figure 6.40). Consider the event E: the center of a circle is covered by an active sensor. It is clear that $p\frac{\alpha}{2} \leq Pr[E] \leq p\alpha$. There are approximately $N = \Theta(1/r^2)$ such circles. It follows that

$$
\begin{aligned}
Pr[\text{Unit square is covered}] &\leq Pr[\text{Centers of all the circles are covered}] \\
&= \prod_{i=1}^{N} Pr[E] \\
&= (1 - Pr[\bar{E}])^N \\
&\leq (1 - (1 - \Theta(p\alpha))^{\Theta(nr^2)})^N \\
&\leq \exp\left(-(1 - \Theta(p\alpha))^{\Theta(nr^2)} N\right) \\
&= \exp\left(-(1 - \Theta(p\alpha))^{\Theta(nr^2)} / \Theta(r^2)\right)
\end{aligned}
$$

It follows from the inequalities above that if $Pr[\text{Unit square is covered}] \geq 1 - 1/n$ then also $\exp\left(-(1 - \Theta(p\alpha))^{\Theta(nr^2)} / \Theta(r^2)\right) \geq 1 - 1/n$, which in turn implies that $\frac{(1-\Theta(p\alpha))^{nr^2}}{r^2} \leq \log\left(\frac{n}{n-1}\right) \leq \frac{1}{n-1}$. It follows that $p = \Omega\left(\frac{\log(n/r^2)}{nr^2\alpha}\right)$. We have proved the following theorem:

Theorem 6.6.3 *Suppose that n sensors with coverage angle $\alpha(n)$ and reachability radius $r(n)$ are thrown randomly and independently in the interior of a unit square. If*

$$Pr[\textit{Sensors provide full coverage of the unit square}] \geq 1 - 1/n,$$

then the probability $p(n)$ that a given sensor is active satisfies $p(n) = \Omega\left(\frac{\log(n/r(n)^2)}{nr(n)^2\alpha(n)}\right)$, asymptotically in n. ■

6.7 Bibliographic comments

The basic theme of this chapter concerns the use of location awareness (or simply, geometric coordinates) for enabling ad hoc network communication. Interestingly enough, the fundamental concept of geographic proximity, described in Section 6.1 is based on the two papers by Toussaint (1980), O'Rourke and Toussaint (1997), was developed not in the context of ad hoc networks but rather as a tool for classifying nodes in statistical pattern recognition. First, Toussaint (1980) describes the RNG and its relation to MST and DT and second O'Rourke and Toussaint (1997) completes the description with the NNG and GG. Similarly, the GG that appeared in citegabriel, provided a statistical approach to variational analysis.

UDGs have been used early on in order to model broadcast networks (Kann (1994)), facility location (Wang and Kuo (1988)) and frequency assignment (Hale (1980)), while Clark et al. (1990) looks at NP-hard problems in such graphs. A useful reference is also Marathe et al. (1995). The recent emergence of ad hoc networks has given such a tremendous impetus to the ideas preprocessing of UDGs for the construction of spanners (Section 6.2) and efficient information dissemination (Section 6.3) that it would be worthwhile to write a separate book about it. Nevertheless, there are two seminal papers by Kranakis et al. (1999) which develops the concepts of compass routing and face routing for *local* routing and Bose et al. (2001) which builds the GG as a as a *distributed, local, planar* spanner of a UDG. Sections 6.2 and 6.3 discuss additional algorithms on the same themes but it is worth mentioning that several problems remain open when transmission ranges of the sensors are not identical or when the setup is three dimensional. The class of UDGs with irregular transmission ranges was first introduced by Barrière et al. (2003b) in order to propose robust position-based routing.

The issue of locality in distributed computations was initiated with the work of Linial (1992). For further studies on what can be computed locally (including a characterization of regular graphs that can be weakly colored locally), the reader is referred to the work of Naor and Stockmeyer (1995).

The next section addresses the localization problem of how a sensor can determine its geographic position. This is, of course, possible when the sensor is GPS enabled but not so if its view is obstructed. Part of Section 6.4 is based on Barbeau et al. (2004). For additional information on building an efficient infrastructure in ad hoc networks, the reader may wish to consult Mauve et al. (2001), the forthcoming survey article Barbeau et al. (2007) and the book Parkinson and (Eds.). Additional literature is related to the concepts of anchors (see Wattenhofer et al. (2005)) and beacons see Ratnasamy et al. (2003)) and Capkun et al. (2002) on establishing *relative* coordinate systems.

Comprehension of the material in Section 6.5 and Section 6.6 depends heavily on a reasonable understanding of probability theory. Chapter 2 provides such an introduction but the reader could also consult the beautiful books of Ross (1996, 2nd edition) as well as the more advanced Ross (2002). Related to this section is the book by Penrose (2003) on random geometric graphs (as well as the related papers Penrose (1999), Penrose (1997)). Section 6.6 is based on Kranakis et al. (2004) which in turn is an extension of the work of Shakkottai et al. (2003).

6.8 Exercises

1. Prove that the proximity graph GG defined in Section 6.1.1 is planar without making use of Theorem 6.1.2.

2. Complete the remaining details of the proof of Theorem 6.1.2.

   (a) Prove that $NNG \subset MST$.
   (b) Prove that $GG \subset DT$.
   (c) Show with examples that all the inclusions in the statement of Theorem 6.1.2 are proper.

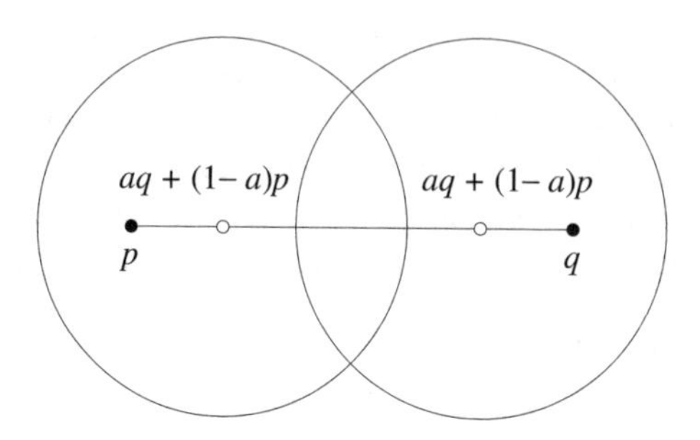

The α-lune defined by points p, q and the parameter α.

3. An extension of the GG is the α-Gabriel Graph (α-GG), which subsumes both the GG and the RNG. Assume $1/2 \leq a \leq 1$. The set of edges of the α-GG is defined by $(p, q) \in E \Leftrightarrow$ the intersection of the discs $D(p(1-a) + aq, ad(p, q))$ and $D(q(1-a) + ap, ad(p, q))$ does not contain any point in the pointset P.

   (a) Show that $1 - \text{GG} = \text{GG}$ and $(1/2) - \text{GG} = \text{RNG}$.
   (b) Extend Theorem 6.1.2 by showing that $RNG \subset \alpha - GG \subset GG$.

4. Which of the graphs MST, NNG, RNG, GG and DT have bounded degree and which do not and why?

5. Prove that the half-space proximal graph is not necessarily planar (Figure 6.17) but it has bounded degree at most 6 (Figure 6.16).

6. Show that in a random UDG, whereby the points of the pointset P are uniformly distributed over a given convex region, compass routing can reach the destination with high probability.

7. Show that in the planar graph depicted in Figure 6.20 compass routing fails to find the destination.

8. Indicate the faces traversed in Figure 6.21 when using face routing.

   (a) Is the route determined by face routing unique?

(b) How does face routing work when the line segment from source s to destination t does not cross the planar graph?

9. In this exercise, we are referring to the $3/2 - NA(k)$ algorithm.

   (a) Prove Theorem 6.4.3.

   (b) What is the size of the memory required by nodes when using the $3/2 - NA(k)$ algorithm?

10. This exercise refers to coverage and connectivity of random UDGs.

    (a) What is the value of s_n derived from Equation 6.5?

    (b) Use the generalization of the coupon collector's problem stated in Exercise 3.11 of Motwani and Raghavan (1995) in order to derive a sufficient condition for k coverage.

11. Complete the details of the proof of Proposition 6.5.4 by examining the trees MaxDLTR and MidDLTR.

12. Consider the trees in Proposition 6.5.4. Show that for any pair among these trees, the expected number of common edges is a constant independent of n. **Hint:** this follows easily from the well-known identity $E[X] = \sum_k Pr[X > k]$, where X is the random variable that counts the number of edges common to the two trees.

13. Complete the details of the proof of Theorem 6.4.3.

14. In the sensor network below the sensors have range 3 units and each subsquare has side equal to 1 unit.

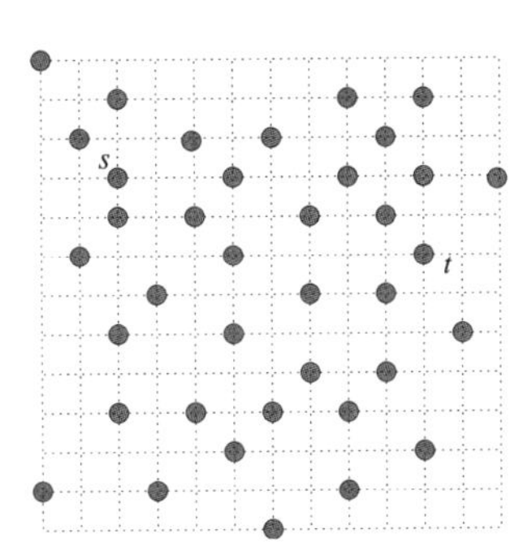

Draw the path traversed starting from node s and ending at node t when using the greedy routing algorithm specified.

(a) If current node is c, move to c's neighbor forming smallest angle with line ct.

(b) If current node is c, move to c's neighbor towards t that is furthest from c.

(c) If current node is c, move to c's neighbor towards t that is closest to c.

(d) Which algorithm would perform best if minimizing the number of hops was the primary concern?

15. The two sensors A, B depicted in the pictures below are transmitting synchronously and at the same time, have reachability range 1, and are at distance $0 < d < 1$ from each other.

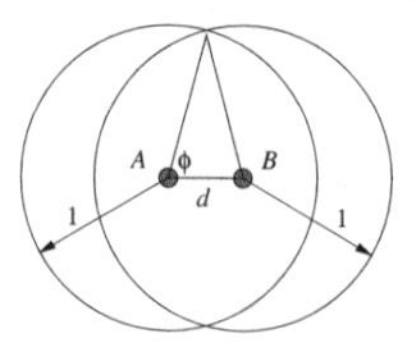

(a) The *hidden* sensor problem occurs when two sensors that lie outside the range of each other transmit at the same time to another sensor. Shade the region(s) containing hidden sensors.

(b) The *exposed* sensor problem refers to the inability of a sensor that is blocked because of the transmission by a nearby transmitting sensor to transmit to another sensor. Shade the region(s) containing exposed sensors.

(c) Compute the area of the region in each case above.

16. In poisson graphs there is a constant $\lambda > 0$ such that the probability distribution of the degree sequence of the network satisfies

$$Pr[D = k] = e^{-\lambda} \cdot \frac{\lambda^k}{k!}. \tag{6.17}$$

Show that $E[D] = Var[D] = \lambda$.

17. In Erdös-Renyi graphs (see Erdös and Renyi (1959)) there is a p such that a random node is linked with another node with probability p and is not linked with probability $1 - p$. Show that the degree distribution of a random node satisfies

$$Pr[D = k] = \binom{N-1}{k} p^k (1-p)^{N-1-k} \tag{6.18}$$

and also $E[D] = (N-1)p$ and $Var[D] = (N-1)p(1-p)$.

18. In the Waxman model (see Waxman (1988)), a set of N nodes are distributed randomly and independently on an $L \times L$ square. If $d(u, v)$ denotes the Euclidean distance between nodes u, v then they are joined by a link with probability

$$P(u, v) = \beta e^{-\frac{d(u,v)}{\alpha L}}, \tag{6.19}$$

where α, β are real numbers such that $0 < \alpha, \beta \leq 1$. This model is clearly inspired from and is also similar to the Erdös-Renyi model except for the fact that the probability of existence of a link between two nodes is inversely proportional to their distance. The parameter α tunes the ratio between short and long distance edges, while β controls the average degree of the network.

19. Explain and clarify similarities and differences between the random UDG model and the three random models presented in Exercises 16, 17 and 18.

20. (⋆) n omnidirectional antennas with reachability radius $r(n)$ are dropped randomly and independently with the uniform distribution over the interior of a unit length segment. Show that asymptotically in n, as $n \to \infty$,

$$Pr[\text{network provides coverage}] \to \begin{cases} 1 & \text{if } r(n) \geq \frac{\ln n}{n} + \frac{\ln \ln n}{n} \\ 0 & \text{if } r(n) \leq \frac{\ln n}{n} - \frac{\ln \ln n}{n}. \end{cases}$$

21. (⋆) As with Exercise 20, assume that the n points are generated using a Poisson process with rate of arrival n. Let the n points be denoted $p_1, p_2, \ldots, p_n$ and $X_i = p_i - p_{i-1}$ be the distance between points p_i and p_{i-1}.

    (a) Show that a necessary and sufficient condition that the sensors provide coverage and k-connectivity is that for all i, $X_{i-k} + X_{i-k+1} + \cdots + X_i \leq r(n)$.

    (b) The probability that $X_i > r$ is $\exp(-nr)$. Calculate the probability that $X_{i-k} + X_{i-k+1} + \cdots + X_i > r$. **Hint:** Prove by induction on k that

$$Pr[X_{i-k} + X_{i-k+1} + \cdots + X_i > t] = \frac{n^k t^k}{k!} e^{-nt}.$$

22. (⋆) As in Exercises 20 and 21, n omnidirectional antennas with reachability radius $r(n)$ are dropped randomly and independently with the uniform distribution over the interior of a unit length segment. Show that asymptotically in n,

    (a) as $n \to \infty$, if $r(n) \leq \frac{(k+1)\ln n - \ln \ln n}{n}$

$$Pr[\text{network provides coverage/}k\text{-connectivity}] \leq n^{-1/k},$$

    (b) and asymptotically in n, as $n \to \infty$, if $r(n) \geq \frac{(k+1)\ln n}{n}$

$$Pr[\text{network provides coverage/}k\text{-connectivity}] \geq e^{-1/k \ln n}.$$

23. (⋆) Conclude by proving the following result. n omnidirectional antennas with reachability radius $r(n)$ are dropped randomly and independently with the uniform distribution over the interior of a unit length segment. Then for $k = o(\ln n)$, asymptotically in n, as $n \to \infty$, $Pr[\text{network provides coverage/}k\text{-connectivity}] \to$

$$\begin{cases} 1 & \text{if } r(n) \geq \frac{(k+1)\ln n}{n} \\ 0 & \text{if } r(n) \leq \frac{(k+1)\ln n}{n} - \frac{\ln \ln n}{n}. \end{cases}$$

24. In Section 6.6, sensors are dropped from an airplane randomly and independently with uniform distribution. To ensure that a given subregion of the unit square contains enough sensors with high probability. In particular, use Chernoff bounds to show that with high probability the expected number of sensors that drop within any given circle C is proportional to the area of the circle.

25. (⋆) Use the approach of Section 6.6 to solve the coverage problem for a unit square region when the sensors can be placed only on its perimeter. In this case, assume that the reachability radius of the sensors is equal to 1. Suppose that n sensors with

coverage angle $\alpha(n)$ and reachability radius 1 are thrown randomly and independently on the perimeter of a unit square. If the probability $p(n)$ that a given sensor is active satisfies $p(n) = \Omega\left(\frac{\log(n/\sin^2(\alpha(n)/4))}{n\alpha(n)\sin(\alpha(n)/4)}\right)$, then

$$Pr[\text{Sensors provide full coverage of the unit square}] \geq 1 - n^{-O(1)}. \quad (6.20)$$

Furthermore, if Inequality 6.20 is valid then $p(n) = \Omega\left(\frac{\log n}{n\alpha(n)}\right)$.

26. The *width* of an inner face of a planar graph is the max w such that the circle with diameter w lies in the interior of the face and the width of a planar graph is the max width of an inner face. Let the random variable w_n be the width of the graph. Prove the following result for any planar graph formed on the pointset P. Let the random variable w_n denote the width of the planar graph. Then

$$Pr\left[w_n < r_n\right] \geq 1 - \frac{1}{n},$$

provided $r_n \geq \sqrt{\frac{9\ln n}{n\pi}}$.

7

AD HOC NETWORK SECURITY

> ...applied mathematics is similar to warfare: sometimes a defeat is more valuable than a victory because it helps us to realize the inadequacy of our arms or of our strategy.
> Rényi (1984)

The renowned mathematician Alfred Rényi in his magnificent imaginary dialogue on the application of mathematics (see Rényi (1984, Vol I, page 260)) has Archimedes, the ancient Greek world's applied mathematician par excellence and Heron, the king of Syracuse, argue about the value of applications of mathematics, while the mighty Roman army is outside the gates of the city. The quotation above is Heron's reply to Archimedes' claim on the importance of constructing even crude mathematical models because they can help us understand a practical situation better, and think over all logical possibilities. One can argue that network security from its very inception has resembled warfare with malicious intruders attacking innocent victims but at the same time with the victims learning from their mistakes in order to enhance security.

Wireless local area networks (LANs) have witnessed an increasing popularity and their spread to locations like airports, cafes, businesses and university campuses has broadened the *unknown boundaries* of existing networked systems. This rapid deployment coupled with the inherent vulnerabilities of deployed protocols exacerbated the security difficulties. In addition to exposure to typical network threats, wireless networks present unique challenges due to the fact that the communication is open access to anyone within range (Table 7.1). In addition, a risk analysis of wireless threats is required (see Barbeau et al. (2006)) in order to evaluate properly the trade-offs between threats and countermeasures.

Wireless network security certainly involves traditional threats, like passive and active attacks, but it must also address the new settings pertaining to mobility and the pervasiveness of the ad hoc environment, like weaknesses in mobile devices and users, computer-enabled vehicles and software components. Some examples can clarify these issues. If mobility is to be supported, then additional issues have to be taken into account. For example, there is the

Principles of Ad hoc Networking Michel Barbeau and Evangelos Kranakis

Table 7.1 Wireless security challenges.

Network type	Challenge
Wireless	Open medium
Mobility	Handover implies change of security parameters
Ad hoc	Infrastructure based security not applicable
Sensor	In-network processing

need to reconfigure security parameters when a node makes a handover in WiMAX/802.16 networks because base station IDs are included in the formation of authorization keys (IEEE et al. (2006)). A handover implies a change of base station and the need to renegotiate all security parameters. In wireless networks that can operate in the ad hoc mode, security is even more challenging because of the difficulty to apply infrastructure based tools such as public-key infrastructure (PKI). Ad hoc nodes can be used to support the communications for a network of sensors, which also present their own challenges. Indeed, end to end security needs special consideration because sensors networks perform data aggregation. This form of data processing inside the network requires inspection of the payload of data packets (Karlof and Wagner (2003)).

This chapter is dedicated to the study of security problems in ad hoc networks. The fundamental issues of authentication are first studied in detail in Section 7.1. Physical layer attacks are addressed in Section 7.2. Section 7.3 discusses two important application protocols, WiFi/802.11 and ZigBee, . Section 7.4 addresses Biometrics-based key establishment. Issues in routing and broadcast security are examined in Sections 7.5 and 7.6. Finally, Sections 7.7 and 7.8 address location based security methods for ad hoc and directional sensor networked systems.

7.1 Authentication techniques

Security protocols are based on communication exchange rules between participating hosts that must be followed in various security applications. Authentication protocols are particularly subtle since even a minor change in the rules or even rearrangement of commands can lead to misunderstandings and even unpleasant consequences. This pertains to generic techniques leading to better understanding of the fundamental security issues in ad hoc network authentication.

7.1.1 Signatures, authentication and hashing

Signatures have long been used as proof of authenticity of a document, or reaching an agreement. Important features of signatures include (1) *unforgeability:* proof that the signer (and nobody else) signed the document, (2) *authenticity:* convincing the recipient of the document's authenticity, (3) *unreusability:* signature "belongs" to the document being signed and cannot be "moved" elsewhere, (4) *unalterability:* the document cannot be changed after signing and still obtain the same signature and (5) *unrepudiatability:* signer cannot later claim (s)he did not sign the document.

A *digital signature* scheme is a method for signing a message in electronic form that can be transmitted over a computer network. Although conventional signatures are physically attached to the document, this is not the case in digital signatures where the algorithm used must somehow bind the signature to the message. A conventional signature is verified by comparing it to an authentic one. However, forging is easy in electronic communication because of the ease of copying. Since electronic copies of digital messages are identical to the original, care is needed so that electronic signatures are not maliciously copied. Nevertheless, digital signatures can be verified by anyone using a public verification algorithm.

Formally, a signature scheme consists of a finite set P of possible messages, a finite set A of possible signatures, and finite set K of possible keys as well as a signing algorithm (a function $P \to A$) and a verification algorithm (a function $P \times A \to \{true, false\}$). The *signing algorithm* associates to each $k \in K$ a function $Sig_k : P \to A$ (which should be easy to sign but computationally infeasible to forge) in such a way that the *verification algorithm* is satisfied, that is, for $k \in K$, the function $Ver_k : P \times A \to \{true, \ false\}$ must satisfy

$$Ver_k(x, y) = \begin{cases} true & \text{if } y = Sig_k(x) \\ false & \text{if } y \neq Sig_k(x). \end{cases}$$

Two widely used signature schemes are based on variations of the well-known RSA cryptographic system and the ElGamal signature scheme (on which the Digital Signature Standard is based), which we outline in the sequel.

The RSA signature scheme is based on selecting an integer $n = pq$ that is the product of two distinct primes p and q, as well as two integers e, d such that $ed \equiv 1 \bmod \phi(n)$, where $\phi(n)$ is the Euler totient function. The integers n, e are made public while p, q, d are kept private. A message M is signed using the function $Sig(M) \equiv M^d \bmod n$ while verification is based on $Ver(M, N) = true \Leftrightarrow M \equiv N^e \bmod n$.

Another one is the ElGamal scheme, which was specifically designed for signing, as opposed to RSA, which was designed for encryption. The scheme is based on selecting integers p, α, β, a such that p is prime, a is a generator of the multiplicative group Z_p^* and $\beta \equiv \alpha^a \bmod p$. The integers p, α, β are made public while a is kept secret. The signing procedure is based on selecting a random $k \in Z_{p-1}^*$ and defining $Sig(M, k) = (\gamma, \delta)$, where $\gamma = \alpha^k \bmod p$ and $\delta = (M - a\gamma)k^{-1} \bmod (p - 1)$ while the verification procedure is based on $Ver(M, \gamma, \delta) = true \Leftrightarrow \beta^\gamma \gamma^\delta \equiv \alpha^M \bmod p$.

It is known that the ElGamal scheme is no more secure than the discrete logarithm problem. For many applications (like smart cards, mobile and cellular telephony), a shorter signature is required. Several methods have been proposed and many more are still under investigation based on elliptic curve implementations of ElGamal because of their processing efficiency.

7.1.1.1 Message authentication

Message authentication is concerned with protecting the integrity of a message, validating the identity of the originator, and non-repudiation of the origin (dispute resolution) and is the electronic equivalent of a signature on a message. In general, it is required that different messages should have different authenticators, and that authentication should be easy to assign and verify.

In order to authenticate a message, an *authenticator, signature*, or *message authentication code* (abbreviated by MAC) is sent along with the message. The MAC is generated via some algorithm that depends on both the message and some (private or public) key known only to the sender and receiver. Typically the message may be of any length, but more often the MAC is of fixed size, requiring the use of a *hash function* to condense the message to the required size if this is not achieved by the authentication scheme.

Authentication is a rather complex topic. However, all known algorithms are based on variants of three fundamental methods: *three-way handshake*, *trusted third party*, and *public key* .

In a three-way handshake, a client C and a server S want to establish a session key K_{session}. They already share handshake keys K_{CH} and K_{SH} and also generate random numbers R_C and R_S, respectively. The four steps of the protocol are as follows.

$$
\begin{array}{ll}
\text{Step 1. } C \rightarrow S: & ID_C, E_{K_{CH}}(R_C); \\
\text{Step 2. } S \rightarrow C: & E_{K_{SH}}(f(R_C)), E_{K_{SH}}(R_S); \\
\text{Step 3. } C \rightarrow S: & E_{K_{CH}}(f(R_S)); \\
\text{Step 4. } S \rightarrow C: & E_{K_{SH}}(K_{\text{session}});
\end{array}
$$

Note that the terms $f(R_C)$, $f(R_S)$ in Steps 2 and 3 are challenge responses, respectively.

In the trusted third party method, two clients A, B want to establish a session key K_{session} via a third party server S. The protocol uses a timestamp T, a *lifetime* L (for the validity of the session key) and keys K_A, K_B that A and B share with S. The four steps of the protocol are as follows.

$$
\begin{array}{ll}
\text{Step 1. } A \rightarrow S: & ID_A, ID_B; \\
\text{Step 2. } S \rightarrow A: & E_{K_A}(T, L, K_{\text{session}}, ID_B), E_{K_B}(T, L, K_{\text{session}}, ID_A); \\
\text{Step 3. } A \rightarrow B: & E_{K_B}(T, L, K_{\text{session}}, ID_A), E_{K_{\text{session}}}(T, ID_A); \\
\text{Step 4. } B \rightarrow A: & E_{K_{\text{session}}}(T + 1);
\end{array}
$$

Note that in Step 3, A merely forwards to B the message $E_{K_B}(T, L, K_{\text{session}}, ID_A)$ and the term $T + 1$ in Step 4 is to indicate that B's reply must arrive within a time unit from the time it was sent, which also cannot exceed the lifetime L.

Finally, the public-key-based scheme is as follows.

$$
\begin{array}{ll}
\text{Step 1. } A \rightarrow B: & E_{\text{public}_B}(R_A); \\
\text{Step 2. } B \rightarrow A: & f(R_A);
\end{array}
$$

In Step 1, A selects a random number R_A that it encrypts with B's public key prior to sending it to B. Then host B replies with a challenge response $f(R_A)$.

7.1.1.2 Hashing

Hash functions are used to condense an arbitrary length message to a fixed size, usually for subsequent signature by a digital signature algorithm. *Hashing* generally applies to the message and does not generally use keys. It appears in the literature with many different names, for example, message digest (MD), Contraction Function, Compression Function, Fingerprint, Cryptographic Checksum, and Message Integrity Check (MIC). Hashing takes a variable-length input and converts it to a fixed-length (generally small) output, called the hash value. A MAC is a one-way hash function whose hash value depends both on the message content as well as a key. A hash function H will always have collisions. This is because the space of hash values is smaller than the message space. Therefore at issue here is, given a hash value y, how difficult it is to find a message x such that $H(x) = y$.

Good cryptographic hash functions h should have the following properties: (1) h should destroy all homomorphic structures, for example, be unable to compute the hash value of two messages combined given their individual hash values, (2) h should be computed on the entire message, (3) h should be a one-way function so that messages are not disclosed by their has values, (4) it should be computationally infeasible, given a message and its hash value, to compute another message with the same hash value, and it should resist *birthday attacks* (finding any two messages with the same hash value). Thus, the length of the hash should be large enough to resist birthday attacks.

The design structure of hash functions is very complex and the basic principles were first proposed by Merkle (1980). Hashing involves the repeated use of a compression function F , which accepts two inputs: one from the previous step (called *chaining value* or CV) and a block of the message. In general, the compression function must be carefully selected to be collision resistant, but it is a very difficult problem to ensure that the resulting (iterated) hash function is collision resistant. Certainly, since collisions will always exist, it is the computational infeasibility that matters. Several hash functions are known in literature, including the family of MDs known by the acronyms MD2, MD4, and MD5, as well Secure Hash Algorithm (SHA) and HMAC.

The collision resistance property can also be analyzed mathematically as follows. Given a hash function $h : X \to Z$ with $m \geq 2n$, where $|X| := m$ and $|Z| := n$, what is the probability of collision? For example, if $X =$ set of all humans, $h(x) =$ birthday of x, and $Z =$ the 365 days of a year. In general, we must compute $Pr[\exists x \neq x'(h(x) = h(x'))]$, which is equal to $1 - Pr[\forall x \neq x'(h(x) \neq h(x'))]$. Let us make the natural assumption that the hash is *uniformly distributed*,that is, $|h^{-1}(z)| \approx m/n$ for all $z \in Z$. What is the probability that k elements $z_1, \ldots, z_k$ of Z chosen at random are pairwise distinct? First of all, observe that

$$Pr[\text{no collision}] = \frac{\frac{n!}{(n-k)!}}{n^k} = \frac{n(n-1)\cdots(n-(k-1))}{n^k} = \prod_{i=1}^{k-1}\left(1 - \frac{i}{n}\right).$$

Therefore

$$\begin{aligned} Pr[\text{no collision}] &\approx \prod_{i=1}^{k-1} e^{-i/n} \\ &= e^{(1+2+\cdots+k-1)/n} \\ &= e^{-\frac{k(k-1)}{2n}}, \end{aligned}$$

where we used the approximation $e^{-t} \approx 1 - t$, for small t. It follows that the probability, say p, of at least one collision is

$$p \approx 1 - e^{-\frac{k(k-1)}{2n}}.$$

Consequently, it follows that

$$e^{-\frac{k(k-1)}{2n}} \approx 1 - p.$$

$$-\frac{k(k-1)}{2n} \approx \ln(1-p), \text{ that is } k^2 - k \approx 2n \ln\left(\frac{1}{1-p}\right)$$

Ignoring the term $-k$, we obtain

$$k \approx \sqrt{2n \ln\left(\frac{1}{1-p}\right)},$$

which provides the values of k as a function of the collision probability p. Thus we see that birthday attacks impose lower bounds on the size of a "secure" MD.

7.1.2 Signatures in networking

When signatures are being used in a networked environment, it is necessary to employ shorter signatures in order to maintain a reasonable communication overhead. Moreover, in some instances one must take into account real-time requirements, such as robustness to packet loss and robustness to denial-of-service (DoS) attacks. In general, the following requirements must be satisfied. (1) Fast generation of authentication information. and fast verification by the receiver. (2) Instant authentication by sender and receiver without unnecessary packet buffering. (3) Robustness to packet loss. (4) Scalability whereby authentication should be independent of the number of users. A variety of protocols is being employed, like *commitment protocols* (e.g. *certificate based*, *cookie based*, and *challenge response*), *self-authenticating values*, *one-way chains*, and *Merkle hash trees*.

The idea used in committing to a single value is illustrated when Alice has a secret s that she does not want to reveal now (e.g. a prediction that she can reveal at a later time). Alice can lock the secret s as follows.

1. Alice takes a one-way function F.
2. Computes $c := F(s)$.
3. Publishes c.

F is a well-known one-way function. but it is very unlikely that someone can compute s from F and c.

Self-authenticating values are used when there are instances where short values are needed that a receiver can authenticate efficiently. These values are the same length as a MAC. Rather than using a separate MAC for each value, values are used that a receiver can authenticate without any additional information. For this purpose, commitment protocols are used: if the receiver obtains a commitment c to a secret s over an authenticated channel, the secret s is a self-authenticating value because the receiver can instantly authenticate it by verifying $c = F(s)$.

Certificate based commitments are an advanced form of authentication, and at this time they are in their infancy with respect to compatibility and ease of use. In using them, one obtains a certificate from a certificate issuing authority and selects a directory one wants to protect. It then enables client certificates for this resource. A cookie, however, is a token containing information about the state of a transaction in a network and are used for interactions between Web browsers and clients. They also minimize storage requirements of servers and in a sense put the burden on the client. Cookies consist of: name, value, associated value, expiration, domain (cookie is intended for), path (to restrict dissemination), secure field (if set cookie will be sent only over secure channel).

7.1.2.1 Lamport one-time signatures

The Lamport one-time signature scheme (see Lamport (1979)) uses a set of $2k$ random numbers $Y = \{y_1, y'_1, \ldots, y_k, y'_k\}$. For any message $M = M_1 \cdots M_k$ (given as a sequence of bits), select a k-element subset of Y in such a way that

$$\text{selection} = \begin{cases} y_i & \text{if } M_i = 0 \\ y'_i & \text{if } M_i = 1 \end{cases}$$

which gives the signature of a message. If $M \neq M'$ then a different k-element subset will be selected.

7.1.2.2 The Bos–Chaum scheme

The Bos–Chaum scheme (introduced in Bos and Chaum (1992)) improves on the size of Lamport signatures by using the idea of set systems. A *Sperner system* (see Bollobas (1986)) on a set B is a set $\mathcal{B}$ of subsets of B such that $\forall B_1, B_2 \in \mathcal{B}(B_1 \subseteq B_2 \Leftrightarrow B_1 = B_2)$. Observe that given k, the set of all k-element subsets of a given set is a Sperner system. Also for a given set of size $2n$, the maximum possible size of a Sperner system is $\binom{2n}{n}$ (Exercise 11).

The signature scheme is as follows. To sign a k-bit message choose n sufficiently large such that $2^k \leq \binom{2n}{n}$, where $n =$ length of signature. Let B be a set of size $2n$ and let $\mathcal{B}$ be the set of n-element subsets of B. Let $\phi : \{0, 1\}^k \rightarrow \mathcal{B}$ be a public injective function (Exercise 13) and $f : Y \rightarrow Z$ a one-way function. Choose $y_1, y_2, \ldots, y_{2n}$ random and put $z_i = f(y_i)$, for $i = 1, 2, \ldots, 2n$. The y_is are secret and z_is are public. For the message $M = M_1 M_2 \cdots M_k$ (given as a sequence of bits) define

$$Sig(M_1 M_2 \cdots M_k) = \{y_i : i \in \phi(M_1 M_2 \cdots M_k)\}$$

and if $a_1, \ldots, a_n$ is the signature then verification can be done using

$$Ver(M, a_1, \ldots, a_n) = true \Leftrightarrow \{f(a_i) : 1 \leq i \leq n\} = \{z_i : i \in \phi(M_1 \cdots M_k)\}.$$

It can be shown using Stirling's formula on $n!$ (Exercise 12) that asymptotically $k \approx 2n$, which gives a 50% improvement over Lamport signatures.

7.1.2.3 One-way chains

Lamport's hashing technique has the advantage of committing to a sequence of values as opposed to a single value. The *one-way chain* is a cryptographic primitive whose need arises in many instances, like in routing protocols, where you want to commit to a sequence of values. Of course, you can commit to each element of the sequence individually but this is usually not very efficient. To generate a one-way chain of length n the following steps are necessary:

1. Randomly pick the last value v_{n-1} of the chain.
2. Generate the chain $v_{n-1}, v_{n-2}, \ldots, v_0$ by repeatedly applying a hash function H.
3. v_0 is the commitment to the entire chain.

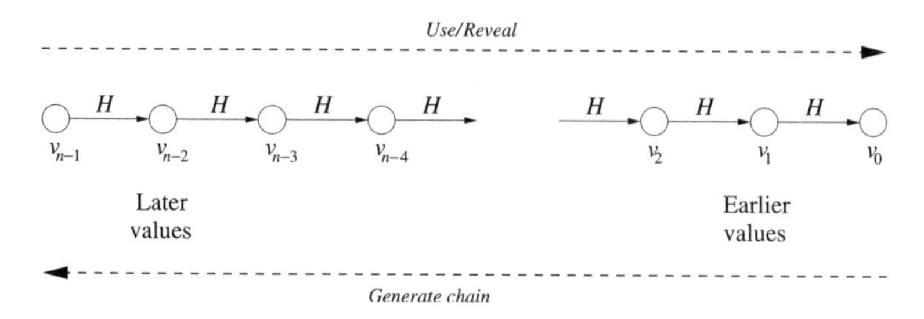

Figure 7.1 Constructing one-way hash chains.

The construction of a one-way hash chain is depicted in Figure 7.1. Take a one-way collision resistant hash function H and construct the chain by iterating Equation 7.1

$$v_i = H(v_{i+1}), \text{ for } i = n-2, n-3, \ldots, 0. \tag{7.1}$$

Authentication with one-way hash chains, is depicted in Figure 7.2. To authenticate that a value v_j belongs to the chain use an earlier value v_i of the chain and verify that Equation 7.2 is true

$$H^{j-i}(v_j) = v_i. \tag{7.2}$$

Given an earlier value v_i can an adversary find a later value v_j, where $i > j$? That is, $H^{j-i}(v_j) = v_i$. This is difficult, even when the value v_{i+1} is released, provided that H is a good collision resistant hash function.

H H H H H H

v_j v_i

$H^{i-j}(v_j) = v_i$

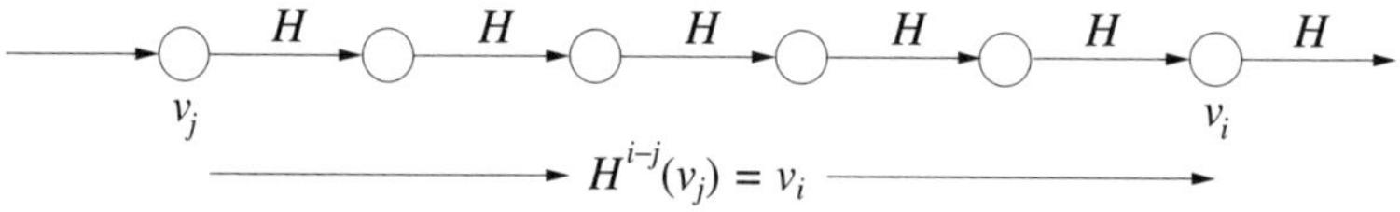

Figure 7.2 Authentication in one-way hash chains.

7.1.2.4 Nonsequential disclosure

Elements of the one-way chain are normally disclosed one after the other in sequence: note that disclosing an element discloses all previous elements as well. There are instances where either we want to open the commitment in any order or open only a few of the commitments. Nonsequential disclosure is a technique developed by Merkle and works as follows (Figure 7.3)

1. Place the secret values as leaves of a full binary tree.
2. Compute the value of an interior node "recursively up the tree" by hashing the values of the leaves of the node.
3. The root of a subtree commits to all the values of its subtree.

The resulting data structure is called *Merkle authentication tree*. To authenticate a sequence $v_0, v_1, \ldots, v_{n-1}$ of values, *blinding* is used in the Merkle authentication tree (the idea is

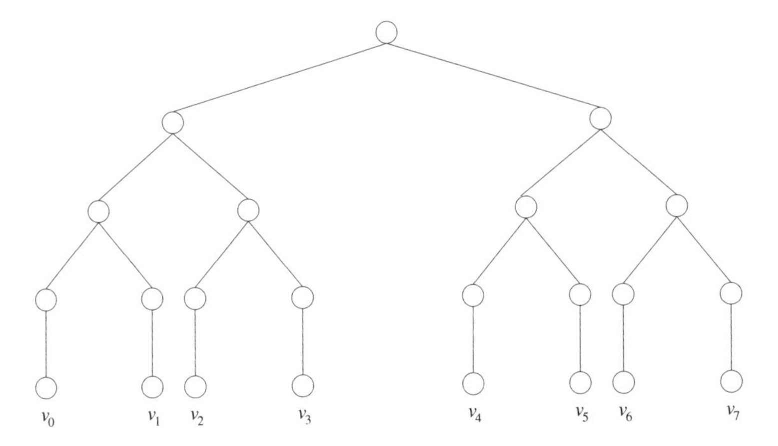

Figure 7.3 Forming a Merkle tree.

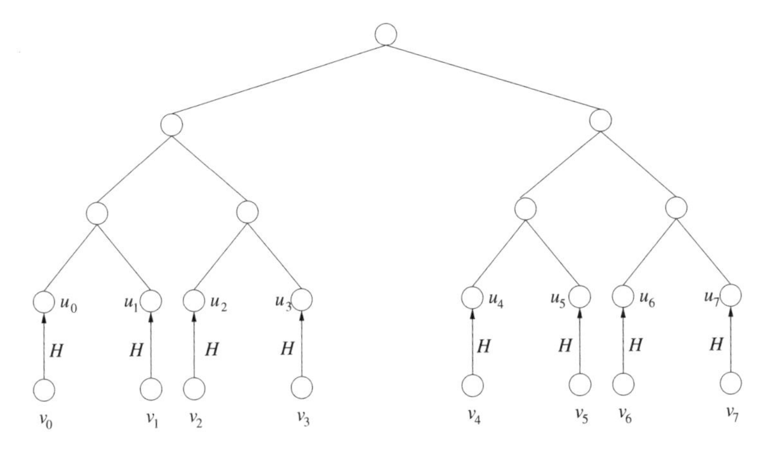

Figure 7.4 Blinding in Merkle authentication trees.

depicted in Figure 7.4). First, we blind the values $v_0, v_1, \ldots, v_{n-1}$ using a hash function H: $u_i = H(v_i)$. To facilitate the construction of the Merkle authentication tree we assume n is a power of 2. To authenticate the values $v_0, v_1, \ldots, v_{n-1}$:

1. Place the values on the leaves of a full binary tree.
2. Blind the values with a one-way hash function H, that is, $u_i = H(v_i)$.
3. Values of internal nodes u_p are derived from values of their two children u_l, u_r after hashing:

   $$u_p = H(u_l \| u_r),$$

   that is, concatenate and hash the left u_l and right u_r children of u_p.
4. Use the resulting hash value of the root of the tree in order to commit the whole tree.

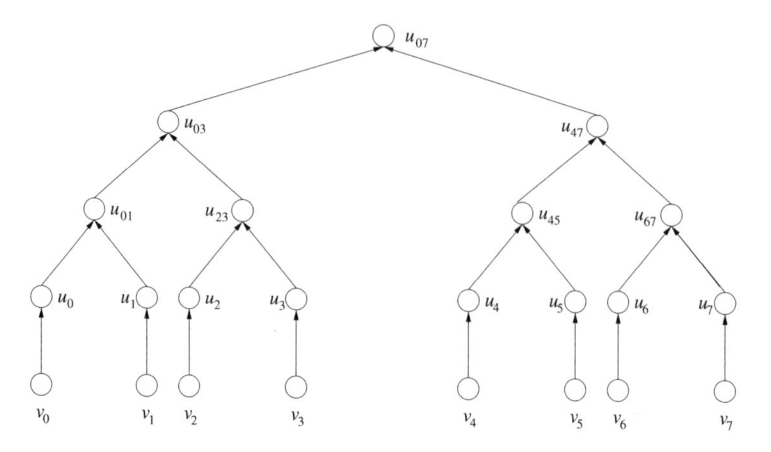

Figure 7.5 Recursive hashing in Merkle authentication trees.

The next step is to use bottom-up recursive hashing in order to build the completed Merkle authentication tree. As depicted in Figure 7.5, the value of a parent node u_p is derived from values of its two children u_l, u_r (left and right, respectively) after hashing $u_p = H(u_l \| u_r)$. In turn, to authenticate a value v_i the sender executes the following sequence of operations:

1. reveals v_i,
2. reveals i,
3. reveals all sibling nodes of the nodes on a path from v_i to the root of the tree,

while the receiver uses these values to verify the path from v_i to the root.

Figure 7.6 depicts an example whereby a receiver with authentic value u_{07} can verify Equation 7.3

$$u_{07} = H\left(H(u_{01} \| H(H(v_3) \| u_2)) \| u_{47}\right). \tag{7.3}$$

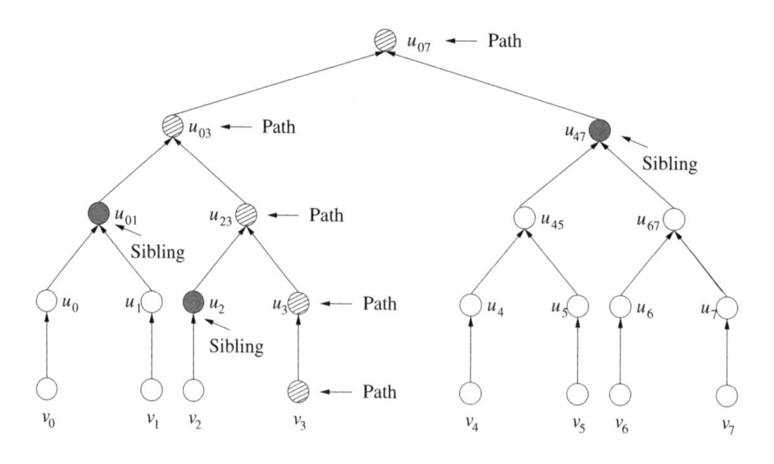

Figure 7.6 Example of Merkle authentication trees.

7.1.3 Distribution of keys

To prevent unauthorized access in streaming broadcasting systems, the data transmitted is encrypted with a symmetric key and only paying subscribers know this key. It is unpredictable which subscribers will decide to pay. Moreover, a user that shares a secret key with the broadcasting system may decide to log out, in which case the streaming supplier must change the key so that the user cannot decode the broadcast any longer. This typically involves a large pool of subscribers, each of whom shares a unique secret with the broadcasting system.

New subscribers that wish to decode a transmission must authorize the broadcasting system to charge their account. The current encryption key K is transmitted to these users and there is no need to change K and inform the remaining recipients during this interaction. However, if some user decides to stop receiving data, then the encryption key has to be changed and all recipients be informed about this update. Typically, a new encryption key needs to be broadcast periodically. Ad hoc systems are sensitive to these interactions because of the limited channel broadcast capacity and the fact that the devices are battery operated. Added to this is the fact that updating the key may leak sensitive information. Therefore the duration of sending and receiving should be kept to a minimum.

7.1.3.1 A High-energy Solution

Consider a set U of wireless users. For each user $A \in U$ let $s(A)$ be a symmetric key shared by user A and the sender. Distribute a new key K by transmitting a ciphertext $E_{s(A)}(K)$. (Asymmetric encryption is less attractive for practical applications because it would increase the size of the record). The problem with this solution is that it has a high energy cost for the users. Indeed, to maintain anonymity, the ordering of the records has to be random and on the average a legitimate user has to receive and decrypt half of the records just to find the key. However, even if there was no demand for anonymity, the set U is unknown to a user (the user only knows that (s)he belongs to U) and therefore a large fraction of the ciphertexts has to be received and decrypted anyway.

7.1.3.2 Facilitating the Search

The cost of the search for the ciphertext can be reduced by placing an identifier in an extra data structure. For example, using a hash function H, the sender may publish a random number r and then define an identifier for user A as a prefix of $H(r, A, s(A))$. By using a sufficiently long prefix the number of collisions can be made small. In general however, at least $\log n$ bits are required for each of the n users. In addition, the payload transmission (which is encrypted with the key K) may be preceded by a test sequence of the form $E_K(N)$, for some known string N (for instance a nonce). Then a user can decrypt this test message with all candidate keys and identify the right one. Nevertheless, the size of the identifiers is a more serious issue since even for a set of 1000, users at least 10 bits are needed. This is a large overhead compared to the ciphertext size. For 64-bit keys, this would make more than 15% overhead of transmission size. It seems that the only good point of this solution is that the number of decryptions is reduced to just one.

More efficient solutions are proposed in Cichon et al. (2005) where the reader can also find additional references.

7.2 Physical layer attacks

In commercial systems, security starts from the link layer and above. Security at the physical layer is not addressed. The physical layer is vulnerable to attacks such as jamming and scrambling. Jamming is achieved by introducing a source of noise strong enough to significantly reduce the capacity of the channel. Jamming is either unintentional or malicious. The information and equipment required to perform jamming are not difficult to obtain. Poisel (2003) has published a book on the topic of jamming alone, that is, how to build jamming systems and to counter systems that are by construction jamming resistant. Jamming attacks are likely to occur. Resilience to jamming can be augmented by increasing the power of signals or increasing the bandwidth of signals using spreading techniques, for example, frequency hopping or direct sequence spread spectrum. Note that a number of options are available to raise the power of a signal, for example by using a more powerful transmitter, a high gain transmission antenna or a high gain receiving antenna. Jamming is easy to detect with radio spectrum monitoring equipment. Sources are relatively easy to locate using radio direction finding tools. Law enforcement can be involved to stop jammers. Jammed segments of bandwidth, once detected, can also be avoided in a spread spectrum scheme. The risk associated with jamming is major, but not critical since its impact can be mitigated.

Scrambling is a sort of jamming that is carried out for short intervals of time and targeted to specific frames or parts of frames. Scramblers can selectively scramble control or management information with the aim of affecting the normal operation of the network. The problem is of greater amplitude for time sensitive messages, which are not delay tolerant, such as the channel measurement report requests or responses. Slots of data traffic belonging to targeted users can be scrambled selectively, forcing them to retransmit, with the net result that they get less than their fair share of the bandwidth. Selectively scrambling the traffic of other users can theoretically reduce the effective bandwidth of the victims and accelerate the processing of the data of the attacker (if it is a user). Scrambling is relatively more difficult to achieve than jamming because of the attacker's need to interpret control information and to send noise during specific intervals. There are technical difficulties for an attacker to address, but they are solvable. The likelihood of occurrence is possible. Scrambling is more difficult to detect than jamming because of the intermittent nature of the attack and the fact that scrambling can also be owing to natural sources of noise. Scrambling and scramblers can be detected by monitoring anomalies in performance criteria. This issue has been studied for WiFi/802.11 systems by Raya et al. (2004).

7.3 Security of application protocols

This section focuses on the security issues of WiFi/802.11 (Section 7.3.1) and ZigBee (Section 7.3.2).

7.3.1 WiFi/802.11 confidentiality

Confidentiality in ad hoc networks means that the contents of every message can be understood only by its source and destination. It is a challenging issue. The wireless nature of the media greatly facilitates the interception of traffic.

An interesting example of a system using symmetric-key-based cryptographic techniques is the Wired Equivalent Privacy (WEP) of WiFi/802.11. The first generation of WiFi/802.11 wireless networks adopted WEP for encryption (IEEE (1999a)).

It is based on the *exclusive or* (XOR) binary operator and RC4 (Sterndark (1994)). Network-wide keys are, however, not so secret keys. The fact that the keys are shared between all the network members is an enabler for attacks from insiders. In WEP, keys are reused for different messages. If the contents of one message is known, then the other messages using the same key can be decrypted. This is illustrated hereafter. Figure 7.7 illustrates how RC4 works. Each rectangle represents a piece of data involved in the process. First, the message text, that is, *LOGIN* in this example, is converted into a binary form using some coding scheme, in this case ASCII. The *exclusive or* of the message in ACSII and a secret key stream is calculated. The message in binary form and secret key stream are of the same length. The calculation is done bit by bit. The result of the calculation is the ciphertext. It is the element of data that travels over the network. It assumed that if the secret key is unknown, then it is difficult to recover the original message. This is a reasonable assumption because there are so many values that the secret key may take.

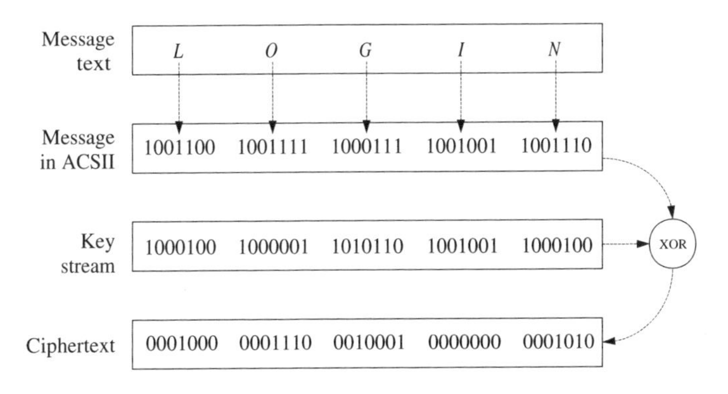

Figure 7.7 The RC4 encryption.

It has been demonstrated, however, that WEP is relatively easy to crack. The idea is illustrated in Figure 7.8. The problem is not due to RC4 per se, but to the way it is used in WEP. Secret key streams are reused for different messages. The key values are secret. They are numbered. The number of the key used for encrypting a given message can be, however, determined by inspecting the contents of its header, which is never encrypted. Figure 7.8 shows two intercepted messages encrypted with RC4 using the same secret key stream, that is, *Ciphertext* 1 and *Ciphertext 2*. The third rectangle shows the result of taking the *exclusive or* of Ciphertext 1 and Ciphertext 2. The key streams being the same in both, they cancel each other out. The result consists of the *exclusive or* of the two original messages in plain text, which is not too different from the actual plain text of the messages. If the contents of one message can be guessed, then the contents of the other message can be determined, as can the secret key stream. For example, it is usual for a system to send the text *LOGIN* at the start of a session with a computer. In the example of Figure 7.8, the *exclusive or* of *the exclusive or of the two original messages in plain text* (third rectangle) and the ACSII of the *LOGIN* message (fourth rectangle) is calculated. The result (fifth

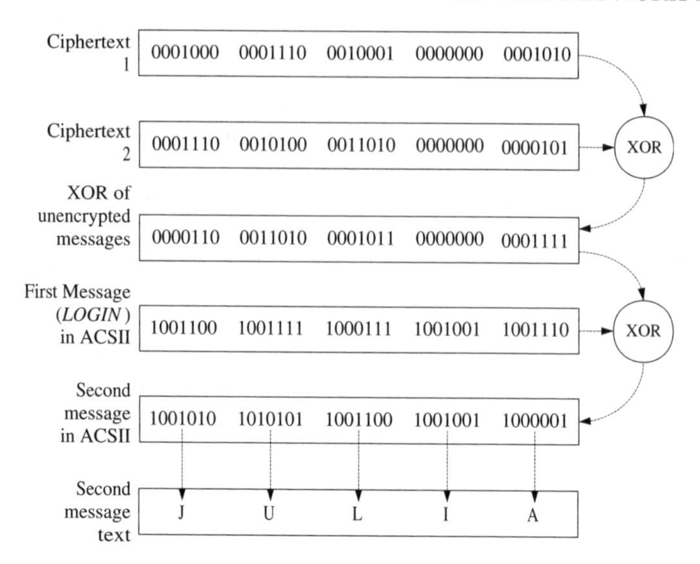

Figure 7.8 Cracking RC4 messages.

rectangle) is converted into ASCII and reveals the user name. With WEP, the frequency of key reuse is relatively high, resulting in a good number of messages encrypted with the same key, that can be combined together to guess their plaintext contents.

Moreover, WEP addresses neither message authentication nor replay protection. Replay protection ensures that messages are freshly generated and are not retransmissions by attackers of previously intercepted messages. For the sake of efficiency, replay protection is often combined with message authentication.

Recent developments, namely the WiFi Protected Access (WPA) (WiFi (2004)) and standard 802.11i (IEEE (2004)), introduce much stronger confidentiality protection mechanisms in WiFi/802.11 networks. Firstly, encryption key establishment uses asymmetric-key-based techniques. Secondly, WPA uses the Temporal Key Integrity Protocol (TKIP), which is RC4-based but with longer non reused keys. TKIP comprises a mechanism to ensure message integrity and avoid replay, the Michael method. 802.11i supports both TKIP and Advanced Encryption Standard (AES) . WiMAX/802.16e uses the Data Encryption Standard (DES) or AES to encrypt data traffic protocol data units (IEEE et al. (2006)). AES includes a mechanism for the protection of the integrity of data messages, their authentication and replay protection. DES does not.

Asymmetric-key cryptographic techniques require that a private key is used for encryption and another public key is used for decryption. Asymmetric-key cryptographic techniques are termed public-key cryptography. The idea of public-key cryptography was originally launched by Diffie and Hellman (1976). Originally abstract, the idea was evolved to a practical approach by Rivest et al. (1978). For key distribution purposes, public-key cryptography uses the concept of a certification authority (CA), which is an entity that everyone trusts and has public keys. The CA issues each participant a certificate, which is an information structure consisting of the participant's identity paired with its public key. The certificate is digitally signed by the CA, using it's own private key. The authenticity of a certificate can be verified by checking its digital signature, using the CA's public key. Public-key cryptography also includes a mechanism to invalidate certificates. This may be achieved by bounding the certificate with an expiration date, publishing signed lists of revoked certificates or online verification of the validity of certificates. X.509 is a well-known format for such certificates (CCITT (1997)).

7.3.2 ZigBee security

ZigBee is a technology proposed by the ZigBee Alliance (2005) for wireless personal area networking and more specifically for sensor networks. It is designed for short distances (e.g. 10 m) and low data rate operation (20–250 kbps). The sequel provides an overview of security in ZigBee.

ZigBee security includes mechanisms for authentication, confidentiality, integrity protection and replay protection. Authentication is a means of defense against the threat of impersonation of a device, in other words a device taking the identity (i.e. address) of a legitimate participant to use its rights and access resources. ZigBee authentication is supported by network keys and link keys. A legitimate participant must demonstrate ownership of a network key to participate in a ZigBee network. The network key is a means of excluding outsiders. It is a network-wide mechanism. Device to device link keys can also be optionally used. A given link key is associated with two devices. The two devices involved must mutually prove ownership of the link key when they communicate together. It is a stronger mechanism than network keys, since it provides protection against both outsiders and insiders attempting to impersonate an endpoint of a link. Confidentiality is subject to the threat of eavesdropping. The AES 128-bit encryption is used to conceal the payload of ZigBee traffic. The integrity of traffic is vulnerable to tampering while it is in transit. The integrity of a message can optionally be protected by a 32, 64 or 128-bit MIC. Replay protection is provided by numbering protocol data units, a mechanism termed *frame counter*.

Security can be addressed at the link layer, network layer or application support layer, in a nonexclusive manner. Figure 7.9 depicts the structure of ZigBee frames, emphasizing the possible placement of an auxiliary header and the fact that the fields following the auxiliary header become part of the encrypted payload. The auxiliary header contains a security control field, a frame counter, a sender address and a key sequence number. If the auxiliary header immediately follows the MAC header as in (a), then the frame payload includes the network and application support headers and is secured across the link until

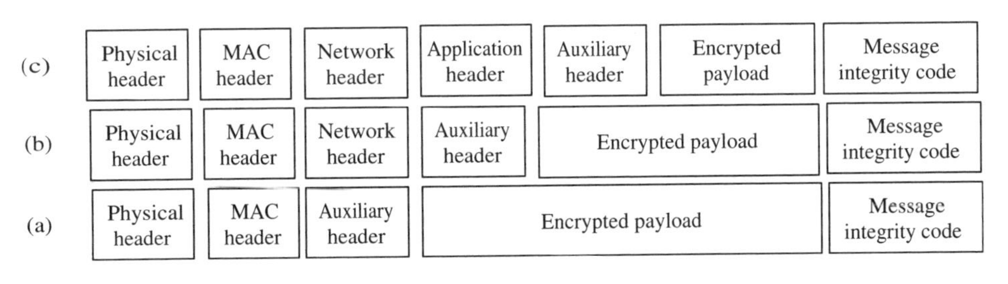

Figure 7.9 ZigBee frame with auxiliary header.

the next hop. If the auxiliary header follows the network header as in (b), then the packet payload includes the application support header and is secured across the network until it reaches its final destination device. If the auxiliary header follows the application support header as in (c), then the application data is secured until it is delivered to the destination application entity. At any of these levels, either the network or the link key can be used to encrypt the payload and calculate the MIC. When available, use of the link key is preferred because it offers stronger security.

ZigBee includes the concept of trust center, which is a role played by a device on a network. The responsibilities of the trust center includes key distribution. Each device can be configured with the address of a trust center with whom it shares a master key. This master key is a long-term key, which can be pre-installed in the device or user-entered. Alternatively, it can be communicated from the trust center to the device, the network being momentarily unsecured. Link keys and network keys can be communicated from the trust center. A link key or a network key can also be established using the master key and Symmetric-Key Key Establishment (SKKE) protocol.

ZigBee defines two security modes: residential and commercial. The residential mode is the most unsecured. The trust center has a network key, which is used to authorize devices to join the network. A list of authorized device addresses, master keys and link keys are optionally stored by the trust center and used in device authorization. In the commercial mode, the trust center maintains a list of authorized devices, master keys, link keys and network keys. All this information is used to authorize devices.

There are several possible network entry scenarios, according to the mode and selection of optional features. As an example, a ZigBee successful network entry scenario is depicted in Figure 7.10. Three devices are involved in the scenario: a joiner, a router and a trust center. The commercial mode is assumed. The joiner (the device that enters the network) is pre-configured with a master key, shared with a trust center. Initially, the joiner sends a

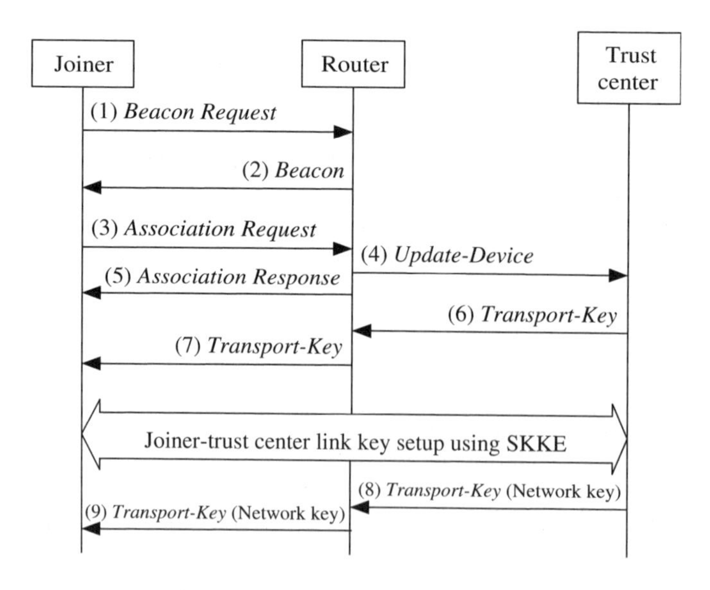

Figure 7.10 ZigBee network entry.

Beacon Request message with the aim of discovering available routers (1). Any available router may reply to the joiner with a *Beacon* message (2). For the sake of simplicity, only one router is depicted. Beacon messages are not secured. The joiner selects a network using the beacons received from the routers. The joiner attempts to join the network using the corresponding router and an *Association Request* message (3). The router learns the address of the joiner. The router accepts the joiner and sends the trust center an *Update-Device* message, which makes the trust center aware of the presence of the joiner and its address. The Update-Device message is secured, at the application support layer, with the router-trust center link key. The router also sends an *Association Response* message to the joiner (5). The joiner is considered *joined*, but not yet authenticated. The joiner is accepted by the trust center. A *Transport-Key* message is sent (with null contents), through the router, to the joiner (6, 7). The SKKE protocol is then used between the trust center and joiner (through the router) to set up a link key between the trust center and the joiner (not detailed in the picture). Finally, the trust center sends the joiner the network key using another Transport-Key message (8, 9). The trust center-joiner link key is used to secure, at the application support layer, this Transport-Key message. The joiner is considered authenticated when it receives this message.

7.4 Biometrics-based key establishment

Networked nodes need to establish keys for authentication, encryption of confidential data, and for secure routing. Key establishment in ad hoc networks may be designed to take into account limited power, memory and computational resources, dynamically varying network topology, scalability and absence of central management. Key establishment in ad hoc networks may be accomplished according to a number of different mechanisms, such as link or network-wide shared symmetric key, asymmetric key (including public-key cryptography, elliptic curve cryptography), random schemes (different key sets are given to each node), location-based keys, and biometrics keys. The shared key model is based on symmetric-key cryptographic techniques, which means that the same key is used for both encryption and decryption. ZigBee security employs symmetric-key cryptographic techniques for protocol data unit authentication and confidentiality, as discussed in Section 7.3.2. The establishment of symmetric keys using biometrics techniques is discussed hereafter.

A biometrics-based key establishment technique was devised by Cherukuri et al. (2003) and Poon et al. (2006) specifically for Body Area Sensor Networks (BASNs). A BASN is a collection of communicating biosensors distributed over a human body. Example devices include electrocardiogram (ECG) sensors (to record the electrical activity of the heart) and photoplethysmogram (PPG) sensors (to estimate the blood pressure). Applications of BASNs include telemedicine and remote patient monitoring. Biometrics features, calculated by distributed synchronous sensors, are used to establish symmetric cipher key encryption keys, that is, keys used to conceal encryption keys while they are transported over the network. Time synchronization, to the millisecond, of the distributed sensors is assumed.

Two devices establish a common symmetric key by agreeing upon a common feature. One device has the role of initiator while the other has the role of responder, see Figure 7.11. The initiator generates the common symmetric key k. The value of a common feature is derived independently on both the initiator and responder, and is used to secretly exchange the key k. The sensed value of the feature plays the role of key encryption key. To exchange

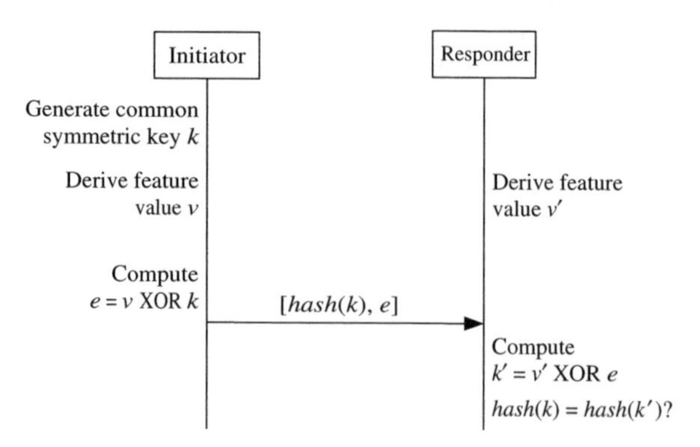

Figure 7.11 Key establishment using the fuzzy commitment protocol.

a symmetric key k with another device, the initiator device first derives the feature value v using an appropriate biometrics sensor. It then computes the one-way hash value $hash(k)$ and the expression

$$e = v \oplus k$$

The $\oplus$ represents the *exclusive or* (XOR) operator. The values $hash(k)$ and e are sent over the insecure communication channel to the responder device. The responder device also computes its own value v' for the same feature and recovers

$$k' = v' \oplus e$$

For this procedure to work properly, both values v and v' must be identical. First, the equality of $hash(k)$ and $hash(k')$ can be tested. If successful, then a challenge response procedure can be run between the initiator and responder to test if both values k and k' match. Otherwise, the key establishment can be performed again until it succeeds.

The biometrics feature used in this process must satisfy properties of uniqueness (high variability from one individual to another) and randomness (high difficulty to guess by an attacker). These issues have been explored by Cherukuri et al. (2003). They suggest the use of blood glucose, blood pressure and body temperature as features satisfying these properties. Poon et al. (2006) use a sequence of Inter Pulse Intervals (IPIs) for that purpose. Crests in ECGs and troughs in PPG occur at the same time points (Figure 7.12). The time points are used as markers for IPIs. A sequence of n IPIs are collected and encoded into

Figure 7.12 ECG with IPI markers.

a binary value of m bits ($n = 33, 67; m = 64, 128$), which constitute the v of the initiator and the v' of the responder. It is important that the collection starts at the same time on both sides. Lack of synchronicity of sensors is the main source of errors according to Poon et al. (2006).

One difficulty is the variation in collected values depending on the locations of sensors, on the human body, calculating the features. The fuzzy commitment protocol of Juels and Wattenberg (1999) addresses that difficulty. The establishment of a pairwise secret c between two peers, an initiator and a responder, is explained with the following example. The peers use a common and abstract reference grid. The secret to be established is a point on that grid. There is a decoding function $f(\_, \_)$ that maps a coordinate (x, y) to the nearest point on the grid, that is $f(x, y) = (round(x), round(y))$. A cryptographic one-way hash function *hash*(), for example SHA-1, is also used (FIPS (2002)).

The calculation done by the initiator is depicted in Figure 7.13. The initiator obtains, by sensing, a pair of values rounded to two decimals, for example,$v = (8.26, 1.37)$. The initiator selects at random a point c on the grid (the secret), for example, $c = (4, 5)$. The offset d, between v and c is calculated, that is, $d = v - c = (8.26, 1.37) - (4, 5) = (4.26, -3.63)$. The initiator sends to the responder the message consisting of $[hash(c[x]|c[y]), d]$, where | represents the concatenation operator. This message is termed the *fuzzy commitment* of c. For this example, the message would be $[hash(4|5), (4.26, -3.63)]$.

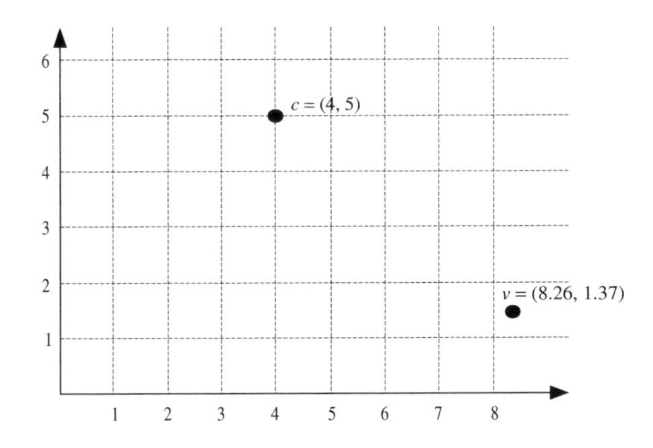

Figure 7.13 Initiator calculation in the fuzzy commitment protocol.

The calculation performed by the responder is depicted in Figure 7.14. The goal of the responder is to recover the value c. The responder obtains, by sensing common features, its own pair of values rounded to two decimals that slightly differs from v, for example, $v' = (7.76, 0.96)$. The difference between v' and d is calculated, that is, $v' - d = (7.76, 0.96) - (4.26, -3.63) = (3.50, 4.59)$. The decoding function $f(\_, \_)$ is applied to $v' - d$, yielding $(4, 5)$. The value for $hash(c[x]|c[y])$ received from the initiator is compared with $hash(f(v' - d))$. If they match, then the responder considers that it has successfully recovered the value of the secret c as $f(v' - d)$.

Juels and Wattenberg (1999) have introduced a biometrics-based encryption scheme, called *fuzzy encryption protocol*, (Figure 7.15). There is a sender and a receiver. The sender gets a message to send, m. The sender generates a symmetric encryption key k, for example,

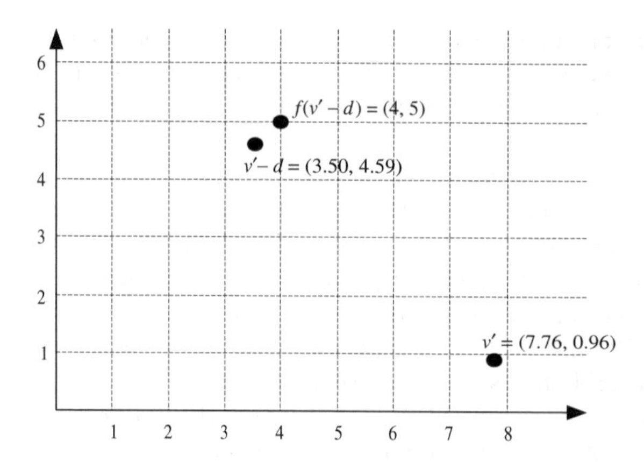

Figure 7.14 Responder calculation in the fuzzy commitment protocol.

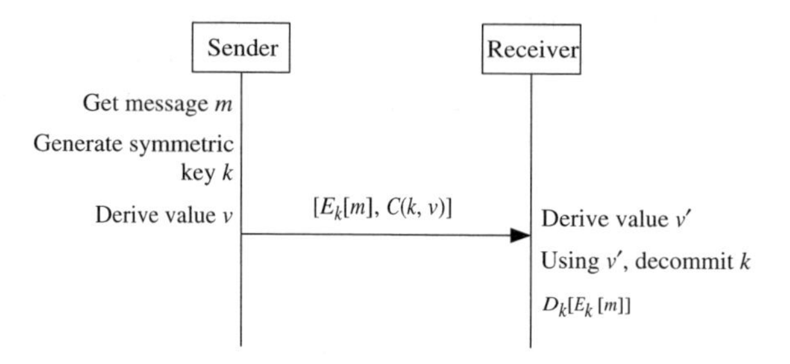

Figure 7.15 Fuzzy encryption protocol.

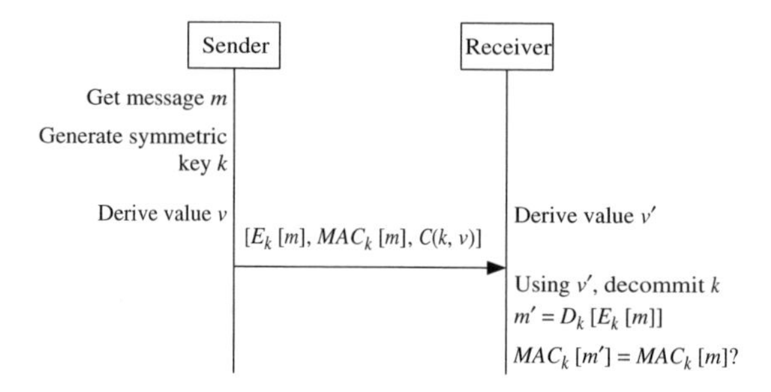

Figure 7.16 Authentication using the fuzzy commitment protocol.

a random value, and then derives a feature value v, using an appropriate biometrics sensor. The sender sends the message m encrypted using key k, denoted as $E_k[m]$, together with the fuzzy commitment of key k using value v, denoted as $C(k, v)$.

The receiver derives its own feature value v', using a biometrics sensor. The receiver uses v' to decommit and to recover the key k concealed in $C(k, v)$. Using k, the message is decrypted, denoted as $D_k[E_k[m]]$.

A biometrics-based encryption protocol including an authentication scheme has been defined by Cherukuri et al. (2003). The procedure is depicted in Figure 7.16. The protocol is similar to the on depicted in Figure 7.15. In addition to the encrypted message $E_k[m]$ and fuzzy commitment $C(k, v)$, the sender transmits the message authentication code $MAC_k[m]$ over message m calculated using key k. In addition to decommitting the key k and decrypting the message m into m', the receiver has to make sure that the message authentication code over m matches the message authentication code over m, that is, $MAC_k[m] = MAC_k[m']$.

7.5 Routing security

Routing security is strongly dependent on the routing protocols and type of network. Of interest here are ad hoc networks, and it is fair to say that there are two categories of routing protocols on such networks: *proactive* and *reactive* (Chapter 5). The former enforces a policy of periodically exchanging information with other nodes to keep routing information up-to-date. The latter attempts to conserve valuable communication resources by exchanging routing information only when needed. There are also several protocols that are hybrids of proactive and on-demand (for example, see Section 5.4).

7.5.1 Routing attacks

Following Hu et al. (2005), *attackers* can be classified into two categories: *passive* and *active*. The former only eavesdrop (e.g. monitor packet content and may also perform traffic analysis) but otherwise do not send any messages, while the latter not only eavesdrop but also take an active part (masquerading by impersonating an authorized user, spoofing, installing rogue access points, replaying and modifying messages) in the attack by injecting packets. Generally, passive attacks are only a threat against privacy and anonymity, while active attacks are a threat against the normal functioning (like data flow) of the network. An attacker may well be a compromised node in the network, in which case it will probably have access to some "critical" cryptographic keys and may also cooperate with other attackers or compromised nodes. Routing attacks can cause either routing disruption or resource consumption. In the former case, an attacker causes packets to be routed in dysfunctional ways, while the latter aims to consume valuable resources like bandwidth, memory and computational power.

The following (not necessarily mutually exclusive) types of attacks have emerged in the security literature: *route advertisement failure*, *packet dropping and injection*, *route modification*, *replay*, *rushing* and *wormhole*. In the sequel we elaborate their functionalities in more detail.

7.5.1.1 Route Advertisement Failure

An attacker can cause *route advertisement failure* by attempting to reduce the amount of routing information that is available to other nodes. This can by done either by failing to advertise or even to update routes, or by destroying and discarding routing packets or parts thereof. Thus by failing to advertise a route, the node becomes unwilling to forward packets to certain selected destinations.

7.5.1.2 Packet Dropping

In *packet dropping*, an attacker could discard data packets sent to certain destinations. By failing to receive routing packets, a node may become ignorant of links available to it and thus fails to pass potentially updated knowledge to neighboring nodes. Similarly, in *packet injection*, extra (control) packets are injected into the network, which may cause data disruptions and even consume more bandwidth and computational resources as other nodes process and forward such packets.

7.5.1.3 Route Modification

In *route modification* attacks, an attacker can modify advertisements by changing (some or all) of the destination, metric or source address, and next-hop information. It follows that either: 1) by advertising a zero metric for all destinations, an attacker can attract data traffic by causing nodes around it to route packets for all destinations toward it rather than toward each actual destination, or 2) by modifying the source address of the advertisement, an attacker can repel data traffic, thus spreading inaccurate route information. The effects can be devastating since an attacker can cause *routing loops* (i.e. packets traverse in cycles), *black holes* (i.e. packets are dropped), *gray holes* (i.e. packets are dropped selectively), *detours* (i.e. packets routed via non-optimal routes), and *network partitions* (thus preventing a certain part of the network from reaching another).

7.5.1.4 Replay Attacks

Replay attacks can be mounted by sending old routing advertisements to some node and thus cause a target node to update its routing table with stale routes.

7.5.1.5 Rushing Attacks

In a *rushing* attack a malicious attacker rushes the dissemination of illegitimate route requests throughout the network. Because of the duplicate suppression mechanism of on-demand routing protocols, affected nodes in the network will ignore later but legitimate route requests.

7.5.1.6 Wormhole Attacks

Wormhole attacks, which are also described in detail in Section 7.8, can be very subtle. A pair of attacker nodes, say X and Y, already connected via a private (typically) fast link (Figure 7.24), forward packets (which either arrive at or are being sent to their local neighborhood) to each other in the ad hoc network. In a sense, X and Y by rushing packets to each other, disrupt routing by short circuiting the normal flow of routing packets. It is important to note that this attack may also create a virtual vertex cut of nodes in the network that they control.

Ideally, a secure routing protocol should be able to prevent all the attacks above but, no doubt, such a task is very difficult indeed. Among possible proposals that could be put forward, it is worth mentioning: (1) securing routing metrics and (2) data packet control. The first proposal arises from the need to exercise some limits on the routing metrics. For example, to prevent routing loops, a routing protocol should enforce a maximum route length policy so that data packets transmitted by the attacker can only cause a limited number of additional transmissions. The second proposal emphasizes the importance of acceptable network performance. For example, by restricting the number of control packets (in response to each data packet), malicious data packets can cause only limited disruptions to individual transmissions.

7.5.2 Preventing malicious packet dropping

Malicious packet dropping is a form of DoS attack. Ad hoc networks are particularly vulnerable to this attack since they lack physical protection and strong access control mechanisms. An attacker either can easily join the network or capture a host and then start disrupting communication by dropping packets. Malicious packet dropping when combined with other attacking techniques, such as shorter-distance fraud, can create even more powerful black hole attacks, which can completely disrupt network communication.

DoS attacks are usually launched by disseminating false routing information in order to make established transmission routes invalid. In distance vector routing, an attacker: (1) can attract traffic in a black hole by advertising shorter distances, (2) by carefully crafting routing update messages can cause network congestion and (3) can mislead other routers to create invalid paths in their routing table by disseminating false information thus causing packets to be dropped because they never reach their destination.

A malicious packet dropping attack is caused by an attacker that either manages to join the network or that somehow compromises a legitimate router and silently drops all or some of the transmitted data packets. The attacker gains access to a router by exploiting a vulnerability (e.g. buffer overflow, or some other authentication weakness). Traditional asymmetric cryptographic techniques (such as public-key signatures) can be used in order to prevent external intruders from entering the network but they are often rendered ineffective since both the signature generation and verification process involve the execution of computationally expensive functions whose execution can disrupt the normal performance of the host. Similarly, symmetric cryptographic primitives (like one-way hash chains, one-time signatures, and authentication trees) for authenticating routing messages. cannot prevent attacks from compromised internal routers. Moreover, although securing distance vector routing protocols by having nodes increase their metric when forwarding routing update messages (see Hu et al. (2003b)) can prevent compromised nodes from claiming shorter distances, a malicious node can still avoid traffic by claiming longer distances.

7.5.2.1 Distributed Probing Scheme

In the sequel, we present a proactive distributed probing technique, due to Just et al. (2003), to detect and mitigate the malicious packet dropping attack. The approach is based on having every node pro-actively monitor the forwarding behavior of other nodes with probing messages that are indistinguishable from normal data packets. Experiments in Just et al. (2003) demonstrate that in a moderately changing network, the probing technique can detect most of the malicious nodes with a relatively low false positive rate. Moreover, the packet delivery rate in the network may increase merely by bypassing the detected malicious nodes.

Given a network with n nodes, there are several parameters that could be taken into account. We could consider either only one probing node for the entire network, or k probing nodes each probing over a distance at most r, for some $k \geq 1$, or even the possibility that all nodes are probing up to infinity. The probing technique is divided into three algorithms: first, the probing path selection, second, the main probing algorithm and third, the diagnosis algorithm, which are described below.

- *Probing path selection*: Probing paths are selected from the routing cache maintained by a mobile node. To reduce overhead, a minimum number of paths is selected that allows for monitoring the forwarding behavior of as many nodes in the routing cache as possible. The probing path selection algorithm returns a set $\mathcal{P}$ of paths with the following properties: (1) for any two paths $p, p' \in \mathcal{P}$ selected, we have that $p \not\subseteq p'$, (2) for any two paths $p, p' \in \mathcal{P}$ if the second farthest node of p is an intermediate node of p' then the farthest node of p will be removed and (3) the length of any path (in terms of number of hops) is greater than 1.

- *Main probing algorithm*: Given a path, the basic idea is for the probing node to send a *probe message* to the farthest node in the path. If an acknowledgment message is received within a certain period of time, all the intermediate nodes are shown to be *good*. Otherwise, a probe message is sent to the second farthest node in the path. This process is repeated until either one node responds to the probe message or else a neighbor nearest node is probed and is found not responsive. In the latter case, the neighbor node in the probed path is either *down* or has moved out to another location. Since the neighbor node is not responsive, there is nothing we can do to monitor the rest of the nodes in the path. Therefore, probing over this path is stopped. If an intermediate node is responsive but a node subsequent to it is not, it is possible that: (1) the intermediate node failed to forward the probe message to the next node; (2) the link between the two nodes is broken by some location change; or (3) the unresponsive node is incapable of responding to the probe message. The diagnosis algorithm will then be called to decide which one is the case.

- *Diagnosis algorithm*: When the probing node in a path detects a responsive node u but whose subsequent node u' is unresponsive, it calls the diagnosis algorithm to determine whether the link $u \rightarrow u'$ is broken at the link level or forwarding level. This is done by first searching the routing cache for another path to u'. If such a path exists, it will probe through this path. If it is still unresponsive, it searches the routing cache for another path. This process repeats until either the routing cache is exhausted or else there is a path $p \in \mathcal{P}$ through which node u' is responsive. In the former case, we declare the node u' as down (since it has probably moved to another location), unless it is feasible to execute a route discovery protocol and reconfirm its presence (but in reality executing a route discovery protocol may not be reasonable as it will cause delay overhead). In the latter case, the diagnosis algorithm appends u to p and sends a probe message to u over p. If an acknowledgment from u is received, then u is declared bad since the link $u' \rightarrow u$ is good but the link $u \rightarrow u'$ is not.

It is evident that the probing and diagnosis algorithms can be improved further and the packet delivery rate can be increased if the node state information is shared with routing cache. The probing technique is of practical significance since it can be implemented independently from routing software and does not require modification to the existing infrastructure. The disadvantage of the probing technique is that it generates relatively high network routing overhead if probe messages do not piggyback on data packets. Additional details can be found in the original paper Just et al. (2003).

7.5.3 Secure ad hoc distance vector routing protocol

The highly dynamic nature of ad hoc networks and the need to operate efficiently with limited resources (e.g. network bandwidth, processing capacity, memory and battery power of individual hosts) make secure ad hoc network routing protocols difficult to design. Added to this is the fact that cumbersome security mechanisms may either delay or prevent routing information exchanges thus leading to reduced routing effectiveness and excessive consumption of network and node resources. It may therefore seem ironic that newly introduced security mechanisms may themselves lead to new opportunities for possible attacks.

This subsection is devoted to securing proactive ad hoc network routing protocols and an outline of Secure, Efficient, Ad Hoc, Distance vector (SEAD). *SEAD* (Secure, Efficient, Ad hoc, Distance vector) is a protocol that was developed by Hu et al. (2003a) in order to secure distance vector routing. Distance vector routing protocols are widely used in networks of moderate size and are rather easy to implement compared to other types of routing protocols. According to Hu et al. (2003a) "SEAD is relatively robust against multiple uncoordinated attackers creating incorrect routing state in any other node, even in spite of active attackers or compromised nodes in the network." Its design uses one-way hash functions and is based in part on the Destination-Sequenced Distance Vector ad hoc network routing protocol (DSDV) (see Perkins and Bhagwat (1994)), which in turn was designed only for trusted environments.

7.5.3.1 DSDV (Destination-Sequenced Distance Vector) protocol

In distance vector routing, each host in the ad hoc network acts as a router and participates in the routing protocol. Hosts maintain routing tables listing distances to all possible destinations within the network. Table entries include destination address, shortest distance to this destination, and next-hop forwarding information. Routing tables are kept up-to-date with updates (essentially executing some form of flooding) that are triggered periodically. To prevent the occurrence of loops, several extensions of distance vector routing protocols have been proposed including split horizon and split horizon with poisoned reverse (for example, see Bertsekas and Gallager (1992, 2nd edition)).

In order to prevent routing loops, DSDV (developed by Perkins and Bhagwat (1994)) adds an *even* sequence number to each routing table entry of the standard distance vector routing protocol and includes it in each routing update being sent. Nodes detecting broken links to a neighbor create new entries with an "infinite" metric and the next *odd* sequence number after the even sequence number in its corresponding routing table entry. When a node receives a route update, for each entry in the route update, it accepts the entry if it has a higher sequence number, or if it has an equal sequence number and a lower metric than the route entry currently in the node's route table for this destination. However, if the sequence number in the update is less than the current sequence number in the table entry, the new update for that destination is ignored (see Perkins and Bhagwat (1994) and Hu et al. (2003a)).

7.5.3.2 Sead

To prevent attacks on route updates both the sequence number and the metric must be authenticated. One-way authentication chains are used (Section 7.1.2). Each node v forms a one-way hash chain $v_{kn}, v_{kn-1}, \ldots, v_0$, and $H(v_i) = v_{i-1}$, where $k-1$ is the maximum hop count, and n is the maximum sequence number this chain allows. These values are used to authenticate routing update entries that specify node v as destination. A certificate authority publishes an authentic seed value v_0 that corresponds to node v. The value v_{ki+j} is used to authenticate a route update entry with sequence number i and metric $k-j$ for the destination node v.

An example is depicted in Figure 7.17. For the destination node v, with max hop count $k=5$ and max sequence number $n=3$. To start the first route update: for entries with node v as destination, node v first sends v_5 as an authenticator for sequence number 0 and metric 0. A recipient would first authenticate v_5 using public seed v_0 and compute v_4 from v_5; the node then advertises sequence number 0 and metric 1 using authenticator v_4. The recipient of this will advertise sequence number 0 and metric 2 using authenticator v_3, and so on. The next time node v starts updates for entries with node v as destination, it uses v_{10} to authenticate sequence number 1 and metric 0. Each update must be authenticated: this can be achieved either with signature, broadcast authentication or pairwise shared keys.

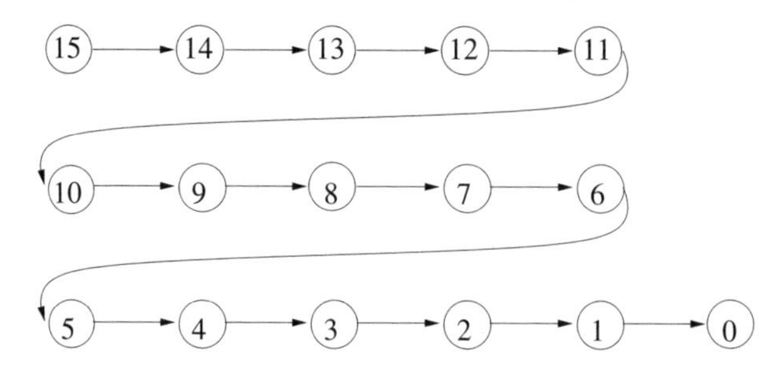

Figure 7.17 Example of SEAD implementation (only indices are depicted).

7.5.3.3 Hash-tree Chains

Hash-tree chains are meant to prevent same distance fraud by preventing attacker from playing the same hash value (i.e. without increasing the metric but replacing the node id with the attacker's node id). A hash-tree chain is a hybrid between hash tree and one-way chain (both ideas discussed in Section 7.1.2). The construction is such that the one-way chain property (Figure 7.18) is used to enforce that nodes cannot decrease the distance metric, while the hash-tree property is used to authenticate node identities (Figure 7.20). Figure 7.19 depicts the Merkle tree used in the construction of the hash tree. A hash tree is constructed between each pair v_{i-1}, v_i of one-way chain values as follows:

1. Derive from v_i a set $b_0, b_1, \ldots, b_n$ of values using a one-way hash function H by $b_j := H(v_i \| j)$, for each j.

2. Next we build a hash tree above these values for authentication as described before.

3. The root of the tree becomes the previous value $b_{0n} := v_{i-1}$.

An example of a hash-tree chain between values v_{i-1}, v_i is depicted in Figure 7.20 whereby each step v_{i-1}, v_i of the chain is replaced with the hash-tree chain.

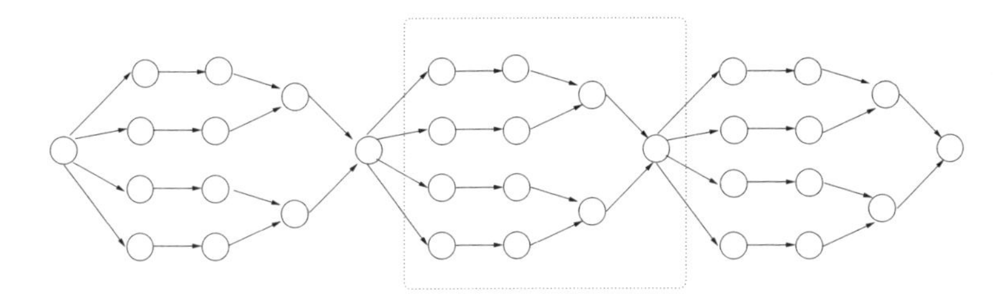

Figure 7.18 Example of hash-tree chain. One-way chain generation.

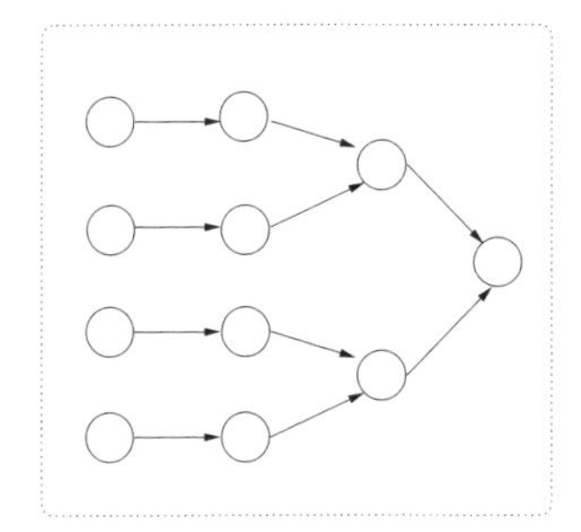

Figure 7.19 Merkle tree.

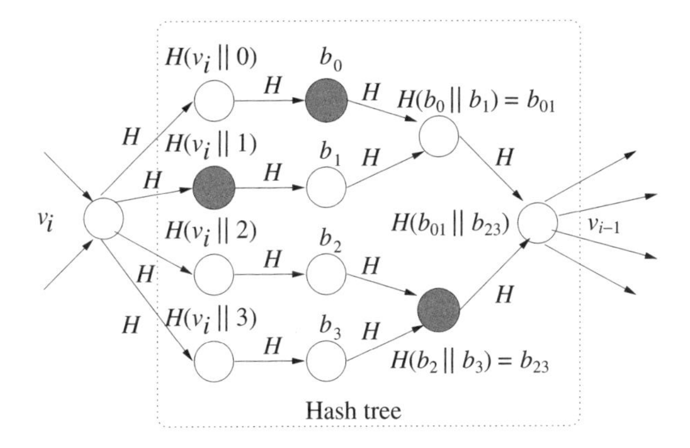

Figure 7.20 Example of using the hash-tree chain.

7.6 Broadcast security

Broadcasting in a communication network is an essential mechanism for scalable digital information distribution and refers to transmitting packets so that they will be received by every host in a network. Typically, the scope of the broadcast is limited to a specific broadcast domain and in practice it is largely confined to LANs, like ethernet and token ring, where its performance impact is not as large as it would be in a wide area network. Standard applications include joining broadcast meetings, global radio streaming and high definition TV. Applications envisioned for wireless networks include real-time monitoring of sensitive content, climate control and seismic safety, personal wireless services, traffic monitoring and vehicular applications.

7.6.1 Issues and challenges

Secure broadcasting involves two sets of participants: *receivers* that want to be certain the content they receive originates from the legitimate sender, and *senders* that may wish to limit the scope of the broadcast to a certain list of receivers (e.g. paying subscribers). It comprises several challenges such as reliability of communication despite receiver heterogeneity, well-behaved congestion control, robustness to packet loss (since a sender may not be able to retransmit lost packets, while in streaming applications at the receiver must use all the data it receives), scalability (involving large numbers of receivers) and security (authentications and confidentiality).

In point-to-point communication, data authentication can be easily achieved through symmetric mechanisms. However this is not the case in broadcasting, where a single packet (that may well be malicious) can reach a large number of recipients. The sought out property in communication networks is authentication via communication asymmetry, since symmetric authentication cannot be secure in a broadcast setting. Digital signatures, where the sender generates a signature with its private key and the receivers can verify it with their corresponding public key, have the required asymmetric property, except for the fact that they have high computation and communication overhead for the sender and the receivers. A casual observer may be led to the conclusion that in order to achieve higher efficiency, shorter keys should be used. Unfortunately whatever gains are achieved in efficiency may be lost because of weaker security.

One can therefore conclude that the requirements of efficient and secure broadcast should include (see also Perrig and Tygar (2002)):

1. low communication overhead with small keys and little or no buffering required by sender and receiver, low computation overhead for signature generation and verification of authentication,

2. reliable key distribution with secure key updating for group key secrecy, forward secrecy, as well as backward secrecy (for newly entering and exiting receivers),

3. robustness to packet loss especially in ad hoc and sensor systems, and

4. scalability for small and large dynamic groups.

Taking these issues into account we discuss in the sequel protocols for securing broadcast communication and we compare their performance.

7.6.2 BiBa broadcast authentication

In this subsection we discuss a signature scheme called *BiBa* (short for bins and balls) that is suitable for broadcast authentication due to Perrig (2001). Its main advantages are fast verification and short signatures. The idea of the BiBa scheme is the following. The signer uses the message M to generate a sufficiently long sequence of values $s_1, s_2, \ldots, s_t$ called "*balls*." The signer randomly throws many balls into a set of bins. Since the signer throws many balls, it has a high probability of observing a collision of balls in bins. The balls that participate in the collision are the signature. An adversary that wants to forge a signature only learns the balls that the signer discloses in signatures, which is a small number compared to the number of balls of the signer. Since the adversary can only throw relatively few balls into the bins, it has very low probability of observing a collision of balls in a bin.

The signer precomputes self-authenticating values (abbreviated *SEALs*) that it subsequently uses to generate BiBa signatures. This amounts to using the message M to generate a sufficiently long sequence of values $s_1, s_2, \ldots, s_t$ called "balls." These are random numbers generated in a way that the receiver can instantly authenticate with the public key. As indicated before, the simplest approach is to use a pseudo random function F as a commitment scheme. (Another approach is to use the leaves of Merkle hash tree.)

To sign a message M, the signer first computes a hash $h := H(M)$ and uses it to compute balls $s_1, s_2, \ldots, s_t$. The signer picks the function

$$G_h : x \rightarrow G_h(x) = MAC(h, x) \bmod n,$$

where MAC is HMAC-MD5, and n is the number of bins. G_h maps balls $s_1, s_2, \ldots, s_t$ into n bins. Each s_i is a self-authenticating value and the signer maps ball s_i to bin $G_h(s_i)$. Then the signer looks for two balls $s_i \neq s_j$ such that $G_h(s_i) = G_h(s_j)$. The signature is $\{s_i, s_j\}$.

The BiBa signature scheme is depicted in Figure 7.21. For verification, the verifier receives message M and signature $\{s_i, s_j\}$. It authenticates s_i, s_j (SEALs) and after computing the hash $h = H(M)$ it checks that

$$G_h(s_i) = G_h(s_j) \text{ and } s_i \neq s_j.$$

In considering the security of this signature scheme recall that the signer has t balls and the range of the hash function is $0..n - 1$. Given a message M, the probability of finding a BiBa signature is equal to the probability of finding at least one two-way collision (at least two balls end up in the same bin when throwing t balls uniformly and independently into n bins. Consider the balls $s_1, s_2, \ldots, s_t$. The first throw is arbitrary. The probability that

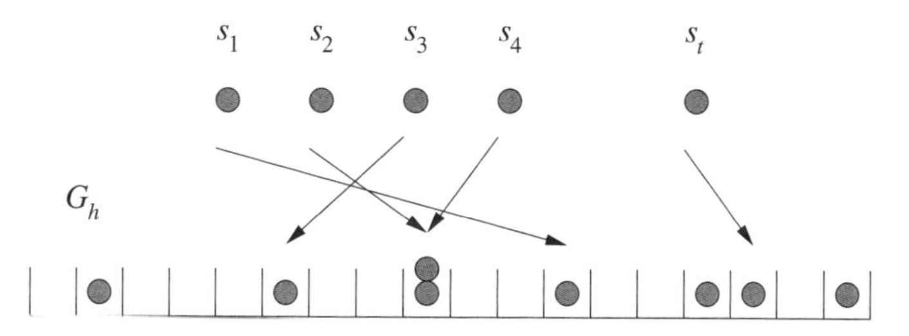

Figure 7.21 The bin-and-balls signature scheme. The function G_h maps the balls $s_1, s_2, \ldots, s_t$ into the n bins. The balls that participate in the collision are the signature. To have collisions there must be more balls than bins.

the second throw occupies a different bin from first throw is $\frac{n-1}{n}$. The probability that the third throw occupies a different bin from first and second throw is $\frac{n-2}{n}$, etc. The probability that the tth throw occupies a different bin from first $t-1$ throws is $\frac{n-t+1}{n}$. Hence, the probability of no collision is

$$\frac{n-1}{n}\frac{n-2}{n}\cdots\frac{n-t+1}{n}.$$

The probability of at least one collision is

$$\begin{aligned}1-\prod_{i=1}^{t-1}\frac{n-i}{n} &= 1-\prod_{i=1}^{t-1}\left(1-\frac{i}{n}\right)\\ &\approx 1-\prod_{i=1}^{t-1}e^{-\frac{i}{n}}\\ &= 1-e^{-\frac{(t-1)t}{2n}}\\ &\approx \frac{(t-1)t}{2n}.\end{aligned}$$

Moreover, just like in the birthday paradox, this last quantity can be made arbitrarily small by choosing n, t appropriately.

Authentication in Unix-like operating systems uses a one-time password method S/key whereby a user's real password is not directly transmitted across the network. Instead the real password combined with a short set of characters and a decrementing counter forms a single-use password. Since the single-use password is used once, passwords intercepted by a password sniffer cannot be useful to an attacker.

In broadcast systems, a sender needs to authenticate a potentially long stream of messages. The sender can only disclose a small number of SEALs before an attacker would have enough information to forge a signature. Since this would limit the number of messages that can be signed, the disclosed SEALs are replenished by using one-way hash chains similar to S/Key.

7.7 Secure location verification

The *secure location determination* problem consists of a set V of *verifiers* (typically a distributed connected system) that wish to determine the location of a *prover*. A restricted version of this is the *in-region verification* and was proposed by Sastry et al. (2003). It consists of a set V of verifiers that wish to verify whether or not a *prover* p is in a given region R (e.g. an area over which physical control can be exercised in order to restrict people's entry). Figure 7.22 depicts a verifier v located at the center of a circular region R. The verifier must decide (according to a protocol) correctly the prover's claim on whether it is inside or outside the given region, where correctly means that the protocol is 1) *complete:* if p and V both behave according to the protocol, and $p \in R$, then V will accept that $p \in R$, and 2) *secure:* if V behaves according to the protocol and accepts p's claim, then p, or a party colluding with p, has a physical presence in R. This is clearly a more restrictive model, since it does not attempt to secure the exact location, but rather the

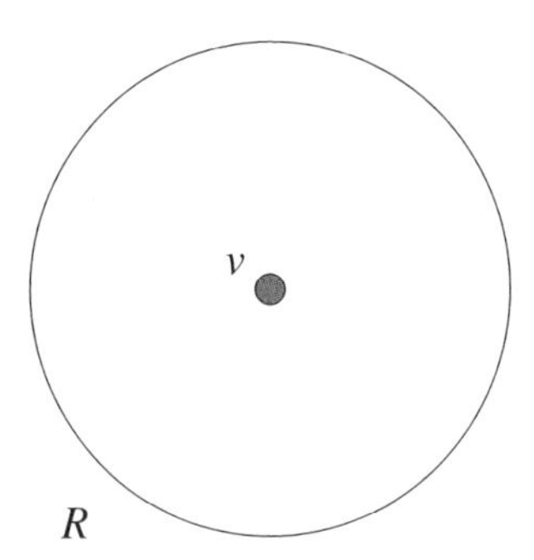

Figure 7.22 A single verifier v (inside region R) and a prover p (not depicted). The prover tries to convince the verifier node v that it is inside the region R (depicted as a circle centered at v).

location in a region. In addition, it is required that both the verifier and the prover are able to communicate using both radio frequency (RF) and sound (typically ultrasound frequencies in order to echo certain values), and the prover must be able to bind its processing delay. The rest of this section is dedicated to a description of the security protocol designed by Sastry et al. (2003).

7.7.1 Simple echo protocol

Consider the circular region depicted in Figure 7.22 with a verifier v located at the center of the circle R. The simple echo protocol uses as parameters the speed of sound $s = 331\ m/s$ and the speed of light $c = 3 \times 10^8\ m/s$ and has the following steps.

Communication Phase:

1. $p \overset{radio}{\longrightarrow} v : \ell$
   //Prover sends claimed location ℓ to verifier
2. $v \overset{radio}{\longrightarrow} p : N$
   //Verifier responds with nonce N
3. $p \overset{sound}{\longrightarrow} v : N$
   //Prover echoes same nonce N

Verification Phase:

4. v accepts iff $\ell \in R$ and elapsed time $\leq d(v, \ell) \cdot (1/c + 1/s)$,

where v is the verifier, p is the prover, N denotes a nonce, that is, an unpredictable random value, and $d(\cdot, \cdot)$ denotes Euclidean distance. According to the protocol, first p sends its location ℓ by radio to v, second, v replies by radio by sending a nonce to p, and third, p immediately echoes the packet back v using ultrasound. The time it takes v to hear the *echo* is the sum of the time it takes using radio plus the time it takes for a return packet using ultrasound, that is, the total elapsed time should be about $d(v, \ell)/c + d(v, \ell)/s$ s. Therefore v can time this process. Suppose that the prover claims to be at some location, say ℓ. Without loss of generality, it can be assumed that $\ell \in R$ (since if the claimed location is not inside R, then the verifier v can reject it immediately). If the elapsed time from the initial transmission to reception of the echo packet is more than this amount, the verifier v

rejects the prover's claim, otherwise, if the elapsed time is at most this expected amount, v accepts it. It follows that in the simple echo protocol the prover convinces the verifier that they are at most a given distance apart.

7.7.2 Echo protocol

The echo protocol, just like the simple echo protocol, consists of a communication and a verification phase. In addition, it is required that the prover can bound its processing delay to be at most Δ_p seconds and can make the verifier node aware of this maximum delay. Then, if the prover claims to be at ℓ, the verifier node can compute the time for a prover actually at ℓ to get the packet back, which is the time for the RF signal to travel from v to ℓ, a processing delay of at most Δ_p, and finally the time for the sound to travel from ℓ back to v. In addition, to prevent malicious users from grossly overstating their true processing delay, a verifier node limits the allowable region over which location claims are accepted. The simple echo protocol is sufficient, if the prover claims a null processing delay. However, as illustrated in Figure 7.23, this is not the case if the prover claims a processing delay of $\Delta_p > 0$, in which case the verifier should not engage in the protocol if the claimed location is within $\Delta_p \cdot s$ of the outside border. As a consequence, one is led to define the *Region of Acceptance (RoA)*, denoted by $RoA(v, \Delta_p)$, to be the area in which the verifier node v is sure that it can correctly verify claims for a prover if the claimed processing delay is Δ_p. Certainly, this region depends on Δ_p and $RoA(v, \Delta_p)$ indicates the region where location claims are permitted by v, if the claimed processing delay is Δ_p.

Communication Phase:

1. $p \overset{radio}{\longrightarrow}$ broadcast: (ℓ, Δ_p)
   //Prover broadcasts claimed location ℓ and processing delay Δ_p
2. $t_s \longleftarrow$ time ()
   $v \overset{radio}{\longrightarrow} p : N$
   //Verifier starts its timer and responds with a nonce N of length n bits
   //We require $\ell \in RoA(v, \Delta_p)$ and $\Delta_p \geq n/b_o + n/b_i$, where b_o and b_i
   //are the outgoing and incoming bandwidth, respectively
   if no such verifier exists or Δ_p is invalid, **abort**
3. $p \overset{sound}{\longrightarrow} v : N$
   $t_f \longleftarrow$ time ()
   //Prover echoes N over ultrasound and verifier records finish time

Verification Phase:

4. **if** sent N differs from received N **return false**
5. **if** $t_f - t_s > d(v, \ell)/c + d(v, \ell)/s + \Delta_p$ **return false**
6. **else return true**

A source of uncertainty is the time it takes to transmit packets. Links have some finite transmission bandwidth. An attacker can exploit the transmission time to launch an attack by guessing the first (or last) few bits of the nonce and send them preemptively. From the point of view of the verifier, let the outgoing and incoming bandwidth be b_o and b_i, respectively. If $b_o > b_i$ then the verifier must stop its timer only after receiving the last bit

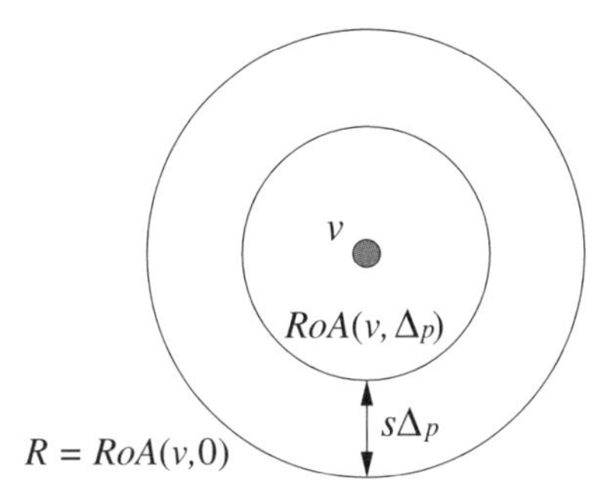

Figure 7.23 A single verifier at the center of a circular region R where there is an upper bound of Δ_p on the processing delay. The diagram illustrates the relationship between $RoA(v, \Delta_p)$ and the region $R = RoA(v, 0)$, which in this case it is obtained for $\Delta_p = 0$.

of the nonce. Indeed, if the verifier stops its timer upon receiving the first bit of the nonce then the attacker could start sending a few randomly guessed bits of the nonce slightly before actually receiving it. The verifier could then conclude the prover to be closer than it actually is. Similarly, if $b_i > b_o$ and the verifier starts timing after the entire nonce has been sent then the attacker could guess and then send the last few bits without having received them. To safeguard against such attacks the verifier must start timing before sending the first bit of the nonce, and stop timing after receiving the entire nonce. This explains why we require $\Delta_p \geq n/b_0 + n/b_i$.

In proving the correctness of the echo protocol, observe that an attacker would not be able to get the sound signal to the verifier in time, and to confirm that the prover is at the claimed location ℓ, a particular verifier node v must verify that the incoming sound signal, which includes the outgoing nonce, is received within $t_{\max} \leq d(v, \ell)/c + d(v, \ell)/s + \Delta_p$ seconds. For additional details of the proof, as well as a modified echo protocol that works for more general regions R, the reader is referred to the original paper by Sastry et al. (2003).

7.8 Security in directional antenna systems

Efficient and robust security in ad hoc networks is an especially arduous task because wireless devices are constrained in memory and processing as well as battery power. The problem is compounded by the fact that such networks have links that are inherently vulnerable to overhearing, eavesdropping, message injection and jamming attacks. Of particular interest is the wormhole attack whereby an attacker can build bogus routing information, selectively drop packets, and even create routing loops by simply embedding two transceivers connected with a high quality, low-latency link in strategically selected locations within the network region. The present section focuses on a "localization" based technique due to Hu and Evans (2004) for preventing wormhole attacks in ad hoc networks when the hosts are composed of directional antennas. On the basis of the paper of Hu and Evans (2004), after explaining the attack in detail in Section 7.8.1, we outline the zoning technique in Section 7.8.2, and finally develop three protocols of successively increasing level of security for preventing such attacks in Section 7.8.3.

7.8.1 Wormhole attacks and their impact on routing protocols

Wormhole attacks can have a devastating effect on ad hoc networks. In its simplest description, it works as follows. Suppose that hosts A and B are not neighbors and that an attacker can receive and send packets at nodes X and Y, respectively, two nodes that it controls. The attacker can make A and B believe they are neighbors by replaying packets. The attacker (Figure 7.24) replays packets received by X at node Y, and vice versa. Thus, in a wormhole attack, an attacker forwards packets through a high quality out-of-band link and replays those packets at another location in the network. Since the attacker replays packets received by X at node Y, and vice versa, if it would normally take several hops for a packet to traverse from a location near X to a location near Y, packets transmitted near X traveling through the wormhole will arrive at Y before packets traveling through multiple hops in the network. The attacker can make A and B believe they are neighbors by forwarding routing messages, and then selectively drop data messages to disrupt communications between A and B.

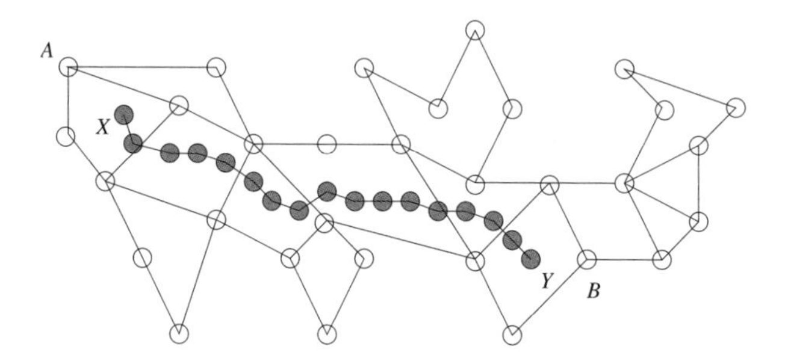

Figure 7.24 Wormhole attack.

7.8.1.1 One-hop Tunneling

It should be noted that for most routing protocols, the wormhole attack has impact on nodes beyond the wormhole endpoints' neighborhoods as well. For example, in Figure 7.25, node A will advertise a one-hop path to B so that C will direct packets towards B through A. As a consequence, the normal operation of standard routing protocols is being disrupted, for example, in on-demand routing protocols (DSR and AODV) or secure on-demand routing protocols (SEAD, Ariadne, Secure Routing Protocol (SRP), the wormhole attack can be

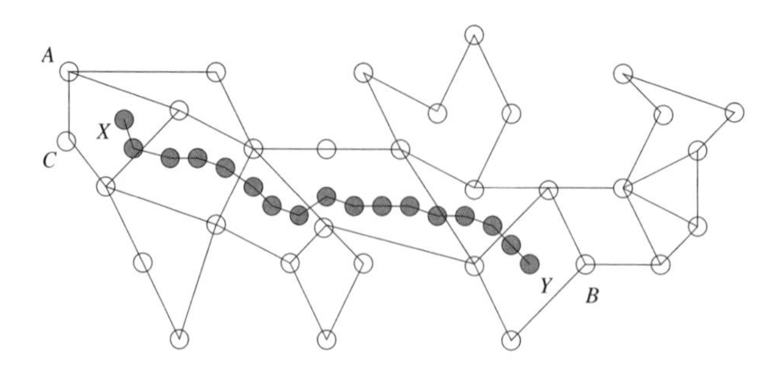

Figure 7.25 Impact on routing protocols: one-hop tunneling

implemented by tunneling ROUTE REQUEST messages directly to target nodes located near the destination node. Moreover, such ROUTE REQUEST messages may arrive earlier if tunneled through high quality channels.

7.8.1.2 Sinkholes, Longer- and Shorter-distance Attacks

Wormhole attacks, by preventing other routes from being discovered, establish full control of the route. Thus, the attacker by altering true distances between hosts can create *longer- and shorter-distance* attacks. The attacker can also discard all messages to create a denial-of-service attack, or more subtly, selectively discard certain messages to alter the function of the network. An attacker with a suitable wormhole can also easily create a *sinkhole* that attracts (but does not forward) packets to many destinations. Moreover, an intelligent attacker may be able to selectively forward messages to enable other attacks.

7.8.1.3 Selecting Disruptions Strategically

An intelligent attacker may be able to place wormhole endpoints at particular locations. Strategically placed wormhole endpoints can disrupt nearly all communications to or from a certain node and all other nodes in the network. In sensor network applications, where most communications are directed from sensor nodes to a common base station, wormhole attacks can be particularly devastating.

7.8.2 Zoning with directional sensors

Directional antennas provide several advantages in efficiency over omnidirectional ones. First of all, transmitting in a particular direction results both in a higher degree of spatial reuse of the shared medium as well as more efficient energy usage. Second, since the transmission range of directional antennas is usually larger than that of omnidirectional antennas, the number of hops for routing to a given destination is reduced and can also connect previously unconnected devices. Finally, directional antennas can increase spatial reuse and reduce packet collisions. Therefore it is not surprising that routing protocols using directional antennas can outperform omnidirectional routing protocols.

Each sensor (or antenna) can be assumed equipped with n directional antennas each spanning an angle of $2\pi/n$ radians. In general, the range of an antenna is divided into n zones, with each zone spanning an angle of $2\pi/n$ radians with a conical radiation pattern. The zones are assumed fixed with non-overlapping beam directions, so that the n zones may collectively cover the entire plane (Figure 7.26). When a node is idle, it listens to the carrier in omni mode but when it receives a message, it determines the zone on which the received signal power is maximal. It then uses that zone to communicate with the sender.

In the discussion of the forthcoming protocols, it is assumed that $n = 6$, that is, there are six zones numbered 1 to 6 oriented clockwise starting with zone 1 facing east (Figure 7.26). This orientation can be established with respect to the earth's meridian regardless of a node's physical orientation. In modern antennas, this is achieved with the aid of a magnetic needle that can remain collinear to the earth's magnetic field and which ensures that a particular zone always faces the same direction.

In *receiving* mode a node can receive messages from any direction, while when in sending *sending* mode a node can work in omni or directional mode. In *omni* mode signals

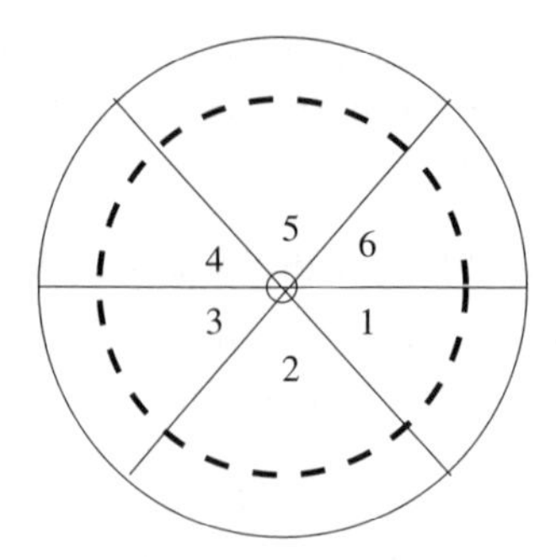

Figure 7.26 Partitioning the range of the sensors into six zones numbered $1, 2, \ldots, 6$ clockwise.

are received with a certain gain G^o, while in *directional* mode with a gain of G^d. Since a node in directional mode can transmit over a longer distance, it is apparent that $G^d > G^o$ (Figure 7.26 where omnidirectional range is depicted with a dashed circle).

7.8.3 Protocols for securing neighbor discovery

Before discussing protocols, we outline the assumptions made concerning security as well as notation. All communication channels are bidirectional (i.e. symmetric, see Figure 7.27), therefore if A can hear B then B can hear A. A mechanism is available for establishing secure links between all pairs of nodes and ensuring that all critical messages are encrypted. Moreover, the sensor network must be "relatively" dense (this density requirement will become apparent in the sequel).

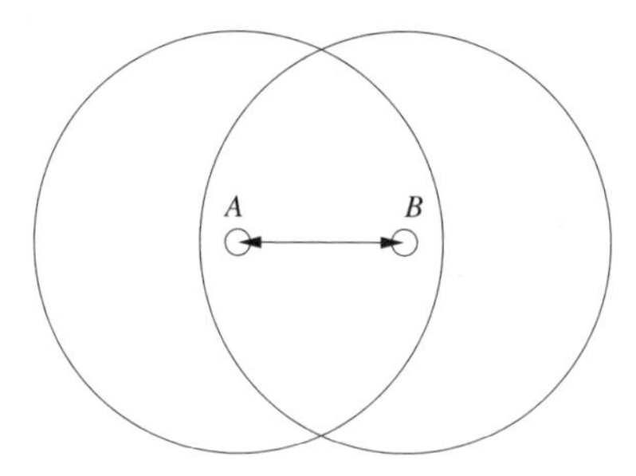

Figure 7.27 Bidirectional communication link.

The following notation will be used in subsequent protocol. Capital letters at the beginning of the alphabet $(A, B, C, \ldots)$ designate *legitimate* nodes, while capital letters at the end of the alphabet $(X, Y, \ldots)$ designate wormhole endpoints. Moreover, the letter R denotes a *nonce*. As usual, $E_{KAB}(M)$ means that message M is encrypted with a key shared by nodes A and B. The notation *zone* denotes a given directional element in the range from 1 to 6, while $\overline{zone}$ denotes the opposite directional element. For example, if $zone = 1$ then $\overline{zone} = 4$. Further, $zone(A, B)$ denotes the zone in which node A hears node B, while *neighbors*$(A, zone)$ is the set of nodes within one (directional distance) hop in direction *zone* of node A.

Equipped with this notation, in the sequel three protocols are presented for enabling secure neighbor discovery. For each protocol given we also discuss deficiencies and how to remedy them in the subsequent protocol.

7.8.3.1 Protocol 1: Secure Neighbor Discovery

The protocol for secure neighbor discovery is as follows.

1. $A \rightarrow Region$: HELLO and ID_A.
2. $N \rightarrow A : ID_N | E_{KNA}(ID_A | R | zone(N, A))$.
3. $A \rightarrow N : R$.

In Step 1, A sends a HELLO message with its ID. In Step 2, all nodes that hear the HELLO message send their node ID and an encrypted message to the announcer. The encrypted message contains the announcer's ID, a random challenge nonce, and the zone in which the message was received. Finally, in step 3, A decrypts the message and verifies that it contains its node ID. It verifies $zone(A, N) = \overline{zone}(N, A)$. If correct, it adds the sending neighbor to its neighbor set for $zone(A, N)$. If the message was not received in the appropriate zone, it is ignored. Otherwise, the announcer transmits the decrypted challenge nonce to the sending neighbor. Upon receiving the correct nonce, the neighbor inserts the announcer into its neighbor set.

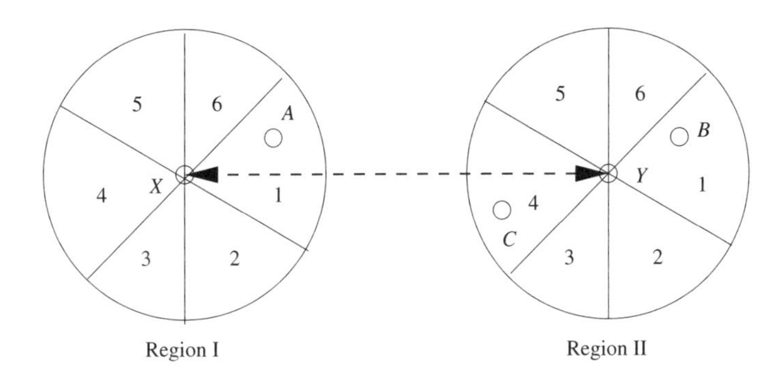

Figure 7.28 Wormhole vulnerability in the first protocol.

There are several weaknesses with the first protocol. The first is a *wormhole vulnerability* (Figure 7.28). An attacker with a wormhole can establish a false distant neighbor. The adversary establishes a wormhole between X and Y, and can trick A and C into accepting each other as neighbors by forwarding messages since they are in opposite zones relative to the respective wormhole endpoints. Another vulnerability is due to *overhearing*. B will hear A and C from the west through the wormhole ($zone(B, A) = zone(B, C) = 4$), and C will hear A directly from the east ($zone(A, C) = \overline{zone}(C, A) = 1$) and C will hear B from the west through the wormhole ($zone(C, B) = \overline{zone}(B, C) = 4$).

7.8.3.2 Protocol 2: Verified Neighbor Discovery

If nodes cooperate with their neighbors they can prevent wormholes since the attacker will only be able to convince nodes in particular regions that they are neighbors. Assume the adversary has one transceiver at each end of the wormhole. An adversary can only trick nodes that are in opposite directions from the wormhole endpoints into accepting each other as neighbors. Hence, nodes in other locations can establish the announcer's legitimacy. Such nodes are called *verifiers*. How do we prevent verifiers from acting through the wormhole? Node C cannot act as a verifier for the link AB since the wormhole attacker could make a node on the other end of the wormhole appear. Node D (Figure 7.29) could act as a verifier, if it satisfies the verifier properties (listed below). This leads to a more precise definition. A *valid verifier* V for the link $A \leftrightarrow B$ must satisfy the following properties:

1. $zone(B, A) \neq zone(B, V)$.
   Node B hears V in a different zone from node A, hence it knows A and V are in different locations, and both cannot be coming through a single wormhole endpoint.

2. $zone(B, A) \neq zone(V, A)$.
   Node B and V hear node A from different directions. A wormhole can deceive nodes in only one direction. So if both B and V are directionally consistent with A in different directions ($zone(B, A) = \overline{zone}(A, B)$ and zone $(V, A) = \overline{zone}(A, V)$), then they know A is not being retransmitted through a wormhole.

Now we are in a position to provide the improved protocol for verified neighbor discovery. First three steps are exactly as in Protocol 1. The remaining three steps are as follows.

4. $N \rightarrow Region$: $INQUIRY|ID_N|ID_A|zone(N, A)$.

5. $V \rightarrow N$: $ID_V|E_{KNV}(ID_A|zone(V, N))$.

6. $N \rightarrow A$: $ID_N|E_{KAN}(ID_A|ACCEPT)$.

In Step 4, all neighbor nodes that hear the HELLO message broadcast an inquiry in all directions except for the received direction and the opposite direction. So, if N received the announcement in zone 1, it will send inquiries to find verifiers to zones 2, 3, 5 and 6.

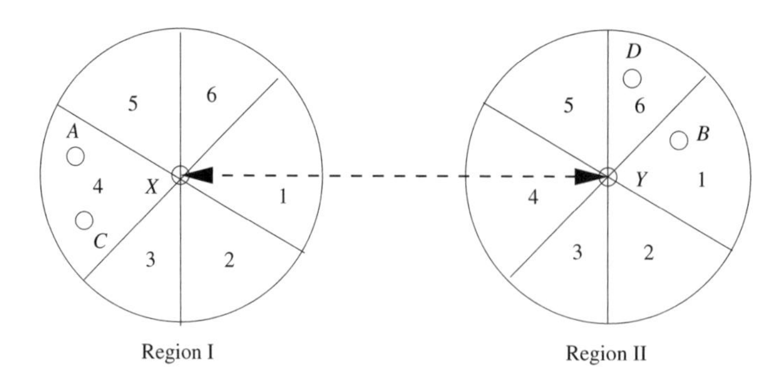

Figure 7.29 Cooperating with neighbors to prevent protocol vulnerabilities.

The message includes $zone(N, A)$, so prospective verifiers can determine if they satisfy the verification properties by having heard A in a different zone. In Step 5, nodes that receive the inquiry and satisfy the verification properties respond with an encrypted message. This message confirms that the verifier heard the announcement in a different zone from N and has completed Steps 1–3 for the protocol to authenticate A and its relative position. To continue the protocol, N must receive at least one verifier response. If it does, it accepts A as a neighbor, and sends a message to A as described in step 6. After receiving the acceptance messages, the announcer adds N to its neighbor set.

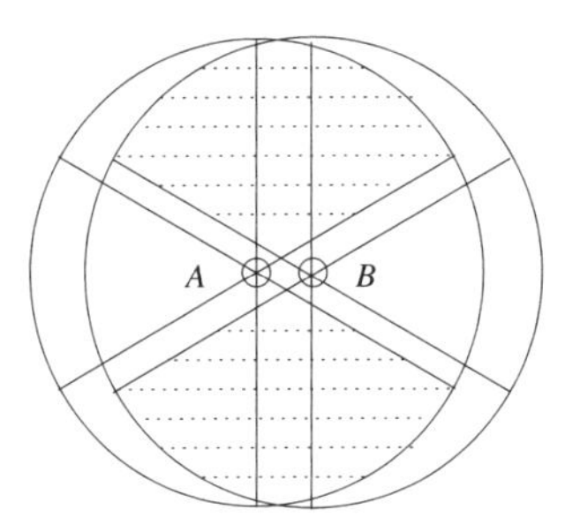

Figure 7.30 Verifier region.

The shaded area, in Figure 7.30, is the verifier region of nodes A and B in the verified neighbor discovery protocol. If there is a node in the shaded region, it can act as a verifier for A and B. Now you can see why you need the sensor network to be dense. It is required that with high probability there is a verifier node in the shaded region to enable A and B to have a successful protocol verification. The shaded region determines a given area. The probability must be sufficiently high that sensors lie within this region and can act as verifiers! The verifier region may still exist when two nodes are slightly out of radio range, and a smart adversary can use this to mislead them into believing they are neighbors.

7.8.3.3 Protocol 3: Strict Neighbor Discovery

Protocol 2 suffers from a wormhole vulnerability known as *Worawannotai attack*. Node B is located just beyond the transmission range of node A. If there is a valid verifier in those areas, the attacker can just put one node in between A and B (node X) and use it to listen to and retransmit messages between A and B (Figure 7.31). Nodes A and B will mistakenly confirm they are neighbors using verifier V, but the attacker will have control over all messages between A and B.

Figure 7.30 depicts the verifier region of two adjacent nodes. Nodes in the shaded region (if any) can act as verifiers for both A and B. Unfortunately, the verifier region may still exist even when two nodes are not within range, in which case an adversary can exploit this to make them think they are neighbors. Figure 7.31 depicts such a Worawannotai attack in which an adversary convinces nodes A and B (two nearby but not neighboring nodes) they are neighbors.

How do we prevent the Worawannotai attack? There are two areas (a, b) where a valid verifier for this protocol could be located. If valid verifier was in those areas, the attacker

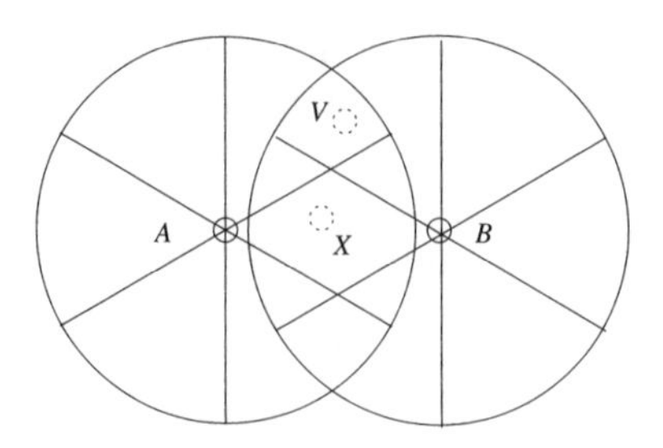

Figure 7.31 Worawannotai attack.

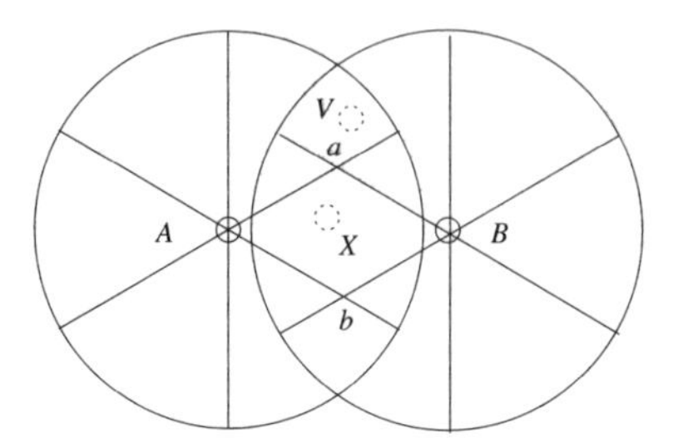

Figure 7.32 Preventing the Worawannotai attack.

could just put one node in between A and B (node X) and use it both to listen to and retransmit messages between A and B (Figure 7.32). A and B mistakenly confirm they are neighbors using verifier V, but the attacker will have control over all messages between A and B.

This strict verification can be enabled when a valid verifier V for the link $A \leftrightarrow B$ satisfies not only the two conditions: 1) $zone(B, A) \neq zone(B, V)$, 2) $zone(B, A) \neq zone(V, A)$ previously defined, but also a third condition: 3) $zone(B, V)$ cannot be both adjacent to $zone(B, A)$ and adjacent to $zone(V, A)$. Thus, in addition to the first two conditions, which guarantee that the adversary cannot replay the confirmation message from verifiers, there is a third condition that ensures that the verifier region is empty when two nodes are out of radio range, so the adversary cannot use this to conduct a Worawannotai attack. The verifier region determined by the previous three rules is depicted by the four regions a, b, c, d (Figure 7.33). These areas are the verifier regions of node A and B in the strict neighbor discovery protocol. Geometric and analytical conditions, as well as a detailed analysis of when a Worawannotai attack is possible, are carried out in the paper by Hu and Evans (2004).

7.9 Bibliographic comments

A good source on cryptographic and security issues of any kind is the seminal handbook of applied cryptography by Menezes et al. (1996). There are four articles dedicated to issues in wireless security in the handbook by Boukerche (2006): topics included range from security architectures, security in ad hoc networks, pervasive computing and mobile commerce. There is a vast literature on authentication techniques. The part from the book of

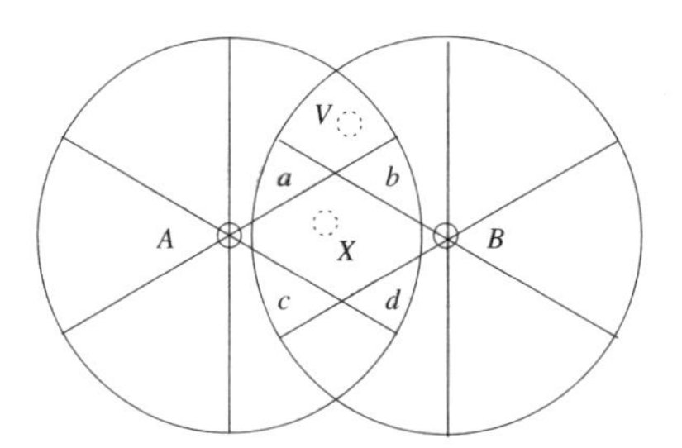

Figure 7.33 Verifier region.

Kaufman et al. (2002) devoted to authentication is still one of the best introductory around. The books by Bishop (2003) and Stallings (2007) also have useful information. A survey of information authentication can be found in Simmons (1988). Clarke and Furnell (2005) provides a survey of attitudes and practices in user authentication on mobile telephones. It is fair to say that it is impossible to understand routing security without a good understanding of routing protocols. A good source for this material is the book by Perlman (2000) while the book by Perkins and Bhagwat (2001)) is more focused on ad hoc networks.

The presentation in Section 7.5.3 on securing ad hoc distance vector routing protocols is based on the work of Hu et al. (2003a). Of related interest is the work of Kumar (1993) (which describes mechanisms to secure distance vector routing protocols using message authentication codes), Smith et al. (1997) (which presents countermeasures that provide security) and Zapata (2002) (which proposes Secure Ad hoc On-demand Distance Vector (SAODV), a protocol that uses a new one-way hash chain for each route discovery to secure the metric field in an RREQ packet) and Wan et al. (2004) (which proposes S-RIP, a protocol with trust mechanisms for securing RIP). Further discussions on rushing attacks can be found in Hu et al. (2003d). Hu and Perrig (2004) provides a survey of secure ad hoc protocols, and Hu et al. (2003b) discusses efficient security mechanisms, while Hu et al. (2005) describe the Ariadne protocol.

The BiBa (Bins-and-Balls) signature scheme is due to Perrig (2001). Mitzenmacher and Perrig (2002) provide improvements and Reyzin and Reyzin (2002) analyze and improve the signature scheme and provides additional one-way schemes and comparisons with BiBa. Perrig and Tygar (2002) is source book on secure broadcast communications for wired and wireless networks.

Section 7.5.2 on preventing malicious packet dropping is largely based on the work of Just et al. (2003). The subject has been considered by several other authors including Awerbuch et al. (2002) that propose a secure routing protocol for resisting byzantine failures in a wireless ad hoc network, Marti et al. (2000) that propose and implement two protocols for detecting and mitigating misbehaving nodes in wireless ad hoc networks by overhearing neighborhood transmissions, and Buchegger and Boudec (2002) who developed the CONFIDANT protocol that enables each node to monitor the behavior and reputation of its neighbors.

In addition to the paper by Sastry et al. (2003) that was cited in Section 7.7, there is extensive literature on the location verification problem in wireless networks. These include localization techniques that can be used to localize devices (Chapter 6 on location awareness and citations thereof). Additionally, on the basis of an idea by Brands and Chaum (1993) of bounding the delay in challenge response schemes, new protocols can be built exploiting a *leash* (essentially a time-to-leave parameter) that specifies how long a packet can stay alive during the protocol communication process (see Hu et al. (2003c)).

Wormholes generally arise because of the ability of certain individual system messages to take shortcuts, and re-emerge safely at the desired point, apparently much faster than what is allowed by the speed of light (see Verissimo (2006)). Wireless systems can be particularly vulnerable to this attack and the "localization" approach described in Section 7.8 is based on the paper by Hu and Evans (2004), where the reader can find additional details on the correctness of the protocols as well as additional literature. Another approach to detecting wormhole attacks, due to Hu et al. (2003c), is to use packet leashes. A temporal packet leash is essentially a *time-to-live* parameter placing an upper bound on the lifetime of a packet thus restricting its travel distance. Packet leashes require robust clock synchronization and precise knowledge of location. The reason is that the sender must include the transmission time and location information in the packet while the receiver should confirm that the packet could have traveled the distance between the sender and itself within the time between reception and transmission. A distributed point of view of wormholes and a discussion of its impact to distributed systems models is available in the survey paper by Verissimo (2006).

7.10 Exercises

1. A challenge response procedure can be run afterward to complete a biometrics-based key establishment procedure between the initiator and responder to test if both values k and k' match. Explain the details of the procedure.

2. The concept of entropy can be used to evaluate the quality (randomness) of a biometrics feature to serve as a key encryption key. Explain how.

3. Biometric derived values v and v' can be used as proof of co-location. In that case, it is sufficient for v and v' to be closed, in some sense. If v, v' are strings of bits $v = v_0, v_1, \ldots, v_{n-1}$ and $v' = v'_0, v'_1, \ldots, v'_{n-1}$ then their Hamming distance is the number of i such that $v_i \neq v'_i$). This can be used to measure the degree of matching. Explain how.

4. Augment the biometrics-based encryption protocol to include an authentication scheme with a replay protection mechanism.

5. Prove the correctness of the verification algorithm of the RSA and ElGamal signature schemes described in Section 7.1.1.

6. Elaborate why when encrypting is done before signing in the RSA scheme (Section 7.1.1), a forger can detach the original message and sign it him(her)self. Show how can this problem be solved by combining signing with encryption.

7. In many applications it is also important to provide a proof of when a message was signed. Suppose we have a hash function h and a key K. The technique, known as

timestamping, timestamps a given message x by selecting a current message *pub* (e.g. a Stock Market quote) and computing $z = h(x)$, $z' = h(x\|pub)$, $y = Sig_K(z')$. The resulting signature is (z, pub, y).

8. There are several complex issues arising from the three fundamental authentication protocols three-way handshake, trusted third party, and public key presented in Section 7.1.1. The reader is advised to study them carefully and analyze their behavior.

9. Discuss the security properties of the following simple MACs.

   (a) Let the integer N be a shared secret known only by the participants and define for any message x,
   $$MAC(x) = x \bmod N$$

   (b) Let $s_1 s_2 \cdots s_n$ be an n-bit sequence that is a secret shared only by the participants. Let a message $x_1 x_2 \cdots x_n$ be n-bit long. Define
   $$MAC(x_1 x_2 \cdots x_n) = y_1 y_2 \cdots y_n,$$
   where $y_i = x_i \oplus s_i$, for $i = 1, 2, \ldots, n$.

   (c) Same as in Part 9b of the exercise, but the output of the MAC is a secret predefined collection of output bits (among the n ones).
   $$MAC(x_1 x_2 \cdots x_n) = y_{i_1} y_{i_2} \cdots y_{i_m},$$

10. Soccer games involve 23 participants, 22 players from the two competing teams plus the referee. Show that with probability more than 50 % a random game has two participants with the same birthday? **Hint:** Show that the probability that two participants have the same birthday is $1 - \frac{365!/342!}{365^{23}} = 0.5073$.

11. The following exercise discusses Sperner systems (see Bollobas (1986)) introduced in Section 7.1.2. Show that

    (a) given k, the set of all k-element subsets of a given set is a Sperner system, and

    (b) for a given set of size $2n$ the maximum possible size of a Sperner system is $\binom{2n}{n}$.

12. The Bos–Chaum scheme presented in Section 7.1.2 gives a 50% improvement over Lamport's signatures. To see this, use Stirling's formula on $n!$ to show that $n \approx \frac{k}{2}$. **Hint:** The parameter k satisfies $2^k \approx 2^{2n} \frac{\sqrt{2}}{\sqrt{\pi n}}$.

13. Show that the injective function $\phi : \{0, 1\}^k \to [B]^n$ required in the Bos–Chaum scheme (introduced in Section 7.1.2) can be constructed with the following algorithm. Given message $M = M_1 \cdots M_k$ choose n minimal such that $\binom{2n}{n} > x := \sum_{i=1}^{k} M_i 2^{i-1}$. Show that the following algorithm represents x as a sum of binomial

coefficients.

$$
\begin{aligned}
&\text{set } \phi(x) = \emptyset, t = 2n, e = n \\
&\textbf{while } t > 0 \textbf{ do} \\
&t = t - 1 \\
&\textbf{if } x > \binom{t}{e} \textbf{ then} \\
&\quad x = x - \binom{t}{e} \\
&\quad e = e - 1 \\
&\quad \phi(x) = \phi(x) \cup \{t + 1\}
\end{aligned}
$$

14. There is no general agreement on what qualifies as a DoS attack in a network. The approach of Hu et al. (2005) specifies the strength of a DoS attack by the ratio between the total work performed by nodes in the network and the work performed by the attacker. Given this model,

    (a) what is the strength of a DoS attack when the attacker sends a single packet that results in a packet flood throughout the network?

    (b) what are upper and lower bounds on the strength of DoS attacks?

15. Consider the distributed probing scheme in Section 7.5.2.

    (a) Why should the probing path selection scheme avoid path redundancy? Are there any disadvantages in doing this?

    (b) Are there other ways to achieve redundancy?

    (c) Discuss advantages and disadvantages of using a spanning tree?

16. Figure 7.34 depicts a sequence number rush attack. Attacker a is 4 hops, and v is 3 hops from destination node d. A usual policy is for a node to use the most recent

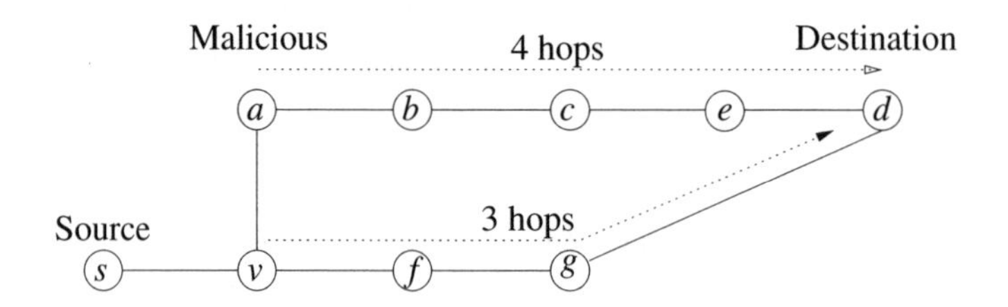

Figure 7.34 Sequence number attacks.

update among the ones it receives. If a hears of new routing updates from d before they reach v then a can rush this update to v. This means that s forwards traffic to d via a until it receives update from f (which contains a shorter route). Elaborate on the following remedies.

(a) Delayed route update adoption policy (i.e. always use the shortest route from the previous sequence number).

(b) Use hash-tree chains.

For additional details see Hu et al. (2003a).

17. Hash-tree chains require large overhead for each step of the chain. Study the following possibilities.

    (a) *Using a single value per node.*
    In a small network, each value b_j corresponds to a node and no two nodes share a value. The tree chain is constructed by encoding the node id.

    (b) *Using one t-tuple of values for nodes.*
    Previous method has overhead $O(n)$ per step of the chain, which may be too big in a larger network.

    For additional details see Hu et al. (2003a).

18. Show how to improve the main probing algorithm in Section 7.5.2 by using binary search.

19. Discuss the correctness of the simple echo protocol that was presented in Section 7.7.1.

    (a) In which way does processing delay and transmission affect its behavior (see Sastry et al. (2003))?

    (b) What can you say about the protocol when the verifier is not at the center of the circular region R (see Sastry et al. (2003))?

20. ($\star$) Prove the correctness of the echo protocol presented in Section 7.7.2 on a circular region (see Sastry et al. (2003)).

21. How can the echo protocol generalize for non-circular regions?

22. Consider wormhole attacks in an ad hoc network as defined in Section 7.8.

    (a) Elaborate on how in a wormhole an attacker can attain longer and shorter distances, DoS and sinkhole attacks.

    (b) How can an attacker select disruptions strategically? Establish and discuss a few scenarios.

23. In sensor networks, traffic is directed from sensors to a base station usually towards a "certain" direction. For the sake of this exercise assume that a set of sensors is located within a square (planar) region Elaborate why (Figure 7.35).

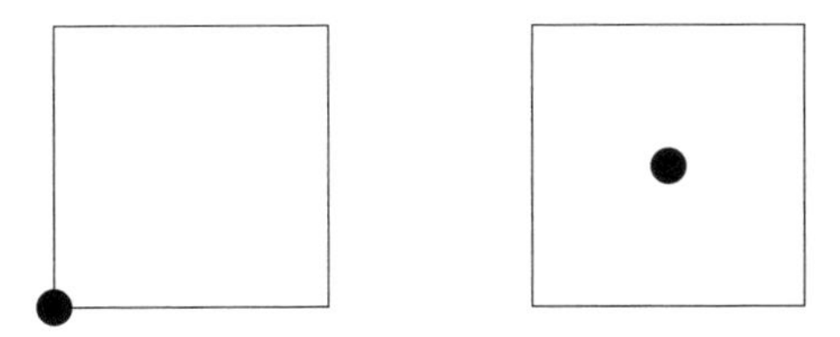

Figure 7.35 Impact of location of base stations on disrupting traffic in a sensor network delimited by a square region.

(a) if the base station is at the corner of the network, a wormhole with one endpoint near the base station and the other endpoint one hop away (from base station) will be able to attract nearly all traffic from sensor nodes to the base station, while

(b) if the base station is at the center of the network, a single wormhole will be able to attract traffic from a quadrant of the network.

24. There are two types of antennas, *omnidirectional* and *directional* (Figure 7.36). Omnidirectional are *isotropic* in the sense that same power (radiation) is transmitted in all directions. *Dipoles* are the simplest type of antennas. The *half-wave* (also known as

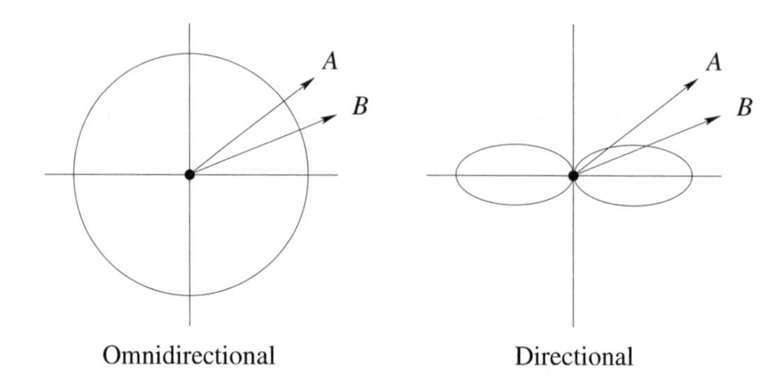

Figure 7.36 Omnidirectional and directional antennas

Hertz) dipole consists of two straight collinear conductors of equal length separated by a small feeding gap. Moreover, the length of the antenna is half of the signal that can be transmitted most efficiently. Similarly, the *quarter-wave* (also known as *Marconi*) is the type used for portable radios. The half-wave dipole has an omnidirectional pattern only in one planar dimension and a figure eight in the other two. For example, the side view along the xy- and zy-plane are figure eight, while in the zx-plane it is uniform (or omnidirectional).

(a) The *beam width* is a measure of the directivity of the antenna. It is the angle within which the power radiated by the antenna is at least half of what it is in the most powerful direction. For this reason it is called *half-power beam width*. When an antenna is used for reception, then the radiation pattern becomes *reception* pattern. How does the beam width affect the protocols in Section 7.8?

(b) Directional antennas have preferred patterns (like an ellipse): For example, in Figure 7.36 B receives more power than A. The radiation patterns of directional antennas may affect the security considerations of the protocols given in Section 7.8. How can these protocols be modified in order to take this into account?

25. The n zones defined in Section 7.8.2 are assumed to cover the entire plane (2π radians). How is the security of the protocol affected given that adjacent directed

antenna beams may have a "certain" overlap? What changes to the protocols may be required to handle this problem?

26. Assume that two nodes A and B are at distance more than their transmission range r, say $r + \epsilon$, for some $\epsilon > 0$. Compute the size of the areas that could contain false verifiers in the Worawannotai attack (For more details see the paper by Hu and Evans (2004)).

Bibliography

Aleksandrov AD 1963 A general view of mathematics In *Mathematics: Its Content, Methods, and Meaning* (ed. Aleksandrov AD, Kolmogorov AN and Lavrentev MA), MIT Press, Cambridge, Massachusetts. Translated from the 1956 Russian edition by S. H. Gould and T. Bartha.

Al-Karaki J and Kamal A 2004 Routing techniques in wireless sensor networks: a survey. *IEEE Wireless Communications* **11**(6), 6–28.

Alonso G, Kranakis E, Sawchuk C, Wattenhofer R and Widmayer P 2003a Randomized protocols for node discovery in ad-hoc multichannel broadcast networks In *2nd Annual Conference on Adhoc Networks and Wireless (ADHOCNOW'03), Montreal, Canada, Oct 09-10, 2003* (ed. Pierre S, Barbeau M and Kranakis E), vol. 2865 of *Lecture Notes in Computer Science*, pp. 104–115. Springer.

Alonso G, Kranakis E, Wattenhofer R and Widmayer P 2003b Probabilistic protocols for node discovery in ad-hoc, single broadcast channel networks In *WMAN (Workshop on Wireless Mobile Adhoc Networks)*. IPDPS Nice, France, April 22–26.

Awerbuch B, Holmer D, Nita-Rotaru C and Rubens H 2002 An on-demand secure routing protocol resilient to byzantine failures *WiSe: Proceedings of the ACM Workshop on Wireless Security*. Atlanta, GA.

Baker F 1995 *Requirements for IP Version 4 Routers* The Internet Society, Request for Comments: 1812, www.ietf.org. June.

Bao L and Garcia-Luna-Aceves JJ 2002 Transmission scheduling in ad hoc networks with directional antennas *Proceedings of the Eighth Annual International Conference on Mobile Computing and Networking (MOBICOM-02)*, pp. 48–58. ACM Press, New York.

Bao L and Garcia-Luna-Aceves JJ 2003 Distributed dynamic channel access scheduling for ad hoc networks. *Journal of Parallel and Distributed Computing* **63**(1), 3–14.

Barbeau M, Hall J and Kranakis E 2006 Detecting impersonation attacks in future wireless and mobile networks In *Workshop on Secure Mobile Ad-hoc Networks and Sensors* (ed. Burnmester M and Yvinec A), vol. 4074 of *Lecture Notes in Computer Science*. Springer.

Barbeau M, Kranakis E, Krizanc D and Morin P 2004 Improving distance based geographic location techniques in sensor networks In *Ad-Hoc, Mobile, and Wireless Networks: Third International Conference (ADHOC-NOW)* (ed. Nikolaidis I, Barbeau M and Kranakis E), vol. 3158 of *Lecture Notes in Computer Science*, pp. 197–210. Springer.

Barbeau M, Kranakis E and Lambadaris I 2007 Establishing a communication infrastructure in ad hoc networks In *Algorithms and Protocols for Wireless Ad Hoc and Sensor Networks* (ed. Boukerche A), John Wiley & Sons.

Barrière L, Fraigniaud P, Narayanan L and Opatrny J 2003a Dynamic construction of bluetooth scatternets of fixed degree and low diameter *Proceedings of the Fourteenth Annual ACM-SIAM Symposium on Discrete Algorithms (SODA-03)*, pp. 781–790. ACM Press, New York.

Principles of Ad hoc Networking Michel Barbeau and Evangelos Kranakis

Barrière L, Fraigniaud P, Narayanan L and Opatrny J 2003b Robust position-based routing in wireless ad hoc networks with irregular transmission ranges. *Wireless Communications and Mobile Computing* **3**(2), 141–153.

Basagni S 1999 Distributed clustering for ad hoc networks *Proceedings of International Symposium on Parallel Architectures, and Networks (I-SPAN'99)*, pp. 310–315. Perth/Fremantle, Australia.

Bashe CJ, Johnson LR, Palmer JH and Pugh EW 1986 *IBM's Early Computers*. MIT Press, Cambridge, MA.

Beech W, Nielsen D and Taylor J 1998 *AX.25 Amateur Packet-Radio Link-Layer Protocol, Version 2.2*. Tucson Amateur Packet Radio Corporation.

Bertsekas D and Gallager R 1992 *Data Networks*, 2nd edition, Prentice-Hall.

Bishop M 2003 *Computer Security: Art and Science*. Addison Wesley.

Blom G and Thoburn D 1982 How many random digits are required until given sequences are obtained. *Journal of Applied Probability* **19**, 518–531.

Bluetooth, SIG, Inc., 2001a *Specification of the Bluetooth System*, www.bluetooth.com.

Bluetooth, SIG, Inc., 2001b *Specification of the Bluetooth System*, www.bluetooth.com.

Bollobas B 1986 *Combinatorics*. Cambridge University Press, Cambridge, MA.

Boone P, Chavez E, Gleitzky L, Kranakis E, Opartny J, Salazar G and Urrutia J 2004 Morelia test: improving the efficiency of the Gabriel test and face routing in ad-hoc networks In *SIROCCO 2004* (ed. Kralovic R and Sykora O), vol. 3104 of *Lecture Notes in Computer Science*. Springer.

Bos JNE and Chaum D 1992 Provable unforgeable signatures In *Advances in Cryptology – CRYPTO '92* (ed. Brickell EF), vol. 740 of *Lecture Notes in Computer Science*, pp. 1–14. Springer.

Bose P, Morin P, Stojmenovic I and Urrutia J 2001 Routing with guaranteed delivery in ad hoc wireless networks. *Wireless Networks* **7**, 609–616.

Boukerche A 2006 Security issues in wireless networks In *Handbook of Algorithms for Wireless Networking and Mobile Computing* (ed. Boukerche A), pp. 905–995. Chapman & Hall/CRC Press, New York.

Brands S and Chaum D 1993 Distance-bounding protocols (extended abstract) In *Advances in Cryptology – EUROCRYPT 93* (ed. Helleseth T), vol. 765 of *Lecture Notes in Computer Science*, pp. 344–359. Springer.

Breeze, Wireless Communications Ltd. 1999 *Radio Signal Propagation*, Tel Aviv.

Buchegger S and Boudec JYL 2002 Performance analysis of the CONFIDANT protocol *MobiHoc*, pp. 226–236. ACM.

Caffery J and Stüber G 1998a Overview of radiolocation in CDMA cellular systems. *IEEE Communications Magazine* **36**, 38–45.

Caffery J and Stüber G 1998b Subscriber location in CDMA cellular networks. *IEEE Transactions on Vehicular Technology* **47**(2), 406–416.

Capkun S, Hamdi M and Hubaux JP 2002 GPS-free positioning in mobile ad hoc networks. *Cluster Computing* **5**, 157–167.

Cavafy C 1976 The morning sea In *The Complete Poems of Cavafy* (ed. Dalven R), Harcourt Brace Jovanovich, A Harvest Book, New York and London.

CCITT 1997 ITU-T, recommendation X.509 (1997 e): Information technology, open systems interconnection, the directory authentication framework.

Chavez E, Dobrev S, Kranakis E, Opartny J, Stacho L and Urrutia J 2004a Route discovery with constant memory in oriented planar geometric networks In *Proceedings of Algosensors 2004* (ed. Nikoletseas S and Rolim J), vol. 3121 of *Lecture Notes in Computer Science*, pp. 147–156. Springer.

Chávez E, Dobrev S, Kranakis E, Opatrny J, Stacho L and Urrutia J 2004b Traversal of a quasi-planar subdivision without using mark bits. *Journal of Interconnection Networks* **5**(4), 395–407.

Chavez E, Dobrev S, Kranakis E, Opartny J, Stacho L and Urrutia J 2005 Half-space proximal: a new local test for extracting a bounded dilation spanner *In OPODIS (workshop on Principles of Distributed Systems), Pisa, Italy, December 2005*, vol. 3974 of *Lecture Notes in Computer Science*. Springer.

Chavez E, Dobrev S, Kranakis E, Opartny J, Stacho L and Urrutia J 2006 Local construction of planar spanners in unit disk graphs with irregular transmission ranges In *Proceedings of LATIN 2006*, March 20-24, Valdivia, Chile (ed. Correa J, Hevia A and Kiwi M), vol. 3887 of *Lecture Notes in Computer Science*, pp. 286–297. Springer.

Cherukuri S, Venkatasubramanian K and Gupta S 2003 Biosec: a biometric based approach for securing communication in wireless networks of biosensors implanted in the human body *Proceedings of 2003 International Conference on Parallel Processing Workshops*, pp. 432–439. Kaohsiung, Taiwan.

Cichon J, Krzywiecki L, Kutylowski M and Wlaz P 2005 Anonymous distribution of encryption keys in cellular broadcast systems In *MADNES* (ed. Burmester M and Yasinsac A), vol. 4074 of *Lecture Notes in Computer Science*, pp. 96–109. Springer.

Clark BN, Colbourn CJ and Johnson DS 1990 Unit disk graphs. *Discrete Mathematics* **86**, 165–177.

Clarke NL and Furnell S 2005 Authentication of users on mobile telephones - A survey of attitudes and practices. *Computers and Security* **24**(7), 519–527.

Clausen T and Jacquet P 2003 *Optimized Link State Routing Protocol (OLSR)*. The Internet Society, Request for Comments: 3626.

Cooper M 2003 Antennas get smart. *Scientific American*, 49–55.

Cormen T, Leiserson C and Rivest R 1990 *Introduction to Algorithms*. MIT Press.

Costas J 1959 Poisson, Shannon and the radio amateur. *Proceeding of the IRE* **47**(12), 2058–2068.

Diffie W and Hellman M 1976 New directions in cryptography. *IEEE Transactions on Information Theory* **22**(6), 644–654.

Dijkstra E 1959 A note on two problems in connection with graphs. *Numerische Mathematik* **1**, 269–271.

Donahoo MJ and Calvert KL 2001 *The Pocket Guide to (TCP)/IP Sockets*. Morgan Kaufmann Publishers, San Francisco, CA.

Ephremides A and Truong TV 1990 Scheduling broadcasts in multihop radio networks. *IEEE Transactions on Communications* **38**(4), 456–460.

Erdös P and Renyi P 1959 On random graphs. *Publicationes Mathematicae* **6**, 290–297.

Feeney L and Nilsson M 2001 Investigating the energy consumption of a wireless network interface in an ad hoc networking environment *Proceedings of Twentieth Annual Joint Conference of the IEEE Computer and Communications Societies (INFOCOM)*, vol. 3, pp. 1548–1557. Anchorage Alaska.

Figuel W, Shepherd N and Trammell W 1969 Vehicle location using signal strength measurements in cellular systems. *IEEE Transactions on Vehicular Technology* **VT-18**, 105–110.

FIPS 2002 *Specifications for the Secure Hash Standard*, PUB 180-2 http://csrc.nist.gov/publications/fips/fips180-2/fips180-2.pdf, August.

Gabriel KR and Sokal RR 1969 A new statistical approach to geographic variation analysis. *Systemic Zoology* **18**, 259–278.

Garcés R and Garcia-Luna-Aceves J March 26-30, 2000 Collision avoidance and resolution multiple access for multichannel wireless networks *INFOCOM 2000*. IEEE.

Garey MR and Jhonson DS 1979 *Computers and Intractability*. WH Freeman, San Francisco, CA.

Gasieniec L, Pelc A and Peleg D 2001 The wakeup problem in synchronous broadcast systems. *SIAM Journal on Discrete Mathematics* **14**(2), 207–222.

Georgiou C, Kranakis E, Marcelin-Jimenez R, Rajsbaum S and Urrutia J 2005 Distributed dynamic storage in wireless networks. *International Journal of Distributed Sensor Networks* **1**(3–4), 355–371.

Haas Z and Pearlman M 2001 ZRP a hybrid framework for routing in ad hoc networks In *Ad Hoc Networking* (ed. Perkins CE), pp. 221–253. Addison-Wesley, chapter 7.

Hale WK 1980 Frequency assignment: Theory and applications. *Proceedings of the IEEE* **68**(12), 1497–1514.

Hata M and Nagatsu T 1980 Mobile location using signal strength measurements in cellular systems. *IEEE Transactions on Vehicular Technology* **VT-29**, 245–251.

Heinzelman WR, Chandrakasan A and Balakrishnan H 2000 Energy-efficient communication protocol for wireless microsensor networks *Proceeding of the Hawaii International Conference on System Sciences*, vol. 2, Hawaii.

Heinzelman WR, Kulik J and Balakrishnan H 1999 Adaptive protocols for information dissemination in wireless sensor networks *MobiCom '99: Proceedings of the 5th Annual ACM/IEEE International Conference on Mobile Computing and Networking*, pp. 174–185. ACM Press, New York.

Hodges A 1983 *Alan Turing: the Enigma*. Simon and Schuster, New York.

Holzmann GJ and Pehrson B 1995 *The Early History of Data Networks*. IEEE Computer Society.

Hu L and Evans D 2004 Using directional antennas to prevent wormhole attacks *Proceedings of the Network and Distributed System Security Symposium, NDSS 2004*. The Internet Society, San Diego, CA.

Hu YC, Johnson DB and Perrig A 2003a SEAD: secure efficient distance vector routing for mobile wireless ad hoc networks. *Ad Hoc Networks* **1**(1), 175–192.

Hu YC, Perrig A and Johnson DB 2003b Efficient security mechanisms for routing protocolsa *NDSS*. The Internet Society.

Hu YC, Perrig A and Johnson DB 2003c Packet leashes: A defense against wormhole attacks in wireless networks *INFOCOM*, San Diego, CA.

Hu YC, Perrig A and Johnson DB 2003d Rushing attacks and defense in wireless ad hoc network routing protocols *Proceedings of WiSe 2003*, pp. 30–40. ACM Press, San Diego, CA.

Hu YC and Perrig A 2004 A survey of secure wireless ad hoc routing. *IEEE Security and Privacy* **2**(3), 28–39.

Hu YC, Perrig A and Johnson DB 2005 Ariadne: A secure on-demand routing protocol for ad hoc networks. *Wireless Networks* **11**(1-2), 21–38.

Huurdeman AA 2003 *The Worldwide History of Telecommunications*. Wiley-Interscience.

IEEE CS 1999a ANSI/IEEE std 802.11 - wireless LAN medium access control (mac) and physical layer (PHY) specifications.

IEEE CS 1999b IEEE std 802.11a - wireless LAN medium access control (MAC) and physical layer (PHY) specifications: High-speed physical layer in the 5 ghz band.

IEEE CS 1999c IEEE std 802.11b - wireless LAN medium access control (MAC) and physical layer (PHY) specifications: Higher-speed physical layer extension in the 2.4 ghz band.

IEEE CS 2004 IEEE Std 802.11i-2004 IEEE standard for information technology- telecommunications and information exchange between systems- local and metropolitan area networks- specific requirements part 11: Wireless LAN medium access control (MAC) and physical layer (PHY) specifications amendment 6: Medium access control (MAC) security enhancements Standard Number IEEE Std 802.11i-2004.

IEEE CS, Theory IM and Society T 2004 IEEE standard for local and metropolitan area networks part 16: Air interface for fixed broadband wireless access systems IEEE std 802.16-2004 (revision of IEEE Std 802.16-2001).

IEEE 2005 What is DSRC? grouper.ieee.org/groups/scc32/dsrc/.

IEEE, LAN MAN Standards Committee of the CS, the IEEE Microwave Theory and Society T 2006 IEEE standard for local and metropolitan area networks part 16: Air interface for fixed and mobile broadband wireless access systems amendment 2: Physical and medium access control layers for combined fixed and mobile operation in licensed bands and corrigendum 1 IEEE Std 802.16e-2005 and IEEE Std 802.16-2004/Cor 1-2005 (Amendment and Corrigendum to IEEE Std 802.16-2004).

Japan Amateur Radio League 2005 The D-STAR project www.jarl.com/d-star/ (in Japanese).

Jaynes ET 2003 *Probability Theory: The Logic of Science*. Cambridge University Press.

Johnson D 1994 Routing in ad hoc networks of mobile hosts *Proceedings of the IEEE Workshop on Mobile Computing Systems and Applications*, pp. 158–163. Santa Cruz, CA.

Johnson DB and Maltz DA 1996 Dynamic source routing in ad hoc wireless networks In *Mobile Computing* (ed. Imielinski T and Korth H), vol. 353 of *The Kluwer International Series in Engineering and Computer Science*, pp. 153–181. Kluwer Academic Publishers, chapter 5.

Johnson D, Maltz D and Hu YC 2004 *The Dynamic Source Routing Protocol for Mobile Ad Hoc Networks (DSR)*. INTERNET-DRAFT, July.

Juels A and Wattenberg M 1999 A fuzzy commitment scheme *CCS '99: Proceedings of the 6th ACM conference on Computer and communications security*, pp. 28–36. ACM Press, New York.

Just M, Kranakis E and Wan T 2003 Resisting malicious packet dropping in wireless ad hoc networks In *Ad-Hoc, Mobile, and Wireless Networks, Second International Conference, ADHOC-NOW 2003 Montreal, Canada, October 8-10, 2003* (ed. Pierre S, Barbeau M and Kranakis E), vol. 2865 of *Lecture Notes in Computer Science*. Springer.

Kann V 1994 Polynomially bounded minimization problems that are hard to approximate. *Nordic Journal of Computing* **1**(3), 317–331.

Karlof C and Wagner D 2003 Secure routing in wireless sensor networks: attacks and countermeasures Proceedings of the First IEEE International Workshop on Sensor Network Protocols and Applications, *Ad Hoc Networks* **1**(2-3), pp. 293–315.

Kasten O and Langheinrich M 2001 First experiences with bluetooth in the smart-its distributed sensor network *Workshop on Ubiquitous Computing and Communications, PACT 01*. Barcelona, Spain.

Kaufman C, Perlman R and Speciner M 2002 *Network Security: Private Communication in a Public World*, 2nd edition, Prentice-Hall.

Kendall D 1953 Stochastic processes occurring in the theory of queues and their analysis by the method of the imbedded markov chain. *Annals of Mathematical Statis* **24**, 19–53.

Kleinrock L 1975 *Queueing Systems, Volume I: Theory*. Wiley-Interscience.

Kleinrock L 1976 *Queueing Systems, Volume II: Computer Applications*. Wiley-Interscience.

Kleinrock L and Tobagi FA 1975 Packet switching in radio channels. *IEEE Transactions on Communications* **COM-23**, Part I, 1400–1416; Part II, 1417–1433.

Klimow GP 1979 *Bedienungsprozesse*. Birkäuser Verlag.

Knuth D 1985 Donald Knuth: interviewed by D. J. Albers and L. A Steen In *Mathematical People: Profiles and Interviews* (ed. Albers DJ and Alexanderson GL), pp. 183–203. Birkhäuser Boston.

Kranakis E, Krizanc D and Urrutia J 2004 Coverage and connectivity in networks with directional sensors In *Euro-Par 2004 Parallel Processing, 10th International Euro-Par Conference*, Pisa, Italy, August 31-September 3, 2004, Proceedings (ed. Danelutto M, Vanneschi M and Laforenza D), vol. 3149 of *Lecture Notes in Computer Science*, pp. 917–924. Springer.

Kranakis E, Singh H and Urrutia J 1999 Compass routing on geometric networks *Proceeding of 11th Canadian Conference on Computational Geometry*, pp. 51–54. Vancouver, BC.

Kumar B 1993 Integration of security in network routing protocols. *SIGSAC Review* **11**(2), 18–25.

Lamport L 1979 Constructing digital signatures from a one-way function. Technical Report CSL-98, SRI International.

Law C, Mehta A and Siu KY 2001 Performance of a bluetooth scatternet formation protocol *The Second ACM Annual Workshop on Mobile Ad Hoc Networking and Computing (mobiHoc 2001)*. ACM Press.

Leeper D 2002 Wireless data blaster. *Scientific American* **286**(5), 65–69.

Li SR 1980 A martingale approach to the study of occurrence of sequence patterns in repeated experiments. *Annals of Probability* **8**, 1171–1176.

Li XY, Frieder O and Wang Y 2003 Localized routing for wireless ad hoc networks *Proceedings of IEEE International Conference on Communications ICC*, vol. 1, pp. 443–447. Anchorage, Alaska, May.

Li XY, Wang Y and Song WZ 2004 Applications of k-local mst for topology control and broadcasting in wireless ad hoc networks *Proceedings of the 23rd Annual Joint Conference of the IEEE Computer and Communications Society (INFOCOM-04)*, Piscataway, NJ.

Linial N 1992 Locality in distributed graph algorithms. *SIAM Journal on Computing* **21**(1), 193–201.

Little JDC 1961 A proof for the queuing formula $L = hW$. *Operations Research* **9**(3), 383–387.

Lyons R 2000 *Quadrature signals: Complex, but Not Complicated*, www.dspguru.com/info/tutor/quadsig.htm, November.

Marathe M, Breu H, Hunt, III HB, Ravi SS and Rosenkrantz DJ 1995 Simple heuristics for unit disk graphs. *Networks: an International Journal* **25**(2), 59–68.

Marti S, Giuli T, Lai K and Baker M 2000 Mitigating routing misbehavior in mobile ad hoc networks *Proceedings of the 6th Annual International Conference on Mobile Computing and Networking (MOBICOM-00)*, pp. 255–265. ACM Press, New York.

Mauve M, Widmer J and Hartenstein H 2001 A survey on position-based routing in mobile ad hoc networks. *IEEE Networks* **1**(6), 30–39.

McLarnon B 1997 VHF/UHF/microwave radio propagation: A primer for digital experimenters *Proceedings of 16th ARRL and TAPR Digital Communications Conference*, Baltimore, MD.

Menezes A, van Oorschot PC and Vanstone SA 1996 *Handbook of Applied Cryptography*. CRC Press.

Merkle R 1980 Protocols for public key cryptosystems *IEEE Symposium on Research in Security and Privacy*, pp. 122–134. IEEE Computer Society Press.

Mitola J 2000 *Software Radio Architecture - Object-Oriented Approaches to Wireless Systems Engineering*. John Wiley & Sons, New York.

Mitzenmacher M and Perrig A 2002 Bounds and improvements for BiBa signature schemes. Technical Report TR-02-02, Harvard University Cambridge, Massachusetts, MA.

Miu A, Tan G, Balakrishnan H and Guttag J 2002 An efficient scatternet formation algorithm for dynamic environments *IASTED International Conference on Communications and Computer Networks (CCN)*. Cambridge, MA.

Motwani R and Raghavan P 1995 *Randomized Algorithms*. Cambridge University Press.

Moy J 1998 OSPF version 2 The Internet Society, Request for Comments: 3561.

Naor M and Stockmeyer L 1995 What can be computed locally? *SICOMP: SIAM Journal on Computing*, **24** 1259–1277.

Ogier R, Templin F and Lewis M 2004 Topology dissemination based on reverse-path forwarding (TBRPF) The Internet Society, Request for Comments: 3684.

O'Rourke J and Toussaint GT 1997 Patterm recognition In *Handbook of Discrete and Computational Geometry* (ed. Goodman JE and O'Rourke J), pp. 797–813. CRC Press, New York.

Pahlavan K and Krishnamurthy P 2002 *Principles of Wireless Networks*. Prentice Hall PTR, Upper Saddle River, New Jersey.

Parkinson B and Spilker J(eds.) 1996 *Global Positioning System: Theory and Application (Volume I, Progress in Astronautics and Aeronautics)*. American Institute of Aeronautics and Astronautics.

Penrose MD 1997 The longest edge of the random minimal spanning tree. *The Annals of Applied Probability* **7**(2), 340–361.

Penrose MD 1999 On k-connectivity for a geometric random graph. *Random Structures and Algorithms* **15**, 145–164.

Penrose M 2003 *Random Geometric Graphs*. Oxford University Press, Oxford.

Perkins C and Bhagwat P 2001 DSDV routing over a multihop wireless network of mobile computers In *Ad Hoc Networking* (ed. Perkins CE), pp. 53–74. Addison-Wesley.

Perkins C, Belding-Royer E and Das S 2003 Ad hoc on-demand distance vector (AODV) routing The Internet Society, Request for Comments: 3561.

Perkins CE and Bhagwat P 1994 Highly-dynamic destination-sequenced distance-vector routing (DSDV) for mobile computers *Proceedings, 1994 SIGCOMM Conference*, pp. 234–244. London, UK.

Perlis AJ 1987 The synthesis of algorithmic systems *ACM Turing Award Lectures: The First Twenty Years, ACM Press Anthology Series*. ACM Press, New York, Addison-Wesley.

Perlman R 2000 *Interconnections, Second Edition: Bridges, Routers, Switches, and Internetworking Protocols*. Addison-Wesley.

Perrig A 2001 The BiBa one-time signature and broadcast authentication protocol *ACM Conference on Computer and Communications Security*, pp. 28–37. Washington, DC.

Perrig A and Tygar JD 2002 *Secure Broadcast Communication in Wired and Wireless Networks*. Kluwer Academic Publishers.

Petrioli C, Basagni S and Chlamtac I 2003 Configuring bluestars: multihop scatternet formation for bluetooth networks. *IEEE Transactions on Computers* **52**(6), 779–790.

Pierce JR 1980 *Signals: The Telephone and Beyond*. W. H. Freeman and Company, San Francisco, CA.

Poisel R 2003 *Modern Communications Jamming Principles and Techniques*. Artech House Publishers.

Poon C, Zhang YT and Bao SD 2006 A novel biometrics method to secure wireless body area sensor networks for telemedicine and m-health. *IEEE Communications Magazine* **44**(4), 73–81.

Preparata FP and Shamos MI 1985 *Computational Geometry: an Introduction*. Springer, Berlin.

Ramanathan R and Hain R 2000 Topology control of multihop wireless networks using transmit power adjustment *Proceedings of the 2000 IEEE Computer and Communications Societies Conference on Computer Communications (INFOCOM-00)*, pp. 404–413. IEEE, Los Alamitos, CA.

Rathi S 2000 Bluetooth protocol architecture. *Dedicated Systems Magazine*, pp. 28–33. http://www.dedicated-systems.com.

Ratnasamy S, Papadimitriou C, Shenker S, Stoica I and Rao A 2003 Geographic routing without location information *Proceedings of MOBICOM*, pp. 96–108. San Diego, CA.

Raya M, Hubaux JP and Aad I 2004 Domino: A system to detect greedy behavior in IEEE 802.11 hotspots *Proceedings of the 2nd International Conference on Mobile Systems, Applications and Service (MobiSys)*, pp. 84–97. Boston, MA.

Rényi A 1984 A dialogue on the applications of mathematics In *Mathematics: People, Problems, Results* (ed. Cambell DM and Higgins JC), pp. 255–263. Wadsworth International, Belmont, California. Reprinted from Rényi A 1967 A dialogue on the applications of mathematics In Dialogue on Mathematics, pp. 28–47. Holden-Day.

Reyzin L and Reyzin N 2002 Better than BiBa: Short one-time signatures with fast signing and verifying *ACISP: Information Security and Privacy: Australasian Conference (Melbourne, Australia, July 3-5, 2002)*, vol. 2384 of *Lecture Notes in Computer Science*. Springer.

Rivest R, Shamir A and Adlemen L 1978 A method for obtaining digital signatures and public key cryptosystems. *Communications of the ACM* **21**, 120–126.

Ross S 1996 *Stochastic Processes*, 2nd edition, John Wiley & Sons.

Ross S 2002 *Probability Models for Computer Science*. Academic Press.

Salonidis T, Bhagwar P and Tassiulas L 2000 Proximity awareness and fast connection establishment in bluetooth *Proceedings of the 1st ACM International Symposium on Mobile Ad Hoc Networking and Computing (MobiHoc)*, 141-142. Boston, MA.

Salonidis T, Bhagwat P, Tassiulas L and LaMaire R 2001 Distributed topology construction of bluetooth personal area networks *Proceedings of the Twentieth Annual Joint Conference of the IEEE Computer and Communications Societies (INFOCOM-01)*, pp. 1577–1586. IEEE Computer Society, Los Alamitos, CA. August.

Sastry N, Shankar U and Wagner D 2003 Secure verification of location claims *Proceedings of the ACM Workshop on Wireless Security (WiSe'03)*, pp. 1–10. ACM Press, San Diego, CA

Segaller S 1998 *NERDS 2.0.1: A Brief History of the Internet*. TV Books, New York.

Shakkottai S, Srikant R and Shroff NB 2003 Unreliable sensor grids: coverage, connectivity and diameter *INFOCOM*.

Shannon CE 1949 Communication in the presence of noise. *Proceeding of the IRE* **37**(1), 10–21.

Simmons GJ 1988 A survey of information authentication. *Proceedings of the IEEE*, **76** 603–620.

Smith SW 1999 *The Scientist and Engineer's Guide to Digital Signal Processing*, 2nd edition, California Technical Publishing, San Diego, CA.

Smith D 2001 *Digital Signal Processing Technology*. The American Radio Relay League, Newington, CT.

Smith BR, Murthy S and Garcia-Luna-Aceves JJ 1997 Securing distance-vector routing protocols *NDSS*. IEEE Computer Society.

Stallings W 2007 *Network Security Essentials: Applications and Standards*, 3rd edition, Pearson/Prentice Hall.

Standage T 1998 *The Victorian Internet*. Walker and Company, New York.

Sterndark D 1994 *RC4 Algorithm Revealed*, http://groups.google.com/group/sci.crypt/msg/10a300c9d21afca0, September.

Stevens WR 1998 *Unix Network Programming, Networking APIs: Sockets and XTI*. Prentice Hall, Upper Saddle River, NJ.

Stroh S 2003 Wideband: multimedia unplugged. *IEEE Spectrum* **40**(9), 23–27.

Tanenbaum AS 4th edition, 1996 *Computer Networks*. Prentice-Hall International.

Tobagi FA and Kleinrock L 1976 Packet switching in radio channels, part III. *IEEE Transactions on Communications* **COM-24**, 832–845.

Tobagi FA and Kleinrock L 1977 Packet switching in radio channels, part IV. *IEEE Transactions on Communications* **COM-25**, 1103–1119.

Toussaint GT 1980 The relative neighborhood graph of a finite planar set. *Pattern Recognition* **12**(4), 261–266.

Turing A 1948 Letter to Jack Good.

Verissimo PE 2006 *Travelling Through Wormholes: A New Look at Distributed Systems Models*, vol. 37, 1, pp. 66–81. ACM Press, ACM SIGACT News Distributed Computing Column 21. New York, March.

Wan T, Kranakis E and van Oorschot P 2004 S-RIP: A secure distance vector routing protocol *International Conference on Applied Cryptography and Network Security (ACNS)*, vol. 3089 of *Lecture Notes in Computer Science*, pp. 103–119. Springer.

Wang DW and Kuo YS 1988 A study on two geometric location problems. *Information Processing Letters*. **28**(6), 281–286.

Wang Z, Thomas RJ and Haas ZJ 2002 Bluenet - A new scatternet formation scheme, *Proceedings of the 35th Annual Hawaii International Conference on System Sciences (HICSS)*, p. 61. Hawaii, January.

Wattenhofer M, Wattenhofer R and Widmayer P 2005 Geometric routing without geometry *Proceedings of International Colloquium on Structural Information and Communication Complexity (SIROCCO)*, vol. 3499 of *Lecture Notes in Computer Science*, pp. 307–322. Springer.

Waxman BM 1988 Routing of multipoint connections. *IEEE Journal Selected Areas in Communications (Special Issue: Broadband Packet Communications)* **6**(9), 1617–1622.

WiFi A 2004 Wi-Fi, prottected access (WPA) enhanced security implementation based on IEEE p802.11i standard, version 3.1.

Wong VWS, Zhang C and Leung VCM 2006 A survey of scatternet formation algorithms for bluetooth wireless personal area networks In *Handbook of Algorithms for Wireless Networking and Mobile Computing* (ed. Boukerche A), pp. 735–754. Chapman and Hall/CRC Press, New York.

Xiao Y 2005 Energy saving mechanism in the IEEE 802.16e wireless MAN. *IEEE Communications Letters* **9**(7), 595–597.

Youngblood G 2002a A software defined radio for the masses, part 1. *QEX* **Jul./Aug.**, pp. 13–21.

Youngblood G 2002b A software defined radio for the masses, part 2. *QEX* **Nov./Dec.**, pp. 10–18.

Youngblood G 2002c A software defined radio for the masses, part 3. *QEX* **Nov./Dec.**, pp. 27–36.

Youngblood G 2003 A software defined radio for the masses, part 4. *QEX* **Mar./Appr.**, pp. 20-31.

Zapata MG 2002 Secure ad hoc on-demand distance vector routing. *Mobile Computing and Communications Review* **6**(3), 106–107.

Zaruba GV, Basagni S and Chlamtac J 2001 Bluetrees: Scatternet formation to enable bluetooth-based ad hoc networks *IEEE ICC'01*, pp. 273–277. Helsinki, Finland.

Zeng QA and Agrawal D 2002 Modeling of handoffs and performance analysis of wireless data networks. *IEE Transactions on Vehicular Technology* **51**(6), 1469–1478.

Zhang B and Mouftah H 2005 Qos routing for wireless ad hoc networks: problems, algorithms, and protocols. *IEEE Communications Magazine* **43**(10), 110–117.

ZigBee Alliance 2005 *Zigbee Specification, Version 1.0*, www.zigbee.org, June.

Index

Principles of Ad hoc Networking Michel Barbeau and Evangelos Kranakis